T0136730

Time Warps, String Edits, Macromolecules

THE DAVID HUME SERIES
PHILOSOPHY AND COGNITIVE SCIENCE REISSUES

The *David Hume Series on Philosophy and Cognitive Science Reissues* consists of previously published works that are important and useful to scholars and students working in the area of cognitive science. The aim of the series is to keep these indispensable works in print.

TIME WARPS, STRING EDITS, AND MACROMOLECULES

DAVID SANKOFF & JOSEPH KRUSKAL

Introduction by JOHN NERBONNE

THE DAVID HUME SERIES
PHILOSOPHY AND COGNITIVE SCIENCE REISSUES

CSLI PUBLICATIONS

Copyright © 1999
CSLI Publications
Center for the Study of Language and Information
Leland Stanford Junior University
Printed in the United States
03 02 01 00 99   1 2 3 4 5

Library of Congress Cataloging-in-Publication Data

Time warps, string edits and macromolecules : the theory and practice of sequence
comparison / edited by David Sankoff and Joseph B. Kruskal
p.   cm. – (David Hume series)
Originally published: Reading, Mass. : Addison-Wesley Pub., Advanced Book Program
1983.
Includes bibliographical references and index.

ISBN 1-57586-217-4 (pbk. alk. paper)

1. Sequences (Mathematics)   2. Pattern perception
3. Computational complexity.   4. Molecular Biology   5. Speech processing systems.
I. Sankoff, David.   II. Kruskal, Joseph B.,
1928–   . III. Series.
QA292.T55   1999
515' .24--dc21   99-42488
CIP

EDIT DISTANCE AND DIALECT PROXIMITY

John Nerbonne
with Wilbert Heeringa and Peter Kleiweg

David Sankoff and Joseph Kruskal's *Time Warps, String Edits and Macromolecules: The Theory and Practice of Sequence Comparison* (hereafter *Time Warps*) is a young (1983) classic which has inspired developments in computer science, pure and applied linguistics, computational biology, and even music and ethnology. CSLI Publications deserves the appreciation of all these scientific subfields for undertaking its republication. Because its first chapter, "Overview," is Joseph Kruskal's gentle introduction to Levenshtein distance (also known as sequence distance and edit distance) (Kruskal 1999), the book will be useful to students as well as researchers. The overview explores the concept of sequence distance first from the perspective of alignment – finding the correspondence between sequences that minimizes distance. It goes on to introduce a dynamic programming algorithm which efficiently calculates the distance, and provide notes on the history of this and related concepts. Later chapters in *Time Warps* go on to explore several interesting extensions and applications in depth, including recognizing relatedness in DNA, matching speech input to lexical hypotheses, tracking the development of bird songs over time, and to error corrections (e.g., of keyboard input). Some of these chapters introduce significant refinements in order to handle transposed elements, continuous input (rather than sequences of discrete elements), and tree-structure.

This introduction to the new edition of *Time Warps* in *The David Hume Series of Philosophy and Cognitive Science Reissues* focuses on linguistic applications. We shall mention some of the developments since 1983 and suggest why the theory of sequence distance still enjoys enormous popularity in the diverse linguistic fields in which it is applied. The examples come from pure and applied linguistics, including computational linguistics, and draw on work done in Groningen. We shall unfortunately ignore the extensive and very successful work on sequence distance in computational biology to explore which DNA sequences are likely mutations of one another (Waterman 1989, Waterman 1995, Crochemore and Gusfield 1994, Farach-Colton 1998.)

1 COMPUTATIONAL LINGUISTICS

Levenshtein distance and various derivatives give the researcher a tool with which to measure the relative similarity of various sequences. The technique is general enough to apply to sequences of all sorts.

Pronunciations consist of sequences of sounds, so it is not surprising that *Time Warps* includes Kruskal and Liberman's explication of dynamic time warping, the application of Levenshtein distance to speech recognition (Kruskal and Liberman 1999). Although Hidden Markov Models have largely replaced dynamic time warping in speech recognition, Levenshtein distance is still used in speech recognition, e.g., in order to score the relative closeness of hypotheses to (annotated) correct answers (Veldhuijzen van Zanten et al. 1999). Without this, it is difficult to distinguish recognition results beyond the level of 'correct vs. incorrect'. Other researchers have used variants of Levenshtein distance to diagnose potentially pathological pronunciation deviation (Connolly 1997). The interest in speech pathology requires that deviant pronunciations be aligned with normal ones so that researchers and therapists can explore the specific "operations" that appear to be responsible for differences.

The last ten years have seen an explosion in the availability of large text corpora (Church and Mercer 1993, Klavans 1996, Nerbonne 1998). Among these, there has been special interest in parallel, bilingual corpora (Veronis to appear, 2000). Careful analysis of parallel corpora yields information such as nearest translation equivalents, common usage of words and phrases, and subtleties of grammatical use, all of which is of course useful to translators and developers of translation support software, to bilingual dictionary compilers, and to developers of related applications such as computer-assisted language learning. Veronis (2000) has papers on all of these applications. The fundamental technical problem in supporting the extraction of such information is the problem of aligning texts so that users can easily identify corresponding parts, and the most successful approaches have been derived from the dynamic programming algorithm for calculating edit distance (Gale and Church 1993). Gale and Church's (1993) algorithm take sentences as fundamental units, and proceed from the assumption that sentence length in characters will correspond probabilistically in parallel texts. Special accommodation is made for the sentences' expanding, contracting and merging in translation, but the authors are explicit about the technique's legacy from *Time Warps*. The topic of alignment is important enough so that alternative methods are still actively sought (Church 1993, Chen 1993), and new applications in multimedia are under investigation in which the alignment of text plays an important role (Braschler and Schauble 1998).[1]

---

[1] The EU OLIVE project `http://twentyone.tpd.tno.nl/olive/` tries to align notes,

2 DIALECTOLOGY

Linguistics studies language variation systematically as it correlates with any number of extra-linguistic variables, but especially with geography and social variables such as class, age, gender, social network, education and trade. The oldest of these is dialectology, the study of how language varies geographically. Dialectology is not only the most senior of the linguistic variation fields, it is one of the oldest branches of linguistics – the early works date from the 19th century as do many of the techniques. See Petyt (1980) and Niebaum (1983).

Traditional dialectology has focused on recording the variety of speech forms within a given area. This has intrinsic, linguistic interest, but it is also interesting because language variety reflects cultural influence, most obviously in the case of migrations, conquests and the creation of states and borders. Dialectology has amassed large amounts of data concerning the specific geographical distribution of individual words, word forms, syntactic constructions, pronunciations and the like. These are normally present in the form of maps showing which forms are used in which areas. In case the same areas spring out repeatedly, one may speak of a "dialect area."

Unfortunately, detailed maps showing the distribution of words and sounds normally fail to distinguish areas, which leads to a scientific puzzle: linguists and lay people alike feel that they recognize varieties, such as Bavarian German, or the Amsterdam variety of Dutch, or the coastal New England variety of English. But these are not the areas consistently picked out by varying forms. There are usually patterns of variation that run criss-cross, ignoring the well-established boundaries of areas.

This puzzle was, as it turned out, a scientific disappointment. Dialectology was originally pursued *inter alia* in order to test the Neo-grammarian hypothesis that sound change is regular. Among its earliest results is that variation is often geographically irregular, i.e., the areas in which different forms are used often turn out to be (perhaps slightly) different areas (Bloomfield 1933, p.322).

Naturally, some dialectologists responded by collecting more data, hoping to find the right distinguishing features. This has been done to such extremes that even friendly reviewers have warned of the dangers of "atomization" (Coseriu [1]1956, 1975, p.50). Lacking analytical techniques with which distributions of different forms can be compared or combined, the result is that hundreds of forms give hundreds of perspectives – in none of which aggregates such as Bavarian German, Amsterdam Dutch, or coastal New England are recognizable combinations of the atoms.

In many cases, however, it often turns out that changes are cumulative, i.e., that more geographically remote areas are linguistically less similar as well so that it is reasonable to speak of a dialect "continuum" (Chambers and Trudgill 1980, § 1.3,§ 8.1-8.6). But this notion – like other aggregate notions – has resisted satisfactory theoretical formulation.

---

speech recognition results and closed captions for the hearing impaired in order to align associated video material (Netter 1998).

This very brief sketch outlines two theoretical questions in dialectology, how aggregate notions such as the variety of a particular area may be approached, and how those aggregates might be seen to exist on a continuum of linguistic variety. We cannot offer theoretical reconstructions of either of these notions, but, with the help of Levenshtein distance, we can measure differences associated with them.

2.1 Measuring dialect difference

When you can measure, [. . .] you know something—Lord Kelvin

Levenshtein distance is employed to obtain a measure of distance in pronunciation. In contrast to the "same/different" classification of earlier dialectology, the measurements are numerical and may therefore be summed and/or averaged to obtain an aggregate characterization of differences between entire varieties (based on larger sets rather than one word at a time). We obtain then a measure of difference rather than a geographic delineation of (discrete) features of individual words or pronunciations. More general characterizations of dialect differences then become available.

A mathematical perspective may be useful. Most linguistic data is naturally nominal, divided into different categories, which allow no natural aggregation. But once the data can be approached numerically – as the use of Levenshtein distance allows, then addition of atomic differences can be interpreted, so that we can sensibly speak of the distance between entire varieties. The measurement also allows us to approach the question of cumulativity more exactly. If dialectologists have been correct in speaking of cumulative differences, then we should likewise find Levenshtein difference increasing with geographical distance – a reflection of the "continuum."

2.2 Data and method

The *Reeks Nederlands(ch)e Dialectatlassen* (Blancquaert and Pée 1925–1982) contains 1,956 Netherlandic and North Belgian transcriptions of 141 sentences. We chose 104 dialects, regularly scattered over the Dutch language area, and 100 words which appear in each dialect text, and which contain all vowels and consonants.

As the introductory chapter of this book best explains, Levenshtein distance in its most basic variant involves calculating the "cost" of changing one word into another using insertions, deletions and replacements. Levenshtein distance (s_1, s_2) is the sum of the costs of the cheapest set of operations changing s_1 to s_2. The example below illustrates Levenshtein distance applied to Bostonian and standard American pronunciations of *saw a girl*. Boston pronunciation inserts an [r] between *saw* and *a*, deletes the postvocalic [r] in *girl*, and replaces the short vowel in *girl* with a fronted rounded vowel [ø], like the first vowels in German *Köln* or French *meuble*.

Standard American	sɔəgIrl	delete r	1
	sɔəgIl	replace I/ø	2
	sɔəgøl	insert r	1
Bostonian	sɔrəgøl		
		Sum distance	4

Kessler (1995) applied Levenshtein distance to Irish dialects.

The example above simplifies the procedures actually used for clarity: the actual measurements are sensitive to the phonetics of the pronunciations of the basic sounds (*t,d,i*, etc.). To obtain a more sensitive measure, costs are refined based on phonetic feature overlap, so that, e.g. replacing a [d] with a [t] is less costly than replacing the [d] with, say [i]. Replacement costs thus vary depending on the basic sounds involved.

This is done in a way that is standard in linguistics. Each sound is represented by a vector of values for phonetic features. The table below suggests how this system gives rise to characterizations of the overall difference between segments: [i] and [e] are much closer than [i] and [u].

	i	e	u	i-e	i-u
advancement	2(front)	2(front)	6(back)	0	4
high	4(high)	3(mid)	4(high)	1	0
long	3(short)	3(short)	3(short)	0	0
lip-rounding	0(none)	0(none)	1(rounded)	0	1

To avoid making the measurements too dependent on one feature system as opposed to another, two feature systems were tested; the results analyzed below are based on Hoppenbrouwers' (SPE-like) features (Hoppenbrouwers and Hoppenbrouwers 1988), but a feature system developed by Vieregge to measure transcriber accuracy also yields very good results (Vieregge et al. 1984). The distance between two vectors of feature values was taken to be the Euclidean distance between the two, $\delta(X, Y)$, where:

$$\delta(X,Y) = \sqrt{\sum_{i=1}^{n} (X_i - Y_i)^2}$$

Several further refinements were the subject of experimentation, including the optimal representation of diphthongs (one segment vs. two), the value of weighting by frequency and/or information gain, and the use of the standard language as a calibration in measurements (Nerbonne and Heeringa to appear, 1999b). We have begun validating the technique, including these aspects, using cross-validation on unseen Dutch dialect data (Nerbonne and Heeringa 1999a).

2.3 Results

Comparing two varieties results in a sum of 100 word-pair comparisons. Because longer words tend to be separated by more distance than shorter words, the distance of each word pair is normalized by dividing it by the mean length of the word pair. This results in a half-matrix of distances, which is best visualized in Figure 1. This fulfills the first desideratum above, obtaining a means of aggregating over the atomistic differences between individuals words. Using the Levenshtein measure, we can now say something about the overall relation between two varieties – even when individual elements may vary and even be inconsistent in the tendencies they display.

A further goal was to explore how dialect areas might be viewed. The half-matrix of distances above was therefore subjected to hierarchical clustering, using Ward's method, which minimizes the error introduces by clustering (Aldenderfer and Blashfield 1984). This classifies the dialects into relatively close groups. The most significant groups which emerge from clustering correspond to those identified in traditional dialectology, viz., Frisian, Low Saxon, and Franconian (Nerbonne and Heeringa to appear, 1999b). See Figure 2. We take this to validate the techniques, but current work seeks more direct validation.

Finally, we explore the degree to which the techniques confirm the view of dialects as existing on a "continuum." The most direct means of testing this is to examine whether phonetic distance increases with geographical distance, to see whether distances are indeed "cumulative." Pearson's product-moment correlation coefficient shows that geographic and phonetic distance correlate at a level of $r = 0.68$. Given the large amount of data, it is not surprising that this is a statistically significant level. But it is also a fairly strong correlation, accounting for nearly 45% of the variance. The heights of children and parents correlate to about this degree. So we obtain an answer: the "continuum" is justified in its generalization that phonetic differences are cumulative. These results suggests that one may indeed combine the detailed dialectology maps, showing discrete dialect areas with the often expressed view that dialects exists on a continuum, in accord with Chambers and Trudgill's (1980:127) (sceptical) wishes. These are then compatible ways of viewing the dialectal facts.

To explore further the view of dialects as a continuum, we exploit techniques Joseph Kruskal developed in multidimensional statistical analysis, viz., multidimensional scaling (MDS). MDS may be applied to distance matrices such as the matrix of average Levenshtein distances in order to identify a small set of most significant dimensions (Kruskal and Wish 1978). The map shown in this book's frontispiece (see inside cover) distinguishes Dutch "dialect areas" in a way which non-numerical methods have been unable to do (without resorting to subjective choices of distinguishing features). The MDS analysis gives mathematical form to the intuition of dialectologists in Dutch (and other areas) that the material is best viewed as a "continuum." The map is obtained by interpreting MDS dimensions as colors. Since this assigns colors only to the sites for which pronunciations are available, we interpolate over other areas using inverse distance weighting.

Figure 1: The average Levenshtein distance between Dutch dialects. The darker the line between two varieties, the closer they are in (phonetic) Levenshtein distance. Frisian (top center) emerges clearly as a relatively distinct, but internally coherent group.

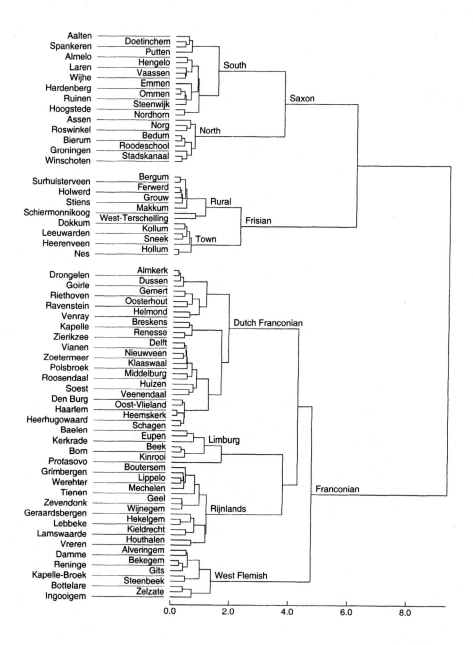

Figure 2: The result of clustering the matrix of Levenshtein distances. The traditional dialect areas (labeled) emerge distinctly. Within the Limburg cluster one finds Protasovo, an emigrant Low German variety, the language of Siberian Mennonites investigated by Nieuweboer (1998) (included out of curiosity).

2.4 Conclusions and prospects in dialectology

We have studied the consistency of the measurements, which, given a set of 100 words, is reliable. In order to validate techniques more rigorously, we have run the techniques on unseen data, and we are currently seeking an objective measure of quality to distinguish the various refinements of Levenshtein distance discussed above. To do this, we shall take measures on a set of variants whose classification into dialect area (Frisian, Saxon, Flemish or Franconian) is uncontested. The question is the degree to which the Levenshtein measure will jibe with the classification. We shall report on this work when it is completed.

Ongoing work applies the technique to questions of convergence/divergence of dialects using dialect data from two different periods. In order to be more generally useful, the distance measure should be applied to the data of other languages. We are also interested in exploring the relation between this (linguistically objective) measure of distance with psycho-acoustical investigations into the perception of varieties (Gooskens 1997).

Further information on the project, also material (data) and some software, is available at http://www.let.rug.nl/alfa/, "Projects."

3 TIME WARPS

Levenshtein distance is still being applied to new areas, more than 15 years after *Time Warps*'s original publication. We may hope that the continued availability of Sankoff and Kruskal's seminal volume will spur further refinement and application of sequence distance measures.

ACKNOWLEDGEMENTS

Wilbert Heeringa and Peter Kleiweg have implemented and maintained all of the dialectology programs described here. Dicky Gilbers, Tjeerd de Graaf, Jack Hoeksema, Wouter Jansen, Brett Kessler, Joseph Kruskal, Hermann Niebaum and Harry Scholtmeier have offered valuable criticism and advice at various points in this project.

REFERENCES

Aldenderfer, Mark S., and Roger K. Blashfield. 1984. *Cluster Analysis*. Quantitative Applications in the Social Sciences. Beverly Hills, CA: Sage.

Blancquaert, E., and W. Pée. 1925-1982. *Reeks Nederlandse Dialectatlassen*. Antwerpen: De Sikkel.

Bloomfield, Leonard. 1933. *Language*. New York: Holt, Rhinehart and Winston.

Braschler, M. and P. Schauble. 1998. Multilingual Information Retrieval Based on Document Alignment Techniques. *Proc., 2nd European Conference, (ECDL)* 183–199.

Chambers, Jack, and Peter Trudgill. 1980. *Dialectology*. Cambridge: Cambridge University Press.

Chen, Stanley. 1993. Aligning Sentences in Bilingual Corpora using Lexical Information. *Proc. of the 31st ACL* 9–17. Columbus, OH: ACL

Church, Kenneth. 1993. char_align: A Program for Aligning Parallel Texts at the Character Level. *Proc. of the 31st ACL* 1–8. Columbus, OH: ACL

Church, Kenneth and Robert Mercer. 1993. Introduction to the Special Issue on Computational Linguistics using Large Corpora. *Computational Linguistics* 19(1):1–24. (See articles in this and in following number 19(2))

Connolly, John H. 1997. Quantifying Target-Realization Differences. *Clinical Linguistics and Phoentics* 11:267–298

Coseriu, Eugenio. [1]1956, 1975. *Die Sprachgeographie*. Tübingen: Gunter Narr.

Crochemore, Maxime, and Dan Gusfield (ed.). 1994. *Combinatorial Pattern Matching: the 5th Annual Symposium*. Berlin and Heidelberg: Springer.

Farach-Colton, Martin (ed.). 1998. *Combinatorial Pattern Matching: 9th Annual Symposium*. Berlin and Heidelberg: Springer.

Gale, William A., and Kenneth W. Church. 1993. A Program for Aligning Sentences in Bilingual Corpora. *Computational Linguistics* 19(1):75–102.

Gooskens, Charlotte. 1997. *On the Role of Prosodic and Verbal Information in the Perception of Dutch and English Language Varieties*. Doctoral dissertation, Nijmegen University.

Hoppenbrouwers, Cor, and Geer Hoppenbrouwers. 1988. De Featurefrequentiemethode en de Classificatie van Nederlandse Dialecten. *TABU: Bulletin voor Taalwetenschap* 18(2):51–92.

Kessler, Brett. 1995. Computational dialectology in Irish Gaelic. In *Proc. of the European ACL*, 60–67. Dublin: ACL.

Klavans, Judith and Philip Resnick, eds. 1996. *The Balancing Act: Combining Symbolic and Statistical Approaches to Language*. Cambridge, MA: MIT Press.

Kruskal, Joseph. [1983] 1999. An Overview of Sequence Comparison. In *Time Warps, String Edits and Macromolecules: The Theory and Practice of Sequence Comparison*, ed. David Sankoff and Joseph Kruskal. 1–44. Reprint, with a foreword by John Nerbonne, Stanford, CA: CSLI Publications.

Kruskal, Joseph, and Mark Liberman. [1983] 1999. The Symmetric Time-Warping Problem: From Continuous to Discrete. In *Time Warps, String Edits and Macromolecules: The Theory and Practice of Sequence Comparison*, ed. David Sankoff and Joseph Kruskal. 125–161. Reprint, with a foreword by John Nerbonne, Stanford, CA: CSLI Publications.

Kruskal, Joseph, and Myron Wish. 1978. *Multidimensional Scaling*. Beverly Hills, CA: Sage.

Nerbonne, John, ed. 1998. *Linguistic Databases*. Stanford: CSLI Publications.

Nerbonne, John, and Wilbert Heeringa. 1999a. Computational Comparison and Classification of Dialects. *Zeitschrift für Dialektologie und Linguistik*. Spec. iss.

ed. by Jaap van Marle and Jan Berens w. selections from 2nd Int'l Congress of Dialectologists and Geolinguists, Amsterdam, 1997.

Nerbonne, John, and Wilbert Heeringa. to appear, 1999b. Computationele Vergelijking en Classificatie van Dialecten. *Taal en Tongval* 51.

Netter, Klaus. 1998. POP-EYE and OLIVE: Human Language as the Medium for Cross-lingual Multimedia Information Retrieval. *Proc. of 2nd International Conference on Languages and the Media.* Berlin.

Niebaum, Hermann. 1983. *Dialektologie.* Tuebingen: Niemeyer.

Nieuweboer, Rogier. 1998. *The Altai Dialect of Plaudittsch (West-Siberian Mennonite Low German).* University of Groningen.

Petyt, K.M. 1980. *The Study of Dialect : An Introduction to Dialectology.* London: André Deutsch.

Veldhuijzen van Zanten, Gert, Gosse Bouma, Khalil Sima'an, Gertjan van Noord, and Remko Bonnema. 1999. Evaluation of the NLP Components of the OVIS2 Spoken Dialogue System. OVIS Technical Report 84. Amsterdam/Eindhoven/Groningen: NWO Language-Speech Technology Programme. avail. at http://odur.let.rug.nl:4321/publijst.html.

Veronis, Jean. to appear, 2000. *Parallel Text Processing.* Dordrecht and Boston: Kluwer Academic.

Vieregge, Wilhelm H., A.C.M.Rietveld, and Carel Jansen. 1984. A Distinctive Feature Based System for the Evaluation of Segmental Transcription in Dutch. In *Proc. of the 10th International Congress of Phonetic Sciences*, ed. Marcel P.R. van den Broecke and A. Cohen, 654–659. Dordrecht: Foris.

Waterman, Michael S. 1995. *Introduction to Computational Biology: Maps, Sequences and Genomes.* London: Chapman and Hall.

Waterman, Michael S. 1989. Sequence Alignments. In *Mathematical Methods for DNA Sequences*, ed. Michael S. Waterman. 53–92. Boca Raton: CRC.

LIST OF CONTRIBUTORS

David W. Bradley	Data Processing Department, California State University, Long Beach	Long Beach, California, 90840
Richard A. Bradley	Department of Biology, University of New Mexico	Albuquerque, New Mexico, 87131
Robert J. Cedergren	Département de Biochimie, Université de Montréal	Montréal, Québec H3C 3J7
Václav Chvátal	School of Computer Science, McGill University	Montreal, Quebec H3A 2K6
James M. Coggins	Computer Science Department, Worcester Polytechnic Institute	Worcester, Massachusetts, 01609
Joseph Deken	Department of General Business, University of Texas	Austin, Texas, 78712
Bruce W. Erickson	Laboratory of Biochemistry, The Rockefeller University	New York, N.Y., 10021
Daniel S. Hirschberg	Information and Computer Science, Department, University of California, Irvine	Irvine, California, 92717
Melvyn J. Hunt	National Research Council	Ottawa, Ontario, K1A 0R6
Joseph B. Kruskal	Mathematics and Statistics Research, Bell Laboratories	Murray Hill, New Jersey, 07974

Matthew Lennig | Bell-Northern Research and INRS—Télécommunications | Montreal, Quebec H3E 1H6

Mark Liberman | Acoustical and Behavioral Research, Bell Laboratories | Murray Hill, New Jersey, 07974

Sylvie Mainville | Département de Mathématiques, Université du Québec à Trois-Rivières | Trois-Rivières, Québec G9A 5H7

William J. Masek | GenRad, Inc. 300 Baker Street | Concord, Massachusetts, 01742

Paul Mermelstein | Bell-Northern Research and INRS—Télécommunications | Montreal, Quebec H3E 1H6

Andrew S. Noetzel | Department of Statistics and Computer Information Systems, Baruch College, C.U.N.Y. | New York, N.Y., 10010

Michael S. Paterson | School of Computer Science, University of Warwick | Coventry, Warwick United Kingdom

David Sankoff | Centre de Recherche de Mathématiques Appliquées, Université de Montréal | Montréal, Québec H3C 3J7

Stanley M. Selkow | Computer Science Department, Worcester Polytechnic Institute | Worcester, Massachusetts, 01609

Peter H. Sellers | Department of Mathematics, The Rockefeller University | New York, N.Y., 10021

Robert A. Wagner | Department of Computer Science, Duke University | Durham, North Carolina, 27706

CONTENTS

CHAPTER ONE

AN OVERVIEW OF SEQUENCE COMPARISON

Joseph B. Kruskal

1. INTRODUCTION

It is often necessary to compare two or more sequences, or strings, or vectors, or continuous functions of time, etc., and measure the extent to which they differ. Comparison is used for identification, for error-correction, and for determining relationships. In many situations there is a natural correspondence between the elements, or components, or coordinates, or points of time, etc., in one sequence and those in the other, and the only sensible comparison is between corresponding elements. In such situations it is easy to make the comparison. *Sequence comparison* deals with the more difficult comparisons that arise when the correspondence is not known in advance, perhaps because some underlying correspondence has been disturbed by the gain or loss of elements in one or both sequences.

Most existing methods for comparing sequences to see how different they are can be applied only to sequences of equal length, and are based on comparing corresponding elements. For example, some well-known methods for comparing sequences **a** and **b** are Euclidean distance $[\Sigma_{i=1}^{n} (a_i - b_i)^2]^{1/2}$, city block distance $\Sigma_{i=1}^{n} |a_i - b_i|$, and Hamming distance, which is simply the number of positions in which the corresponding elements are different. Here a_i corresponds to b_i, and the comparison between them is expressed in the term $(a_i - b_i)^2$ or the term $|a_i - b_i|$ or, in the case of Hamming distance, a term that is 1 if $a_i \neq b_i$ and 0 if $a_i = b_i$.

In some situations the appropriate correspondence to use is not known in advance, either because a_i has no special connection with b_i or simply because the sequences have different lengths. The general approach used in sequence comparison is to seek the appropriate correspondence by optimizing over all possible correspondences that satisfy suitable conditions, such as preserving the order of the elements in the sequence. The optimization procedures used fall within the framework of dynamic programming, though in several cases they were invented without explicit use or apparent awareness of that framework. (Knowledge of dynamic programming will not be required in this volume.) The central themes of sequence comparison are:

1. distance functions that are appropriate to use when there is no natural correspondence of elements;
2. optimum correspondences between sequences;
3. dynamic programming algorithms for calculating distances and optimum correspondences;
4. applications; and
5. the properties of distances, optimum correspondences, and algorithms.

The applications fall into two major categories: discrete and continuous. In the discrete case, the objects under study are ordinarily sequences (typically of different lengths), though limited attention has also been given to more complicated structures. An individual sequence can be written $a = (a_1, \cdots, a_n)$, where the elements a_i are drawn from some *alphabet*. The alphabet might consist of all numbers or of all vectors in some low-dimensional vector space, but more often it is a special alphabet characteristic of the application, e.g., the twenty amino acids, or the four nucleotides, or the symbols on some particular computer terminal keyboard, or the 40 phonemes of English, or the 113 syllables (properly, *"morae"*) of Japanese. In the continuous case, the objects under study are ordinarily continuous functions of a variable t, usually time, over intervals of different lengths. An individual function can be written $a(t)$, over the interval 0 to T, where the function values may be numbers, e.g., the amount of material emerging from a gas chromatograph column at time t, or may lie in a vector space, e.g., a vector of coefficients that describes the frequency content of a speech waveform at time t. Continuous situations are dealt with by converting them into discrete ones, by sampling the time interval at discrete times $t_1, t_2, \ldots$.

In discrete situations, sequence deletions and insertions may result from processes that lead to actual physical deletion and insertion. In situations that have been converted from continuous to discrete by sampling, it is also possible for physical deletion and insertion to occur. In addition, however, compression and expansion along the time axis can give rise to changes that closely resemble deletions and insertions and have often been so described. There are, however, subtle differences between the methods used to handle deletion–insertion and compression–expansion.

1.1 Applications to molecular biology

Proteins and nucleic acids are macromolecules central to the biochemical activity of all living things, including chemical regulation and genetic determination and transmission. These macromolecules may be considered as long sequences of subunits linked together sequentially in a chain. In the case of proteins, the units are drawn from an alphabet of twenty amino acids, and the sequences are typically a few hundred units long. In the case of nucleic acids, the units are drawn from an alphabet of four different nucleotides (or bases), and the sequences are typically from tens to thousands of units long (for ribonucleic acid, i.e., RNA) or millions of units long (for deoxyribonucleic acid, i.e., DNA).

The methods of sequence comparison are helping to answer several important questions in biology. One set of questions revolves around *homology* of macromolecules: Two structures are said to be homologous if they have a common evolutionary ancestor, and are said to have a high degree of homology if the differences that have developed between them are relatively minor. Are certain macromolecules homologous? Which parts of one molecule are homologous to which parts of the other? If we compare many different molecules in one species with the corresponding molecules in another species, how high is the typical degree of homology? For what pairs of species is this typical degree of homology high? Knowledge of homology sheds light not only on evolution, but also on structure and function in organisms. Another set of questions revolves around the structural configuration of individual RNA molecules, i.e., the way in which different parts of the chain join to each other, since the methods of sequence comparison can help determine this also.

In this volume, Part I is devoted to molecular biology, and various other chapters are relevant to molecular biology, as listed in the introduction to Part I. There is a voluminous literature on sequence comparison in molecular biology, though much of it unfortunately is based on cruder methods than the dynamic programming methods described in this book. For access to this literature, the reader may consult Beyer *et al.* (1974), Wilbur (1983), and the entire issue of *Nucleic Acid Research,* Volume 10, No. 1 (1982), particularly Goad and Kanehisa (1982), Kanehisa and Goad (1982), Nussinov *et al.* (1982a, 1982b), and Auron *et al.* (1982).

1.2 Applications to human speech

A very active area for sequence comparison methods is speech research, where they are often referred to as *(dynamic) time-warping* or *dynamic programming* or elastic matching. A recent volume of selected reprints (Dixon and Martin, 1979) contains many papers explicitly making use of time-warping, and dynamic time-warping is now routinely incorporated into most modern systems concerned with recognition of speech or speaker. As the terminology above suggests, speech processing performs sequence comparison almost exclusively by dynamic-programming methods.

Physically, speech is transmitted as a pressure wave through the atmosphere. Conceptually, a preliminary step converts the received pressure wave into a vector function of time, $\mathbf{a}(t)$, where the multi-dimensional vector at t describes the pressure wave in a short interval surrounding t. (For example, the coefficients of the vector might give the energy in several frequency bands, or give the "linear predictor coefficients.") Before actual use, however, $\mathbf{a}(t)$ is converted into a sequence of vectors or phonetic symbols, which is the information actually dealt with, by sampling at regular times or by more elaborate means such as dividing the wave into natural segments, possibly followed by classification into phonetic categories.

Consider the much-studied problem of recognizing an isolated word selected from a limited vocabulary. (Note that this is far from the entire speech-recognition problem.) A very common approach is to provide the recognizer with one or several versions of each vocabulary word in sequence form. To recognize an unknown word, the recognizer converts it to sequence form, compares it to many stored sequences, and selects the one that matches best. Of course, a perfect match is unlikely. Compression and expansion may occur, because the rate of speaking increases and decreases substantially in normal speech from instant to instant, even within short words, in a manner that varies from occasion to occasion. This results in the need to stretch or squeeze the time axis locally, which explains the term "time-warping." In addition, deletion and insertion may occur, as when a speaker slurs "probably" to "prob'ly" or even "pro'lly" or expands the dictionary pronunciation "offen" into the spelling pronunciation "often" by inserting "t".

In this volume, speech processing is covered in Part II (see particularly Chapter 5 by Hunt, Lennig, and Mermelstein). For an introduction to the voluminous literature on sequence comparison in speech processing, the reader may consult the collection of reprints by Dixon and Martin (1979). For further information about applications to recognition of isolated words, see papers such as Itakura (1975), White (1978), Sakoe and Chiba (1978), Myers, Rabiner, and Rosenberg (1980), Rabiner, Rosenberg, and Levinson (1978), and White and Neely (1976). For information about recognition of connected speech, see papers such as Myers and Rabiner (1981a,b), Rabiner and Schmidt (1980), and Bridle and Brown (1979).

1.3 Applications to computer science

Another context where strings of symbols are of great importance is computer science, since they are so prominent during input and output (where the symbols may be letters, digits, etc.) and during internal functioning (where the symbols may be bits, bytes, words, records, etc.). A good review in the field of computer science citing many applications is given by Hall and Dowling (1980), though their interests are less sharply focussed than ours. One practical application in this field, often referred to as *string-correction* or *string-editing*, is the correction of human errors made at the input stage (the keyboard of a computer terminal or a keypunch machine). The earliest known work of this kind is Damerau (1964), Alberga (1967), and Morgan (1970), though these forerunners included neither the distance concept nor the dynamic-programming method. Wagner and Fischer (1974) sparked much interest in this problem among computer scientists, and there is now considerable literature. Another application, now in routine use, is the comparison of two computer files, perhaps slightly different versions of the same basic information, where each line or record of the file may be treated as a single symbol. A full description is provided in Hunt and McIlroy (1976), but published material contains only very brief references, as in Kernighan, Lesk, and Ossanna (1978) and Johnson (1980). A related application of a more elaborate algorithm is given in Gosling (1981).

One unusual feature has been introduced by Wagner and colleagues in connection with string-correction: the extension to include what are called *swaps* or *transpositions* (i.e., the interchange of adjacent elements in the sequence), a common human error. Their inclusion increases realism, though at the cost of complicating the algorithms and the mathematics.

1.4 Applications to codes and error control

Strings of symbols are used extensively to code information for transfer over a noisy channel such as radio, telegraph, microwave transmission, and so forth. Frequently just two symbols are used, which may be denoted by 0 and 1 regardless of physical implementation. Since noise in the channel introduces error into the received signal, many questions arise about detection and correction of error. The systematic study of different coding schemes and how well they work is referred to as (mathematical) coding theory (see, e.g., MacWilliams and Sloane, 1977), a large field which closely intertwines at some points with information theory.

Though the primary focus of coding theory has always been on substitution errors (in which the wrong symbols are received), it is also true that insertion and deletion errors (in which too many or too few symbols are received) have long been studied under the heading of synchronization error. However, it was only in the 1960s that attention was given to reconstructing the transmitted sequence (i.e., decoding) in the neighborhood of an insertion or deletion error (e.g., F. F. Sellers (1962), Levenshtein (1965a,b) and Ullman, 1966). Levenshtein (1965a) presented the earliest known use of a distance function that is appropriate in the presence of insertion and deletion errors. His distance function and generalizations of it play a major role in sequence comparison.

Computational methods like those described here seem to be of limited interest in coding theory, probably because a single received sequence (which is only a small part of one message) is virtually never important enough to merit the kind of extended attention that is given to a single sequence in other fields of application. Perhaps as a result, coding theory has not developed these ideas very much, despite Levenshtein's invention of a distance function. See, however, work by F. F. Sellers (1962), Calabi and Hartnett (1969a,b), Levenshtein (1965a,b; 1971), Mills (1978), Tanaka and Kasai (1976), Ullman (1966, 1967), and Kunze and Thierrin (1982).

1.5 Applications to gas chromatography

In gas chromatography (GC), the components of a gaseous mixture are separated from one another when the mixture is swept by a continuous stream of nonreactive carrier gas through a long, densely packed column of special material, from which the components emerge at different times over a period of minutes or hours. The mixture itself enters the column all at once, and progress of different components through the column varies according to the duration and frequency of attachment of the different gas molecules to the packing material. In pyrolysis GC, the gaseous mixture is formed by rapidly heating input material, possibly microorganisms or cells, to a high temperature in the stream of carrier gas.

The amount of material (other than carrier gas) emerging from the column may be measured by a flame ionization detector and plotted as a function of time, to yield a chromatogram. Alternatively, a mass spectrometer may be used to provide detailed information about the emerging material every few seconds. Reiner *et al.* (1979, 1978, 1969) have applied these methods to identification of microorganisms and tissue samples.

However, comparison of chromatograms is sometimes hindered by nonlinear distortions of the time axis, due to nonuniform packing of the column material and nonuniform flow of the carrier gas. In some situations it is feasible to compensate for this distortion by sliding one chromatogram relative to the other as different portions are compared, but "such successes notwithstanding, visual comparison of profiles over a light box proved to be a tedious experience" (Reiner *et al.*, 1979, p. 491). Furthermore, if the multidimensional information provided by the mass spectrometer is to receive effective use, this simple method is not possible.

Time-warping offers an approach to compensating for time distortion which is automatic and can handle as much of the mass spectrogram as the experimenter wishes. Reiner *et al.* (1979) includes preliminary results that illustrate this use of time-warping.

1.6 Applications to bird song

Bird song has been the subject of hundreds of papers in recent years. One reason for interest is that in some bird species, song is an important means of communication, it is learned by the young from their elders, and it has dialect-like variation from place to place. Such a phenomenon, which shares some important characteristics with human language, is rare among nonprimates. As explained in Section 8, the concept of "distance" between songs permits the useful application of quantitative methods in this field, and avoids serious difficulties encountered in a more straightforward approach to quantitative analysis. Chapter 6 by Bradley and Bradley illustrates the use of such distances, and compares simple linear time-warping with dynamic time-warping, to the advantage of the latter.

1.7 Other discrete-time applications: geological strata, tree rings, varves, and text collation

In geology, stratigraphic sequences can be obtained from drill holes and outcroppings. Each stratum may simply be classified as to type, or the classification may be supplemented by additional information about thickness, geochemical or mineral assay, fossils, etc. "Correlation" of different sequences is frequently necessary; i.e., it is necessary to determine which strata in one such sequence correspond geologically to which strata in another. In the simplest case, this can easily be accomplished by a method that geologists call "cross-association." However, the sequences sometimes contain gaps or repetitive parts, due to local nondeposition, eroded strata, faulting, and folding, and the problem becomes more difficult. Smith and Waterman (1980) describe how to apply sequence-comparison techniques to achieve correlation of sequences, and present two brief applications.

In principle, a great variety of other applications may exist similar to the one above, since strata are laid down by many natural processes. Two possible applications are dendrochronology and varve chronology. Dendrochronology refers to the science of dating based on tree rings (see, e.g., Ferguson, 1970). It requires the comparison of sequences obtained from different logs, most notably from logs of the bristle-cone pine. Dates possibly accurate to within one year have been obtained by this means further back than 5000 B.C. Varves are annual layers of sediment, generally clay, in which it is possible to count the years, typically due to variation between the summer and winter sediment (e.g., see Hörnsten (1970), Fromm (1970), and Tauber (1970). Varve chronology requires comparison of sequences from different cores or different locations. It has been extended further back than 10,000 B.C.

Collation of different versions of the same text offers another possible application, as discussed in Cannon (1976, 1973).

1.8 Other continuous-time applications: handwriting and evoked potential waves

Time-warping can be used in computer processing handwritten material such as signatures and line drawings. Typically, the penpoint is tracked during the act of writing or drawing, with pressure of pen on paper or height of pen as a third coordinate, so that the record consists of the penpoint trajectory as a function of time. Time-warping is important to compensate for nonlinear variations in writing speed. Handwriting recognition is addressed by Fujimoto *et al.* (1976) and Burr (1980), verification of signatures by Yasuhara and Oka (1977), and other related topics by Burr (1979, 1981).

There are many other potential applications of the same general kind, since many processes slow down and speed up irregularly. For example, evoked potentials from the brain ("brain waves" in response to a stimulus) are known to vary somewhat in timing from one occasion to another, and might constitute an application of time-warping.

1.9 Visual space-warping for human depth perception

While not an application in the usual sense, the automatic, nonconscious process by which the human visual system fuses images from the two eyes during binocular depth perception is a form of sequence comparison. For simplicity, consider only black and white images that lie along a narrow strip in the horizontal direction (i.e., parallel to the line connecting the eyes). Each image is then a pair of functions $f_i(x_i)$, where the value of f indicates darkness, i indicates left or right eye, and x_i indicates horizontal position. It is known that the process of fusion is achieved by matching values of x_1 with those of x_2 so that corresponding values of f_1 and f_2 agree. Suppose, as is usually true, that the two images differ only because the eyes view a three-dimensional scene from slightly different positions. Consider that which is visible as a single surface of varying darkness whose depth (i.e., distance from the eyes) also varies.

If depth varies only gradually, then fusion requires only continuous "space-warping" by means of compression and expansion. If depth has abrupt variations, as at the edge of a foreground object, then deletion and insertion are also required in order to deal with background regions blocked from one eye or the other by the foreground. Julesz (1971, 1978) has demonstrated that most people can experience a compelling sensation of depth when viewing "random-dot stereograms" that contain no clues to depth other than those depending on fusion, and has used such stereograms to investigate many aspects of vision.

2. HOW SEQUENCES DIFFER

Different exemplars of what is nominally the same sequence are often subject to differences, such as:

substitutions (also called replacements),
deletions and insertions (also called indels),
compressions and expansions,
transpositions (also called swaps).

For example, on some occasions molecules of the same protein differ slightly among organisms of the same species: such differences are sometimes part of evolutionary change. If the same message is transmitted repeatedly on the same communication channel, it will be received differently on different occasions due to noise on the channel. The same word or computer instruction typed on different occasions may differ occasionally, due to human error. In speech, different utterances of the same word differ far more than most people realize, due not only to regional accent, personal idiosyncrasy, function in the sentence, and stress placed on the word, but also speed of speech, mood of the speaker, and the inevitable unexplained factors that are lumped together under the heading of "random variation."

Differences should not be surprising. Rather, in many examples, it is remarkable that the differences are so small. That many individual exemplars of the "same" macromolecule or repeated utterances of the "same" word are sufficiently similar to be recognized as the same, *that* is remarkable. Such close similarity can be maintained only by complex and subtle mechanisms.

Nevertheless, differences are universal. The mechanisms that underlie these differences vary greatly from one situation to another, but there is a common thread. In all our application areas, similar sequences have only *local* differences. For example, the beginning of one sequence corresponds roughly to the beginning of the other sequence, despite the differences, and does not correspond to the end of the other sequence.

If we compare two sequences, the most obvious type of difference between them is the substitution of one element (or term, or component, or unit) for another at the same position in the sequence. Such differences are called *substitutions* or *replacements* and are very common in most of the application areas. There are other important types of differences, however. These include *deletion* of elements and *insertion* of elements, and also *compression* (of two or more elements into one element) and *expansion* (of one element into two or more elements). *Dealing with differences between sequences due to deletion–insertion, compression–expansion, and substitution is the central theme of sequence comparison.* Both deletion–insertion and compression–expansion can conceal the precise correspondence between elements of nominally identical sequences.

Compression–expansion arises primarily in connection with sequences obtained by time-sampling from continuous functions of time, as in the processing of human speech. Compression–expansion is closely similar to deletion–insertion—so similar, in fact, that the differences have often been overlooked, and are clearly described for the first time in Chapter 4. We will

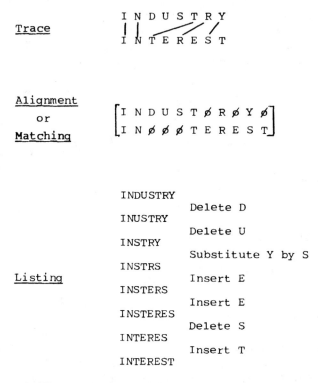

Figure 1. Three modes for analyzing differences between sequences.

describe the differences briefly at the end of Section 4, when we have the machinery to discuss them. As a result of the similarity, it is possible to extend much of the material dealing with either of these topics to the other. In fact, a synthesis incorporating both types of difference at the same time is not difficult and is potentially useful. For historical reasons, however, much of the following material is phrased only in terms of deletion–insertion or only in terms of compression–expansion, rather than in terms of both.

Another type of local change, which is treated in Chapter 7 by Wagner and in Lowrance and Wagner (1975), is a *transposition* or *swap*, which means the interchange of two adjacent elements of a sequence. These are particularly relevant to human keyboard errors, but are also observed in linguistics under the name "metathesis."

3. ANALYSIS OF DIFFERENCES: TRACE, ALIGNMENT, AND LISTING

A basic approach of this field is to analyze the total difference between two sequences into a collection of individual elementary differences. The following discussion is limited to the case where the elementary differences are substitutions, deletions, and insertions. It will be helpful to have a single name to cover both insertions and deletions, so we coin the word *indel* from *in*sertion–*del*etion for this purpose. It is sometimes convenient to treat the elementary differences as *elementary operations,* and to think of the operations as actively changing a *source* sequence into a *target* sequence, step by step. Indeed, the very names "substitution," "insertion," etc., connote this active operational viewpoint.

At least three different modes of presentation for such analyses are in use: a *trace*, an *alignment,* and a *listing.* These are illustrated in Fig. 1 and explained below. It is not generally realized that these are actually different modes of *analysis* as well as different modes of presentation, based on distinctions that will be made clear below. The distinctions are not usually exploited, however; this may be why they are not more commonly noticed.

In addition, for any given mode, many analyses of the same total difference are available: this is illustrated in Fig. 2, using traces. The multiplicity of

```
W  A  T  E  R
|        |
W  I  N  E
```

```
W  A  T  E  R
|  |     |  |
W  I  N  E
```

```
W  A  T  E  R
 \     / /
W  I  N  E
```

```
W  A  T  E  R

W  I  N  E
```

```
W  A  T  E  R
 / / / /
W  I  N  E
```

Figure 2. Different analyses of the same pair of sequences.

alternative analyses is one of the central difficulties of this field. A parsimony principle will be introduced in the next section to select the "shortest" or "least cost" analysis. Part of the multiplicity is an elaboration of the simple fact that

substitution of *a* by *b*

can also be analyzed as a deletion–insertion pair,

deletion of *a* and insertion of *b*.

Which analysis seems more plausible depends on the context.

In parts of at least two major application areas, macromolecules and computer science, listings correspond directly to the natural mechanisms by which sequences are believed to change, but alignments (for macromolecules) and traces (for computer science) are typically used for analysis and presentation. One reason is that the different kinds of analysis yield essentially the same results in many situations, but computation based on alignments or traces is far faster than computation based on listings. Thus listings are primarily of theoretical interest, while alignments and traces are used in practical work.

A *trace* from **a** to **b**, as illustrated in Fig. 1, consists of the *source* sequence **a** above and the *target* sequence **b** below, usually with lines from some elements in the source to some elements in the target. An element can have no more than one line, and the lines must not cross each other (specifically, the source elements with lines must correspond *in order* to the target elements with lines). The lines provide a correspondence, often partial, between source and target, and thereby supply a possible correspondence to fill the gap mentioned in the first paragraph of the chapter. If the elements connected by a line are the same, we refer to the pair of elements as an *identity* or a *continuation* (because the target element "continues" the source element); if they are different, the pair constitutes a *substitution*; in either case we call the pair a *match*. A source element having no line shows a *deletion*; a target element having no line shows an *insertion*; either of these is an *indel*. See Fig. 3 for a summary of this terminology. In the trace of Fig. 1,

I, N, T, and R of the source are continued,
Y of the source is substituted by S of the target,
D, U, and S of the source are deleted,
E, E, and T of the target are inserted.

An *alignment* (or *matching*) between **a** and **b**, as illustrated in Fig. 1, consists of a matrix of two rows. The upper row consists of the *source* **a**, possibly interspersed with *null characters* (or *nulls* for short), which are

TYPE OF OPERATION

DESCRIPTION	Match		Indel (Elementary operations)	
	No action	Substitution or Replacement	Deletion	Insertion
Trace	Identity or Continuation / Line joining identical elements	Line joining different elements	Element in source with no line	Element in target with no line
Alignment or Matching	Column with same element above and below	Column with different elements above and below	Column with ∅ below	Column with ∅ above
Listing	Equal source and target elements which can be traced to each other	Unequal source and target elements which can be traced to each other	Source element which cannot be traced to target	Target element which cannot be traced to source

MODE OF ANALYSIS

Figure 3. Description of operations.

represented here by ϕ. (Some authors use λ or — or a blank for the same purpose.) The lower row consists of the *target* sequence **b**, possibly interspersed with null characters. The column $[^\phi_\phi]$ of null characters is not permitted. A column $[^x_\phi]$ having ϕ below indicates *deletion*; a column $[^\phi_y]$ having ϕ above indicates *insertion;* either of these is an *indel*. A column $[^x_y]$ without ϕ is called a *match*; if $x \neq y$ it is a *substitution;* if $x = y$ it is called a *continuation.* See Fig. 3 for a summary of this terminology. The matches supply a correspondence, often only partial, between source and target, thereby supplying a possible correspondence to fill the gap mentioned in the first paragraph of the chapter. The alignment in Fig. 1 corresponds to the trace in the figure, and has the corresponding continuations, substitutions, etc.

As a mode of analysis, alignments are richer than traces in the sense that an alignment makes some order distinctions between adjacent indels which a trace does not, as Fig. 4 illustrates. There is a simple construction for transforming alignments into traces by abandoning the order restrictions (which transforms all the alignments of Fig. 4 into the trace there). The construction has two steps. First, copy the source and the target from the alignment, one above the other, i.e., copy the top and bottom rows of the alignment but omit the nulls. Then, for each match in the alignment, draw a line in the trace to connect the top and bottom elements it provides. The resulting trace has the same source, target, continuations, substitutions, deletions, and insertions as the alignment. (In mathematical terminology, the mapping from alignments to traces is many-to-one and onto.)

In a great many situations the order of adjacent insertions and deletions is unimportant. If so, traces can be used for analysis, or alignments can be used in a way that ignores the irrelevant aspects of the ordering. On the other hand, there are some situations in which the order of adjacent indels is important, and traces are not adequate. For example, this applies to the constraint that no more than F consecutive deletions and no more than G consecutive insertions are permitted. It also applies to the constraint that no more than K strings of indels are permitted, where each string consists either of consecutive deletions or of consecutive insertions (with no limit on the length of each string). These and other constraints are treated in Chapter 10.

As modes of presentation (rather than modes of analysis), traces and alignments each have their own advantages. Diagrams that are basically alignments, though often improved by inclusion of auxiliary information, have been favored in macromolecular applications. (For such practical use, a more inconspicuous null character than ϕ should be used, e.g., — or a blank). Traces have been favored in some computer science applications.

A *listing* from **a** to **b**, as illustrated in Fig. 1, consists of an alternating series of sequences and elementary operations, starting with the *source* sequence **a** and ending with the *target* sequence **b**, which satisfies the following consistency requirement:

<u>Trace</u>

$$
\begin{array}{cccc}
g & a & b & h \\
| & & & | \\
g & c & d & h
\end{array}
$$

$$
\begin{bmatrix}
g & a & b & - & - & h \\
g & - & - & c & d & h
\end{bmatrix}
$$

$$
\begin{bmatrix}
g & a & - & b & - & h \\
g & - & c & - & d & h
\end{bmatrix}
$$

$$
\begin{bmatrix}
g & a & - & - & b & h \\
g & - & c & d & - & h
\end{bmatrix}
$$

<u>Alignments</u>

$$
\begin{bmatrix}
g & - & a & b & - & h \\
g & c & - & - & d & h
\end{bmatrix}
$$

$$
\begin{bmatrix}
g & - & a & - & b & h \\
g & c & - & d & - & h
\end{bmatrix}
$$

$$
\begin{bmatrix}
g & - & - & a & b & h \\
g & c & d & - & - & h
\end{bmatrix}
$$

Figure 4. Six alignments which correspond to the same trace.

Listing consistency property. Two adjacent sequences in the listing must differ only as provided by the intervening elementary operation.

Thus a listing is, in effect, an algorithm that describes how to change the source into the target. Listings are primarily of theoretical interest, rather than for direct use in applications, and are a much richer mode of analysis than alignments or traces. Listings are also called *derivations* (see Chapter 14), and the same concept is invoked by *sequence of edit operations* (see Chapters 7, 8, 13, and 14).

For any two successive sequences in a listing, consider their elements, but exclude the element, if any, that was inserted or deleted by the intervening elementary operation. Then there is an obvious correspondence between the elements in one sequence and those in the other. By using such correspondences, it is possible to trace forwards and backwards through the listing. Any source element that cannot be traced all the way to the target sequence is called an *overall deletion;* any target element that cannot be traced all the way to the source is called an *overall insertion*; both, of course, are *overall indels*. (The modifier "overall" is used to indicate a distinction, which did not exist for alignments and traces, between elementary operations that are components of the listing and those that are the net result of it.) For any element in the source sequence which can be traced all the way to the target sequence, consider the pair consisting of the source element and its corresponding target element. This pair is called an *overall match*; if the two elements are equal, it is called an *overall identity* or an *overall continuation*; if they are unequal, it is called an *overall substitution.* See Fig. 3 for a summary of this terminology. The overall matches supply a correspondence, often only partial, between source and target, and thereby supply a possible correspondence to fill the gap mentioned in the first paragraph of the chapter. The listing in Fig. 1 corresponds to the alignment and the trace shown there, and has the corresponding continuations, substitutions, etc.

The operations found formed by tracing forward and backward from any single element involve at most one insertion, an unlimited number of substitutions, and at most one deletion. The order of these operations within the listing is important, of course: None of them may precede the insertion or follow the deletion, and substitution $A \to B$ can immediately precede but not immediately follow the substitution $B \to C$. On the other hand, if the order of operations that act on different positions is changed, then it is a simple matter to adjust the intervening sequences so that the listing-consistency property is preserved.

As a practical mode of presentation, listings are awkward and have been little used. As a mode of analysis, listings have theoretical importance because it is possible to generalize them much more broadly than alignments and traces, and because they correspond to plausible underlying mechanisms in several major applications. They are richer than alignments because listings permit many successive changes to be made in a single position and alignments permit only one. In addition, listings make distinctions based on the order in which changes are made, and alignments do not. There is a simple natural construction for transforming listings into alignments by abandoning these distinctions. The first step is to copy the source and target from the listing, spacing them so that each overall match is lined up vertically and forms a column. Then a null character is placed in the blank space below each overall deletion and in the blank space above each overall insertion. The resulting two rows are an alignment which has the same source, target, continuations, substitutions, etc., as the listing. (In mathematical terms, the mapping from listings to alignments is many-to-one and onto.) Listing distance underlies the approach to transposi-

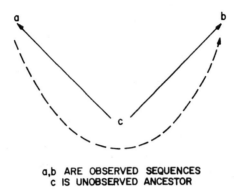

a,b ARE OBSERVED SEQUENCES
c IS UNOBSERVED ANCESTOR

Figure 5. Usual evolutionary situation.

tions used in Chapter 7, though in practice a simpler equivalent distance based on generalized traces is used.

In practice part of the richness of listings is seldom available, namely, the possibility of making several successive changes in one position. Under assumptions that are usually applicable, listings with more than one change in a position will not be selected for use, in accordance with the parsimony principle to be introduced in Sec. 4.

A listing resembles a description of the individual changes that might have taken place during some tangible process of change, such as evolution from one sequence to another, or processing by the nervous system of a planned sequence of motor activities, through several stages from formation to execution. Although this resemblance is legitimately exploited in some applications, it is important that great caution be used in any attempt to interpret listings in tangible fashion, for several reasons:

1. Listings may be formed and used without regard to whether the situation makes it sensible to think of changes as taking place in a series of listing-like steps. For example, in speech recognition it is not easy to make such an interpretation, yet distances (see the next section) based on listings or other analyses are indubitably very useful.

2. During an historical process such as evolution, more than one change in a position is a very real possibility, but (as explained above) the parsimony principle will usually prevent the listing from having this.

3. The operations occur in some definite order in the tangible process of change, and they occur in a definite order in the listing, but there seldom is evidence which permits the listing order to reconstruct the tangible process order.

4. In the usual evolutionary situation, sequence **a** is not the ancestor of sequence **b**, but instead both **a** and **b** are descendants of some unknown

common ancestor **c** (see Fig. 5). If a tangible interpretation is given to the steps from **a** to **b** (dotted path in figure), some of them presumably are the steps from **a** to **c** *in reverse* (insertions become deletions, and deletions insertions, e.g.) while others are the steps from **c** to **b** (in normal manner).

4. LEVENSHTEIN AND OTHER DISTANCES

Levenshtein (1965a) introduced two closely related concepts of distance $d(\mathbf{a}, \mathbf{b})$ from one sequence **a** to another sequence **b**. One of these is the smallest number of substitutions and indels required to change **a** into **b**. The other is the smallest number of indels (no substitutions permitted) required to change **a** into **b**. For example, it is possible to change **a** = INDUSTRY into **b** = INTEREST using seven substitutions and indels, as shown in Fig. 1. However, it is also possible to change **a** to **b** using only six. Two of the many ways in which this can be done are shown in Fig. 6, as listings, alignments, and traces. It can be shown that six is the minimum, so one kind of Levenshtein distance from INDUSTRY to INTEREST is six. It turns out that the minimum number of indels to make the same change is eight, so the other kind of Levenshtein distance is eight.

These concepts are only two of a wide class of meanings that are given to the word "distance" in sequence comparison. In preparation for discussing these many meanings, we rephrase the first preceding definition in terms of four parts.

1. Let the elementary operations be substitutions and indels.
2. Consider all listings from **a** to **b** based on the elementary operations.
3. Let the length of each listing be the number of elementary operations it contains.
4. Then the distance is the minimum length of any listing.

The most obvious way to achieve the other definition in this framework is to reduce the elementary operations to indels only, and this is very much within the larger spirit of the field. Alternatively, we can leave the elementary operations unchanged, but redefine the length of a listing to be

$$(\text{number of indels}) + w \,(\text{number of substitutions}), \qquad \text{with } w \geq 2.$$

If $w > 2$, then it is always shorter for a listing to use a deletion–insertion pair in place of a substitution, and if $w = 2$ it is as short.

Many meanings of distance in sequence comparison can be described using four parts like those above.

1. *Elementary differences.* Each definition is based on some set of *elementary differences* (also called *elementary operations*). In the

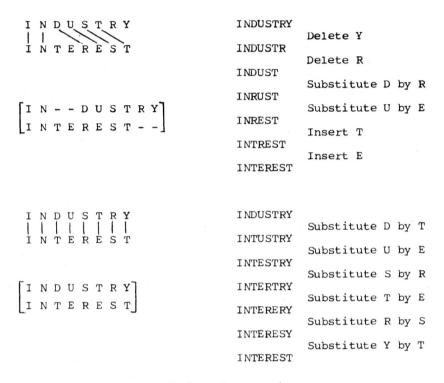

Figure 6. Some shortest analyses.

following discussion, the elementary differences are substitutions and indels. Elsewhere, other elementary differences are added to the list, such as transposition, or compression and expansion.

2. *Acceptable analyses.* Each definition is based on some class of *acceptable analyses* of the difference between **a** and **b**, where the acceptable analyses are based on the elementary differences. In the following discussion, acceptable analyses may be all listings (from **a** to **b**), or all alignments, or all traces. Other possibilities are used elsewhere, such as a constrained set of alignments (e.g., those in which no more than two consecutive deletions and no more than two consecutive insertions occur), or all time-warpings.

3. *Length.* Each definition is based on some function for calculating the *length* of an analysis. The function may incorporate a system of weights or parameters. The "simple length" function (described below) is used here and in most applications based on substitutions and indels. Other

length functions, including those used with compression–expansion, are discussed below.

4. *Parsimony principle.* In general, the distance $d(\mathbf{a},\mathbf{b})$ is defined to be the minimum length of any acceptable analysis of the difference between $\mathbf{a}$ and $\mathbf{b}$.

Algorithms for performing the minimization referred to in (4) are a central theme in sequence comparison. As different meanings of length are considered, different algorithms may be needed, and algorithms to do the minimization are almost always an important consideration. The computing time required may be important, and often computer storage requirements are too. Algorithms are discussed later in the chapter.

Suppose the elementary differences consist of substitutions and indels. One widely used length function, which we call *simple length,* is based on weights for the elementary differences. Specifically, we assume that each possible substitution and indel has a given weight ≥ 0. The simple length of an analysis is the sum of the weights of the elementary operations it contains (in case of a listing, this refers to the operations in the individual steps, not the overall operations). If all the weights are 1, then the simple length function simply counts the number of elementary operations, and we obtain one of Levenshtein's original distances. If the weight of every indel is 1 and the weight of every substitution is $w \geq 2$, we obtain the other. We use the following notation for the weights:

Substitution $x \rightarrow y$	$w(x, y)$
Deletion of x	$w(x, \phi)$ or $w_{\text{del}}(x)$ or $w_{-}(x)$
Insertion of y	$w(\phi, y)$ or $w_{\text{ins}}(y)$ or $w_{+}(y)$

When the null character ϕ is permitted as an argument of w, we refer to *extended* w and use the convention $w(\phi, \phi) = 0$. If the alphabet of possible sequence elements is finite, the weights may be arranged into a matrix (see Fig. 7) where x corresponds to the row and y to the column.

Any given extended w (together with a class of acceptable analyses) leads to a distance function d. For any elements x and y, it is possible to compare

$$w(x, y) \quad \text{with} \quad d(x, y),$$

$$w(x, \phi) \quad \text{with} \quad d(x, \phi)$$

$$w(\phi, y) \quad \text{with} \quad d(\phi, y),$$

since any element is also a length 1 sequence and the null character is the length 0 sequence. It may turn out that w and d are equal in all such comparisons: If so,

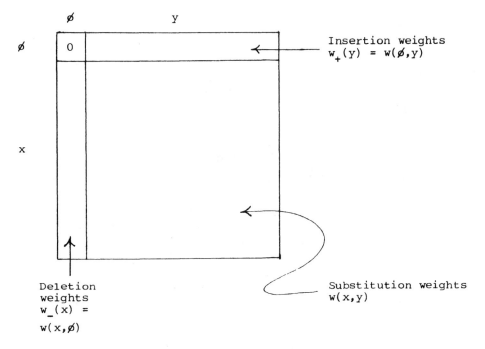

Figure 7. Extended matrix of weights.

we say that *d agrees with w.* When this happens, we can regard *w* as being part of *d,* namely, *w* is *d* restricted to sequences of length 1 or 0. The process of deriving *d* from *w* can be regarded as a process of *extending* the definition of distance from the restricted case of sequences of length $\leqq 1$ to all sequences. We shall refer to this approach as *extension of d from short sequences,* while the approach we have used will be called *definition of d from weights.* When the extension approach is used, the same letter (often *d*) is used for both *w* and *d,* and agreement of *d* with *w* becomes a form of self-consistency for *d.* In the extension approach, it is necessary to make assumptions that ensure this self-consistency. Further discussion may be found in Chapter 10.

We shall use *Levenshtein distance* to cover any distance based on the simple length function. (Existing usage is not entirely clear about the meaning of this phrase, but this definition seems to match usage pretty well.) If the simple length function is used with all listings, or with all alignments, or with all traces as the acceptable analyses, in principle we obtain three different distance functions. However, alignments and traces yield the same distance function in this context, so we only refer to *simple listing distance* and *simple alignment distance.* Even these often turn out to be the same (see Chapter 10).

In the mathematical literature, the word "distance" is ordinarily used to indicate a function *d* which satisfies the *metric axioms:*

1. *Nonnegative property.* $d(\mathbf{a}, \mathbf{b}) \geq 0$ for all $\mathbf{a}$ and $\mathbf{b}$;
2. *Zero property.* $d(\mathbf{a}, \mathbf{b}) = 0$ if and only if $\mathbf{a} = \mathbf{b}$;
3. *Symmetry.* $d(\mathbf{a}, \mathbf{b}) = d(\mathbf{b}, \mathbf{a})$ for all $\mathbf{a}$ and $\mathbf{b}$;
4. *Triangle inequality.* $d(\mathbf{a}, \mathbf{b}) + d(\mathbf{b}, \mathbf{c}) \geq d(\mathbf{a}, \mathbf{c})$ for all $\mathbf{a}, \mathbf{b}, \mathbf{c}$.

Although we do not restrict the word "distance" in this way, we are interested in these important properties. It is common to make suitable assumptions on w in order to achieve the metric axioms for the distance d which is derived from w. Of course, the assumptions needed for this purpose depend on which type of sequence distance is being used. We remark that the triangle inequality always holds for simple listing distance d, whether or not it holds for w, but the same is not true for simple alignment distance.

Although it is generally desirable to use a function d satisfying the axioms above, there are exceptions. A significant one occurs in speech recognition: When comparing a known "template" utterance to an unknown target utterance, there is empirical evidence and a rationale for use of an asymmetric function d (see, for example, Myers *et al.,* 1980). For another, consider transmitting text corrections over a telecommunication channel, as discussed in Goldberg (1982). The length of the corrections signal rather than the number of corrections is to be minimized, and the correction scheme may need to mark which sequence elements are correct in addition to indicating changes. If so, the space required by the marks is represented by positive values for $w(x, x)$ and $d(\mathbf{a}, \mathbf{a})$.

Other length functions. The simple length function described above covers most applications based on indels. There are, however, at least three other types of length functions in use. The most important one is used in connection with time-warping, where compression–expansion plays a central role. It is described last here, because it requires the most space. Another directly generalizes the simple length function, to meet an applied need. Finally, one very unusual length function, used for an application that does not fit the framework above, may be found in Chapter 3.

One reason for using distance in macromolecular applications, as described below in Sec. 9, is that it bears an approximate inverse relationship to the likelihood of one sequence changing into another in some circumstances. In some applications it may be possible for whole strings to be deleted or inserted, with the likelihood decreasing only slowly as the number of elements in the string increases (so that a single deletion of two adjacent elements is far more likely than two deletions of separated elements). To achieve a reasonable correspondence between distance and likelihood in this situation, it is appropriate to use a length function in which the dominant term is a weight for each consecutive string of deletions or insertions, and in addition much smaller weights used (as in simple length) for each single-element deletion or insertion.

Further information about this length function may be found in Sec. 7 of Chapter 10.

In applications using compression–expansion, where sequence comparison is generally referred to as time-warping, the concept of simple length is, of course, not meaningful. A full discussion of the length functions used with time-warping is given in Chapter 4. Here we briefly illustrate the ideas. It was noted in Section 2 that compression–expansion is extremely similar to deletion–insertion. It turns out that the most important differences lie in the nature of the length functions that are appropriate to use. In particular, let us consider expansion of one element in **a** (say, a_6) into two adjacent elements of **b** (say, b_6, b_7). It is possible to achieve the same change with substitutions and indels, e.g., insertion of b_7 together with substitution of a_6 by b_6. (The substitution may be omitted if it happens that $a_6 = b_6$.) The most natural cost to assign to the combination of insertion and substitution is the simple length involved, namely,

$$w_{\text{ins}}(b_7) + w(a_6, b_6).$$

On the other hand, the most natural cost to assign to the expansion is

$$\tfrac{1}{2} w(a_6, b_6) + \tfrac{1}{2} w(a_6, b_7) + w_{1,2},$$

where the former terms are a penalty depending upon how different a_6 is from the elements it expanded into, and the latter term is a penalty for distorting the time axis by matching 1 element with 2 elements. Extensive applications of the time-warping concept may be found in the speech-processing literature: see Sec. 1.2 and Chapters 4 and 5.

5. CALCULATING DISTANCE: THE BASIC ALGORITHM

Distance from **a** to **b** is defined as the minimum length of any acceptable analysis of the difference between **a** and **b**. Efficient methods for performing the optimization to find the distance and the corresponding optimum analysis are a central topic of sequence comparison. In some applications it is the distance, and in others it is the analysis, that is of primary interest. If there happen to be several different optimum analyses, we may wish to know them all.

One concept for finding the minimum, which can be described as a form of dynamic programming, has a remarkable history of multiple independent discovery and publication:

Vintsyuk, 1968, U.S.S.R.; speech processing;
Needleman and Wunsch, 1970, U.S.A.; molecular biology;

Velichko and Zagoruyko, 1970, U.S.S.R.; speech processing;
Sakoe and Chiba, 1970, 1971, Japan; speech processing;
Sankoff, 1972, Canada; molecular biology;
Reichert, Cohen, and Wong, 1973, U.S.A.; molecular biology;
Haton, 1973, 1974a, 1974b, France; speech processing;
Wagner and Fisher, 1974, U.S.A.; computer science;
Hirschberg, 1975, U.S.A.; computer science;

and possibly others. (While these items are listed in chronological order, the question of priorities has many complications, since some of the descriptions are marred in one way or another, and since some discoveries were cited or submitted years prior to the publication shown here.) While the algorithms described in these papers do differ somewhat, they can all be seen as flowing out of the same concept. Different algorithms remain useful today for different situations. We describe here the simplest, most basic algorithm of this kind. Many other related algorithms appear in Chapters 2, 3, 4, 7, 8, 9, 10, 14 and throughout the literature.

One essential common element in all the papers cited is reliance on distances akin to alignment or trace distance, as opposed to listing distance, since use of the latter would require far slower calculations. In many cases, listing distance is the conceptual starting point, though this fact may not be clearly brought out; in such cases, the mathematical equivalence of listing distance with alignment and trace distance is a vital foundation for practical sequence comparison. By way of contrast, consider Chapter 7, in which transpositions are introduced as elementary operations. Listing distance is the conceptual starting point, and in that context listing distance is *not* mathematically equivalent to trace distance as defined above, but to a complicated generalization of trace distance. This leads to great complications in the algorithm, which illustrates the importance of our comment.

The first stage of the following algorithm finds simple alignment distance, and an optional second stage can be used to find an optimum alignment if desired. If the two sequences have length m and n, then the dominant portion of the computing time for this algorithm is proportional to mn, and the dominant portion of memory space is also. If $m = n$, this is n^2, so the algorithm requires computing time that is quadratic in the length of the sequences.

For the sake of concreteness, the following description includes some details that are intended as an aid to understanding, not necessarily for actual use. Let the two sequences be **a** and **b**, with entries a_i and b_j, and having lengths m and n. Substitution weights are given by $w(x, y)$, deletion weights by $w(x, \phi)$, and insertion weights by $w(\phi, y)$. Let $\mathbf{a}^i$ and $\mathbf{b}^j$ indicate the initial segments $a_1 \cdots a_i$ and $b_1 \cdots b_j$ (including the case where the initial segment has length 0, namely $\mathbf{a}^0$ or $\mathbf{b}^0$, which is simply the empty sequence). Of course, $\mathbf{a}^m = \mathbf{a}$ and $\mathbf{b}^n = \mathbf{b}$.

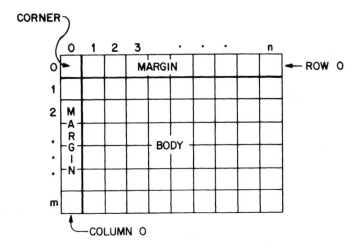

Figure 8. Array used to perform dynamic programming algorithm.

In the first stage, we work forward recursively, and find distances $d(a^i, b^j)$ for successively larger i and j, finally reaching $d(a^m, b^n) = d(a, b)$, which is the distance desired. One procedure is to record these values in an array like that shown in Fig. 8, where we start from the $(0, 0)$ cell in the upper left-hand corner, and move toward the (m, n) cell in the lower right. If the optional second stage is to be carried out, a necessary preparation during the first stage is to record certain pointers. The pointers are described in abstract form in the equation below, and portrayed visually as arrows in the array. The optional second stage, often referred to as "backtracking," consists of tracing along the pointers or arrows.

The recursion of the first stage starts from the obvious value $d(a^0, b^0) = 0$, which provides the entry for the $(0, 0)$ cell. The recurrence equation for a typical cell (i, j) is based on the values in three predecessor cells, $(i - 1, j)$, $(i - 1, j - 1)$, and $(i, j - 1)$. However, if $i = 0$ or $j = 0$, two predecessor values involving negative values of i or j are not used. No cell can be processed until its predecessors have been processed, but otherwise the order of calculation makes no difference.

Now we give the recurrence equation that is used to calculate the (i, j) value, and the corresponding pointer equation that serves as preparation for stage two. The intuitive meaning will be described shortly:

$$d(a^i, b^j) = \min \begin{cases} d(a^{i-1}, b^j) & + w(a_i, \phi) & (\text{``deletion of } a_i \text{''}), \\ d(a^{i-1}, b^{j-1}) & + w(a_i, b_j) & (\text{``substitution of } a_i \text{ by } b_j \text{''}), \\ d(a^i, b^{j-1}) & + w(\phi, b_j) & (\text{``insertion of } b_j \text{''}), \end{cases}$$

$$\text{pointer } (i, j) = \begin{cases} (i - 1, j) & \text{or} \\ (i - 1, j - 1) & \text{or} \\ (i, j - 1) \end{cases}, \quad \begin{array}{l} \text{where choice depends on} \\ \text{the minimizing entry above.} \end{array}$$

The choice in the pointer equation simply records which line was selected as the minimum in the recurrence. In the minimization, any line containing a negative value of i or j is simply ignored. The intuitive meaning of the recurrence formula is that the shortest alignment between $\mathbf{a}^i$ and $\mathbf{b}^j$ may be formed by choosing the shortest of the following three alternatives:

i. use the shortest alignment between $\mathbf{a}^{i-1}$ and $\mathbf{b}^j$, and also delete a_i;
ii. use the shortest alignment between $\mathbf{a}^{i-1}$ and $\mathbf{b}^{j-1}$, and also substitute a_i by b_j,
iii. use the shortest alignment between $\mathbf{a}^i$ and $\mathbf{b}^{j-1}$, and also insert b_j.

One procedure for evaluating the recurrence equation is to start by recording some auxiliary values in the cell, as shown in Figs. 9 and 10. For a body cell, these are the three w values, namely, $w(a_i, \phi)$, $w(a_i, b_j)$, and $w(\phi, b_j)$. For a margin cell, only one of these entries is used, of course. Next the d values from the preceding cell(s) are added to the corresponding w values, as indicated by double arrows in Fig. 11, to form sums. For body cells, there are three preceding cells, three w's, and three sums, while for margin cells there is just one of each. Finally, the minimum of the sums is recorded as the new d value. The pointer equation records which one of the sums is minimum. Pictorially, the pointer is shown as a heavy single arrow in Fig. 11; the other two possible heavy single arrows are shown in dotted form. If more than one of the sums provides the minimum value, this can be indicated in the pointer equation by using a set of two or three pairs on the right-hand side, and pictorially by using two or three arrows.

To prove the validity of the recurrence equation, consider a shortest alignment between $\mathbf{a}^i$ and $\mathbf{b}^j$, whatever it may be. Since an alignment may not contain a column of null letters, the right-hand column must be either $[\begin{smallmatrix} a_i \\ \phi \end{smallmatrix}]$ or $[\begin{smallmatrix} \phi \\ b_j \end{smallmatrix}]$ or $[\begin{smallmatrix} a_i \\ b_j \end{smallmatrix}]$. (If $i = 0$ or $j = 0$, the entry a_i or b_j does not exist, and only one of these cases is possible.) Consider the first case. If we delete the last column from the alignment, the remaining alignment between $\mathbf{a}^{i-1}$ and $\mathbf{b}^j$ must be a shortest possible one, for if there were a shorter one it could be substituted, and this would yield a shorter alignment between $\mathbf{a}^i$ and $\mathbf{b}^j$. The simple length of the entire alignment is the length of the first portion, which must be $d(\mathbf{a}^{i-1}, \mathbf{b}^j)$, plus the simple length of the last column, which is $w(a_i, \phi)$. This gives us the first line

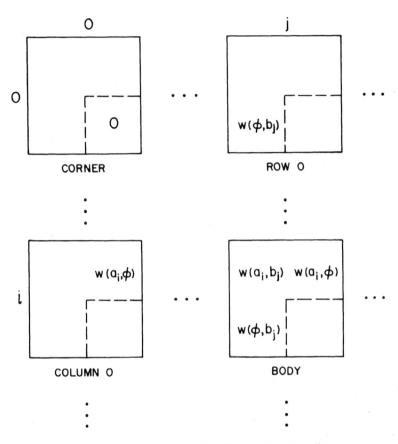

Figure 9. Arrangement of auxiliary values in cells.

in the formula. Similar reasoning in the other two cases gives the other two lines in the formula. Thus $d(a^i, b^j)$ must be one of the three alternatives shown, and since d was defined to be the minimum possible alignment length, it must be the minimum of the three alternatives.

The second-stage "backtracking" calculation, which finds the optimum alignment(s), is very simple. In pointer form, this consists of starting with the final cell (m, n), and using the pointer information repeatedly to obtain a path all the way to cell $(0, 0)$. Pictorially, as illustrated in Fig. 12, this means following along the arrows. In either case, an alignment is read off from the path in reverse

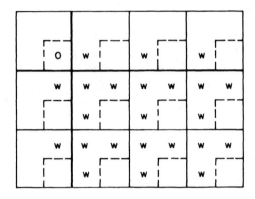

Figure 10. Array before recursion.

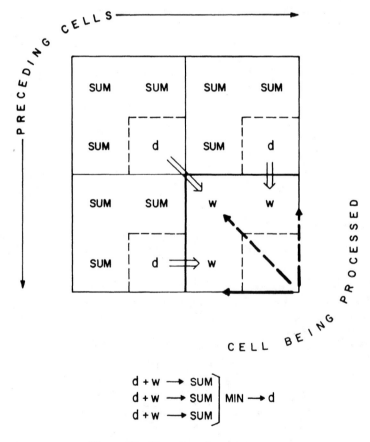

Figure 11. One step of recurrence.

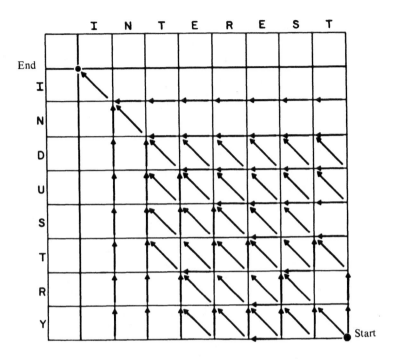

Figure 12. Array used for tracing backwards.

order in the obvious way, which we phrase in terms of the arrows: A diagonal arrow in cell (i, j) means that a_i is substituted by b_j and $[^{a_i}_{b_j}]$ is placed in the alignment; a horizontal arrow in column j means insertion of b_j and $[^{\phi}_{b_j}]$ is placed in the alignment; and a vertical arrow in row i means deletion of a_i and $[^{a_i}_{\phi}]$ is placed in the alignment. If all optimum alignments are desired, all possible alternative paths should be traced out and used. Two of the alignments that can be formed from Fig. 12 in this way are shown in Fig. 6.

6. COMPUTATIONAL COMPLEXITY OF SEQUENCE COMPARISON

Computational complexity has been investigated for several problems in sequence comparison. (The computational complexity of a problem means the *minimum possible computation time* that is needed to solve a worst-case example of that problem.) Consider the problem of calculating the optimal alignment between two sequences of length n. Since the algorithm given in the preceding section requires time proportional to n^2, we know that the computational complexity is not greater than proportional to n^2. Using fairly general assumptions, Wong and Chandra (1976) proved that the computational complexity is in fact proportional to n^2.

However, computational complexity is often very sensitive to the precise assumptions. Chapter 14 by Masek and Patterson presents an algorithm that requires only time proportional to $n^2/\log(n)$, based on two mild assumptions: that the sequence elements come from a finite alphabet, and that the weights (of substitutions and indels) are all rational. Asymptotically, this method is the fastest one currently available, but it has not been proved optimal. As a practical matter, the authors point out that their method does not become faster than a simple method similar to that in the preceding section until $n \geqq 262,419$. When applied to sequences of length m and n with $m \leqq n$, this method uses time proportional to $mn/\min(m, \log(n))$.

Consider the same problem (i.e., calculating an optimal alignment) in the case where indels all have weight 1 and substitutions either are not permitted or have weight >2 (so that it is always more economical to delete and insert than to substitute). Then it is well known and easy to see that calculating an optimal alignment is equivalent to finding the longest common subsequence of the two sequences. Consider algorithms for the latter purpose that are restricted in the type of comparisons they can make between sequence elements. Aho, Hirschberg, and Ullman (1974) have shown that the number of comparisons of the "equal–not equal" type must be proportional to at least n^2 if the alphabet has unrestricted size, and proportional to at least ns if the alphabet has size s and $s < n$. Hirschberg (1978) has shown that the number of comparisons of the "less than–equal to–greater than" type must be proportional to at least $n \log(n)$. Related problems, such as finding the longest common subsequence of N sequences, finding the shortest common supersequence, and finding the longest common *consecutive* subsequence (i.e., substring), are surveyed in Chapter 12 by Hirschberg.

Some computational complexity results are also known for situations in which a sequence is compared not with another sequence, but with a large class, possibly infinite, of sequences defined by a "grammar." The problem is to find the grammatical sequence that is closest to the given sequence and/or to find the distance involved. A survey of computational-complexity results for this problem is given in Chapter 13 by Wagner.

7. HOW SEQUENCE COMPARISON IS USED

Sequence comparison can be used to answer certain clearly drawn questions, as we describe below. As with other methods, however, the most interesting current uses of sequence comparison are hard to describe in that crisp a fashion. One reason for this is that sequence comparison is involved in drawing up questions as well as in answering them. Another reason is that interesting new uses tend to spawn new methods, whose implications, potentialities, and limitations are not immediately clear.

Sequence comparison yields two kinds of information, distances and optimum analyses (e.g., alignments, traces, or listings). In some applications emphasis is on the distances, and in others on the optimum analyses. Sometimes the two kinds of applications are closely intertwined. A simple basic application emphasizing distance is a decision that two molecules (or other sequences generated by an evolutionary process) are homologous. If the distance is small enough, we may conclude that such closeness could not have occurred by chance. (However, a large distance does not imply that they are nonhomologous; it may merely be that their common ancestry was very long ago.) Of course, to set the threshold requires information about how large a distance might come about by chance. This can be gained both by Monte Carlo experiments, which are often used, and by mathematical investigation. (Monte Carlo investigations may be found in Sect. 6 of Chapter 2, Sect. 5 of Chapter 6, and Sect. 5 of Chapter 10. Mathematical investigations are given in Chapter 16 by Deken and Chapter 15 by Chvátal and Sankoff.) A simple basic application emphasizing optimum analysis is to reconstruct as well as possible (i.e., to estimate) the homology between two homologous molecules **a** and **b**. In other words, we want to identify which portions and which elements of **a** correspond to which portions and which elements of **b**, in the sense that corresponding portions or elements descend from a single portion or element of the ancestor molecule. Not only is this information of considerable evolutionary interest, but it can also have direct biochemical value, since a portion with a particular type of activity on one molecule is likely to correspond to a portion with a similar activity on another. One way of solving this clearly drawn problem is to use the homology provided by the shortest alignment (or trace) from **a** to **b**. For reasons explained in Section 9, this is akin to maximum-likelihood estimation of the homology.

Sometimes a given molecule **a** is believed not to be homologous as a whole to another molecule, but merely to contain somewhere a portion **a'** that is homologous to a specified portion **b'** (of some molecule **b**). Then a new problem arises, that of finding which portion **a'** of **a** has the smallest distance to **b'**, what the distance between **a'** and **b'** is, and what the optimum (say) alignment between them is. (A similar goal is referred to as "word-spotting" in speech research.) Chapter 2 by Erickson and Sellers provides two elegant and practical answers, and obtains three interesting applications. In one the authors discover that two plant proteins are approximately cyclical permutations of each other, a phenomenon that had not been previously observed. In another, they tentatively find in the middle of a protein the "translocation signal," i.e., the sequence of approximately 15 amino acids which signals that this protein is to be translocated through the endoplasmic-reticulum membrane. In the third, they

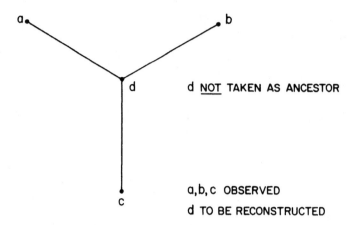

Figure 13. Non-time-oriented family tree.

discover that two repeating sequences of "satellite DNA" contain very similar pieces that were not previously noted. This work illustrates the fact mentioned earlier, that sequence comparison helps in drawing up questions as well as in answering them. If Erickson and Sellers had not been actively working with sequence-comparison methods, it is not at all clear that they would have conceived the *question* described in this paragraph.

A similar but more general question arises when neither portion **a′** nor **b′** is specified in advance, i.e., the search is for a portion somewhere in **a** and a portion somewere in **b** that are homologous. This problem differs significantly from the one above, because if we simply search for **a′** and **b′** having smallest distance, we will surely end up with a meaningless comparison between very short portions, such as a single element sequence **a′** and a single element sequence **b′** that happen to be identical. To cure this difficulty, it is necessary to evaluate a homology not only by the distance between the portions but also by the number of elements it contains, so that a homology containing many elements will be favored over one containing few. Methods are presented in Chapter 10, Smith and Waterman (1981), and Sellers (1980).

Suppose we wish to establish the homology among three molecules **a, b, c**. Since any three homologous molecules must be connected by a non-time-oriented family tree like that of Fig. 13, a satisfying answer would include reconstructing the internal sequence **d** as well as the homology of **d** with each of **a, b, c**, which indirectly provides the homology among **a, b, c**. (In Fig. 13, the common ancestor is *not* assumed to be **d**, but may occur anywhere on the three edges. Wherever it occurs, time proceeds outward from that point along the edges of the tree. We do not concern ourselves with where the common ancestor is located.) One approach to this problem (due to Sankoff, 1975, and Sankoff *et* *al.*, 1973, 1976) is to define a generalized alignment among **a**, **b**, **c**, and **d**. This means a matrix, such as

$$
\begin{bmatrix}
a_1 & a_2 & \phi & a_3 & a_4 \\
\phi & b_1 & b_2 & b_3 & \phi \\
c_1 & c_2 & \phi & \phi & c_3 \\
d_1 & d_2 & \phi & d_3 & d_4
\end{bmatrix},
$$

in which each row consists of one of the four sequences, possibly padded by null characters ϕ, and in which no column consists entirely of null characters. The length of a generalized alignment is defined to be

$$\text{length}(\mathbf{a}, \mathbf{d}) + \text{length}(\mathbf{b}, \mathbf{d}) + \text{length}(\mathbf{c}, \mathbf{d}),$$

where each term refers to the simple length of the ordinary two-row alignment formed by the appropriate two rows of the matrix. The generalized distance among **a**, **b**, and **c** means the minimum length, using any **d**, of any generalized alignment among **a**, **b**, **c**, and **d**. The sequence **d** and generalized alignment, which provide the minimum length, form a satisfying solution to the question.

An algorithm to find the optimum generalized alignment is practical though time-consuming. It is similar to the algorithm described in Sec. 5 above, but a three-dimensional array is used in place of the two-dimensional array. Specifically, define D_{ijk} as the generalized distance among the partial sequences $a_1 \cdots a_i$, $b_1 \cdots b_j$, and $c_1 \cdots c_k$. The recursive step to calculate D_{ijk} from the preceding values contains inner minimizations that have no counterpart in the algorithm of Sec. 5:

$$
D_{ijk} = \min \begin{cases}
D_{i-1,\,j-1,\,k-1} & + \min_{d}\,[w(a_i, d) + w(b_j, d) + w(c_k, d)], \\
D_{i,\,j-1,\,k-1} & + \min_{d}\,[w(\phi, d) + w(b_j, d) + w(c_k, d)], \\
D_{i-1,\,j,\,k-1} & + \;\cdots, \\
D_{i-1,\,j-1,\,k} & + \;\cdots, \\
D_{i,\,j,\,k-1} & + \min_{d}\,[2w(\phi, d) + w(c_k, d)], \\
D_{i,\,j-1,\,k} & + \;\cdots, \\
D_{i-1,\,j,\,k} & + \;\cdots\;.
\end{cases}
$$

The optimum sequence **d** and the optimum generalized alignment can be found by backtracking after the recursion has been completed.

Suppose there are more than three known homologous molecules for which we wish to obtain the corresponding information. Then the actual evolutionary

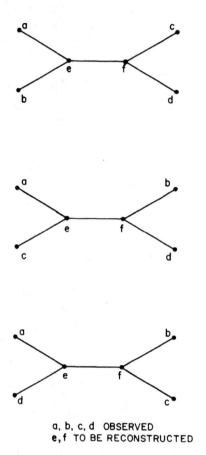

a, b, c, d OBSERVED
e, f TO BE RECONSTRUCTED

Figure 14. Three different non-time-oriented family trees.

history may correspond to any one of many non-time-oriented family trees, as illustrated in Fig. 14 for four molecules, so it is necessary to reconstruct the tree as well as the sequences at the internal nodes and the homologies. In one approach, reconstruction of the tree is taken as a separate initial step, which we discuss in the next paragraph. Reconstruction of the internal sequences and homologies along each branch, for a known tree like that in Fig. 15, is a separate subsequent step. An algorithm for this purpose was introduced in Sankoff *et al.* (1973), further elucidated in Sankoff (1975), applied to a large set of data in Sankoff *et al.* (1976), and is most succinctly described in Chapter 9. It is a direct generalization of the algorithm just described for the three-sequence case.

One approach to reconstruction of the non-time-oriented family tree is based on the pairwise distances among the molecules (Fitch and Margoliash,

1967, and many later papers). Of course, this tree is of great interest in itself, so reconstructing the tree is a prime application where emphasis is on the distances.

Leaving molecules and moving to bird song, the pairwise distances between many songs are a tool that can be used to investigate many questions, such as the historical relationships among songs. One plausible approach would be to analyze the distances by multidimensional scaling. (Information about multidimensional scaling is available from many sources, such as Kruskal and Wish, 1978, or Carroll and Kruskal, 1978.) Chapter 6 by Bradley and Bradley uses another approach. Suppose there are several songs from each of several different bird populations (of a single species in a single general area). If songs change by cultural evolution, as believed, then song distances within each population should tend to be smaller than distances between populations. The authors demonstrate that this is so, and then go on to investigate whether song distance among populations increases with geographic distance among populations, as would occur under a variety of plausible models.

Distance can also be useful in solving problems that are familiar except for the possibility of deletion–insertion or compression–expansion in addition to substitution. For example, the Acceptance–Rejection Problem is to decide whether a newly observed sequence belongs to a given population. (In most applications, the characteristics of the population must be inferred from a sample of known members.) In one acceptance–rejection problem, an automatic speaker-verification system must verify whether a speaker is who she claims to be (perhaps in connection with automated banking operations over the telephone) by comparing a sample of her speech with pre-recorded samples in its memory. In another, a macromolecule must be accepted or rejected as a

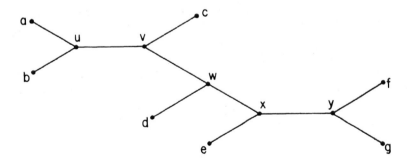

a, b, c, d, e, f, g OBSERVED

u, v, w, x, y TO BE RECONSTRUCTED

Figure 15. Non-time-oriented family tree.

new member of a class of molecules that are all homologous with each other. In many such problems, including those above, distances within the population are expected to be much smaller than those between members and nonmembers, so distances provide a natural approach.

Closely related is the Identification Problem, which is to decide from which of several populations a newly observed sequence has been drawn. One identification problem occurs in coding theory, of course, and Levenshtein invented distances to help solve it. Here each population consists of the sequences that may be received over the noisy channel when a single sequence is transmitted. In coding theory the identification problem is referred to as decoding, and the populations generally have known probabilistic structure. Another important identification problem occurs in speech research, where one of the tasks of an automatic speech recognition system is to identify which of many words, syllables, phonemes, or other units in its memory corresponds to a specific bit of speech it has isolated. Chapter 5 by Hunt, Lennig, and Mermelstein considers syllable identification as one step of a larger problem, namely, identifying an entire sentence that satisfies a specified syntax. Another identification problem is to decide which grammatical computer-language statement was intended in place of the ungrammatical statement that was typed in. In this application, each population consists of sequences that may result from human error when a particular sequence is intended. There is one population for each grammatical statement (in some contexts, the number of grammatical statements may not be finite). Generally, distances within each population are expected to be smaller than those between populations, so distances provide a natural approach.

8. WHY SHOULD DISTANCES BE USED?

Even though we started this chapter with comparison of sequences as the goal, and with such widely used concepts as Euclidean distance, city-block distance, and Hamming distance as the jumping-off place, it is worth reviewing the reasons for using distances. Ordinary variables, such as population or altitude (of a city), height or weight (of a person), and pitch or loudness (of a tone), characterize a *single* unit of study (such as a single city, a single person, or a single tone). Variables like distance (between two cities), perceived similarity (between two tones or other stimuli), and correlation (between two variables) characterize a *pair* of units of study. We refer to variables of the first type as *monadic* and variables of the second type as *dyadic*. Some methods of quantitative analysis explicitly require dyadic input. These include many methods of clustering (more information available from many sources, e.g., Sneath and Sokal, 1973, Wallace, 1978), some methods for reconstructing evolutionary family trees, and multidimensional scaling (see references cited above in Sec. 7). One reason for using distances is in order to make use of such methods. There are many potential applications of this kind using distances provided by sequence comparison, notably in the reconstruction of family trees from macromolecules.

Most methods of quantitative analysis, however, have been developed for monadic variables, which are far more common. The usual approach to quantitative study of a topic is to select several relevant monadic variables (e.g., to study human growth we might select features such as height, weight, circumference of head, length of trunk, etc.) and then to analyze these variables, by methods that range from such simple but important techniques as cross-tabulation, scatter plots and other graphical displays, to more complicated methods, such as regression, analysis of variance, and time-series analysis.

Of course, it is of vital importance that the variables selected not only should be relevant to the topic, but also that they should cover all important aspects of the topic. For some topics, however, including the study of variable-length sequences, it is difficult to find adequate monadic variables. Such topics include:

1. evolutionary study of macromolecules,
2. study of the cultural evolution of bird song, and
3. human perception of stimuli in many domains, such as words, faces, odors, musical compositions, etc.

Where monadic variables are inadequate, dyadic variables such as distance and similarity may provide an important new source of usable information. There are several different approaches to using the dyadic variables. One is direct common sense, illustrated for topic (1) in the applications in Chapter 2. Another is to adapt methods usually used with monadic variables, illustrated for topic (2) in Sect. 6 of Chapter 6. A third is to use methods specially intended for use with dyadic variables, like those mentioned above, also illustrated in Chapter 6. Such a method may be developed or improved specifically for the intended applications. Multidimensional scaling was improved in the early 1960s specifically for use with topic (3), and there has been great growth in such applications since that time. Reconstruction of family trees from sequence-comparison distances goes back at least to the 1960s, and is still undergoing development, as illustrated by Chapter 9.

9. WHY LEVENSHTEIN DISTANCE?

Given that a dyadic variable may be useful, why is Levenshtein distance in particular an appropriate variable to use? We present three rationales. Our first rationale applies in a narrow range of situations, where the differences between **a** and **b** may reasonably be attributed to a series of physical changes. Most notable is the evolution of DNA and RNA, where the changes are called *mutations.* Another is in the comparison of computer files by programs such as *diff* in the UNIX* system. When several slightly different computer files of the same basic information (text, or computer source code, etc.) are in computer storage, these programs can be very helpful in keeping track of the differences, since they compare two files and describe the differences in terms of lines to be inserted or deleted, to get one version from another.

Of course, we do not know which listing from **a** to **b** is the right one but, since mutations are unlikely events, a listing that is longer (even by just one elementary operation) is more unlikely as an explanation of how **a** and **b** arose from some common ancestor than a listing that is shorter. The most plausible listing to use as an explanation is the shortest possible one, so the Levenshtein distance between two sequences is a plausible indicator of the amount of actual historical change between them. The weights that enter into the definition permit us to treat some mutations as more unlikely than others. There are sometimes convincing reasons to do this, particularly as between substitutions and indels.

The second rationale arises in many situations where it would be natural to make use of the likelihood of obtaining **b** from **a**, or of obtaining both **a** and **b** from an unspecified common ancestor. This approach has been tried, but developing a suitable probability model for the underlying process of change can be difficult, and the calculations that result are likely to be slow and expensive. The second rationale for Levenshtein distance is as an elegant, computationally, practical, general-purpose approximation, not to likelihood itself but to a decreasing function of likelihood. For many uses, a monotonic function of likelihood is all that is needed.

Here is an extremely brief indication of an argument that supports the second rationale in one context. Suppose we start with a sequence **a** and permit elementary operations ("mutations") to occur by a simple, Poisson-like random process. The likelihood of obtaining **b** starting from **a** is a sum of terms. Each term corresponds to one listing from **a** to **b**, and the value of that term is a product of factors. Each elementary operation in the listing contributes one factor to the product; in addition each letter that did *not* undergo any change contributes one factor, and each position (between letters, and at the beginning and end of the sequence) at which an insertion did *not* take place contributes one factor. Under suitable assumptions, the following constitute adequate approximations:

1. Each of the factors mentioned has a fixed value, which depends only on the type of factor and on what letter or letters are involved in the substitution, deletion, insertion, non-changed letter, or non-inserted position.

---

*UNIX is a trademark of Bell Laboratories.

2. The likelihood of obtaining **b** starting from **a** is approximated by its largest term, i.e., the likelihood of one "dominant" listing.
3. Less important, the factors that correspond to non-changed letters and to non-inserted positions are approximately equal to 1.

The dominant listing will never have more than one elementary operation for any position in the sequence. Now, transform by the negative of the logarithm function. This transforms the

$$\text{approximate likelihood} \approx \text{likelihood of the dominant listing}$$

into a sum of terms ≥ 0, which we may call weights, i.e., into the simple length of the listing. The weights depend only on the letters involved, and if we are willing to accept approximation (3), we need consider only weights corresponding to substitutions, deletions, and insertions, since the remaining weights are zero; not accepting (3), however, introduces only a minor complication. The approximate likelihood of obtaining **b** starting from **a** transforms into the minimum length from **a** to **b**, namely, Levenshtein distance. Thus to an approximation,

$$\text{distance} = -\log(\text{likelihood}),$$

and $-\log$ is a monotonic decreasing function.

To actually work out this rationale in detail however, is a little more involved than it appears. We have not seen this done in the literature, and our version (which we hope to publish in the future) indicates the need for assumptions and precautions that do not seem to have been stated elsewhere. For example, suppose **a** contains a run of k_1 repetitions of the same letter, and **b** contains a corresponding run of k_2 repetitions, with $m = \max(k_1, k_2)$ and $j = |k_1 - k_2|$. Then the likelihood of obtaining **b** from **a** contains approximately m-choose-j largest terms of the same size, so the approximation in step (2), which is based on a single largest term being dominant, would be improved by incorporation of the factor m-choose-j. If **a** and **b** contain several pairs of corresponding runs, then the combined correction factor, which is the product of the individual factors for each run, can be quite large. This consideration is of particular importance in applications where the sequences tend to have runs, such as certain speech-research applications.

The third rationale is simply an application of the universal *post hoc* justification process. If we use any particular definition for distance, and find that this kind of distance supplies the information we want—that "it works" when we check its performance—then the satisfactory performance justifies the definition. Every well-made application of distance contains such checking and supports this rationale. Of course, this rationale leads naturally not only to checking on how well a given definition works but also to comparing different definitions to see which one works better. Chapter 6 makes just such a comparison, between distances based on linear time-warping and Levenshtein distance, to the advantage of the latter.

REFERENCES

Aho, A. V., Hirschberg, D. S., and Ullman, J. D., Bounds on the complexity of the longest common subsequence problem, *Journal of the Association for Computing Machinery* 23:1–12 (1974).

Alberga, C. N., String similarity and misspellings, *Communications of the Association for Computing Machinery* 10(5):302–313 (1967).

Auron, P. E., Rindone, W. P., Vary, C. P. H., Celentano, J. J., and Vournakis, J. N., Computer-aided prediction of RNA secondary structures, *Nucleic Acids Research* 10(1):403–419 (1982).

Barker, W. C., Wallace, D. C., and Dayhoff, M. O., 5S Ribosomal RNA, pages 95–98, in Dayhoff, M. O. (editor), *Atlas of protein sequence and structure,* Vol. 4, National Biomedical Research Foundation: Silver Spring, Maryland, U.S.A., (1969).

Bertelè, U., and Brioschi, F., *Nonserial Dynamic Programming,* New York: Academic Press, (1972).

Beyer, W. A., Stein, M. L., Smith, T. F., and Ulam, S. M., A molecular-sequence metric and evolutionary trees, *Mathematical Biosciences* 19:9–25 (1974).

Bridle, J. S., An efficient elastic-template method for detecting given words in running speech. *British Acoustical Society Spring Meeting, April 1973,* Chelsea College, London, Paper 73SHC3 (1973).

Bridle, J. S., and Brown, M. D., Connected word-recognition using whole-word templates, *Proceedings of the Institute for Acoustics,* pages 25–28 (1979).

Burr, D. J., A technique for comparing curves, pages 271–277, in *Proceedings of the IEEE Conference on Pattern Recognition and Image Processing, 1979, Chicago,* New York: IEEE (1979).

Burr, D. J., Designing a handwriting reader, pages 715–722, in *5th International Conference on Pattern Recognition, 1980,* Miami Beach, Florida, New York: IEEE (1980).

Burr, D. J., Elastic matching of line drawings, *IEEE Transactions on Pattern Analysis and Machine Intelligence* PAMI-3:703–713 (1981).

Calabi, L., and Hartnett, W. E., A family of codes for the correction of substitution and synchronization errors, *IEEE Transactions on Information Theory* 15(1):102–106 (1969a).

Calabi, L., and Hartnett, W. E., Some general results of coding theory with applications to the study of codes for the correction of synchronization errors, *Information and Control* 15:235–249 (1969b).

Cannon, R. L., An optimal text-collation algorithm, *ICCH: International Conference on Computers in the Humanities, July 20–22, 1973, University of Minnesota: Final Program* (1973).

Cannon, R. L., OPCOL—Optimal text-collation algorithm, *Computers and the Humanities* 10(1):33–40 (1976).

Carroll, J. D., and Kruskal, J. B., Scaling, Multidimensional, pages 892–907, in Kruskal, Wm. and Tanur, J. M. (eds.) *International Encyclopedia of Statistics,* New York: The Free Press and Macmillan Publishing Co., (1978).

Damerau, F. J., A technique for computer detection and correction of spelling errors, *Communications of the Association for Computing Machinery* 7(3):171–176 (1964).

Dixon, N. R., and Martin, T. B. (eds.), *Automatic speech and speaker recognition*, New York City: IEEE Press (1979).

Ferguson, C. W., Dendrochronology of bristlecone pine, *Pinus aristata*. Establishment of a 7484-year chronology in the White Mountains of eastern-central California, U.S.A., pages 237–259, in Olsson, Ingrid U. (ed.), *Radiocarbon variations and absolute chronology*, New York: Wiley Interscience, and Stockholm: Almqvist and Wiksell (1970).

Fitch, W. M., and Margoliash, E., Construction of phylogenetic trees, *Science* **155**:279–284 (1967).

Fromm, E., An estimation of errors in the Swedish varve chronology, pages 163–172, in Olsson, Ingrid U. (ed.), *Radiocarbon variations and absolute chronology*, New York: Wiley Interscience, and Stockholm: Almqvist and Wiksell (1970).

Fujimoto, Y. *et al.*, Recognition of handprinted characters by nonlinear elastic matching, pages 113–119, in *3rd. International Joint Conference on Pattern Recognition, 1976, Coronado, California*, New York: IEEE, (1976).

Goad, W. B., and Kanehisa, M. I., Pattern recognition in nucleic acid sequences. I. A general method for finding local homologies and symmetries, *Nucleic Acids Research* **10**(1):247–263 (1982).

Goldberg, R. N., Minimal string difference encoding, *Journal of Algorithms* **3**:147–156 (1982).

Gosling, J., A redisplay algorithm, *SIGPLAN Notices* **16**(6):123–129 (1981).

Hall, P. A. V., and Dowling, G. R., Approximate string matching, *Computing Surveys* **12**(4):381–402 (1980).

Haton, J. P., Contribution à l'analyse, paramétrisation et la reconnaissance automatique de la parole, Thèse de doctorat d'état, Université de Nancy, Nancy, France (1973).

Haton, J.-P., Une méthode dynamique de comparaison de chaînes de symboles de longeurs différentes; application à la recherche lexicale, *Comptes Rendus Hebdomadaires des Séances de l'Academie des Sciences*, Serie A. **278**(23):1527–1530 (1974a).

Haton, J. P., Practical application of a real-time isolated-word recognition system using syntactic constraints, *IEEE Transactions on Acoustics, Speech and Signal Processing* ASSP-**22**(6):416–419 (1974b).

Hirschberg, D. S., A linear space algorithm for computing maximal common subsequences, *Communications of the Association for Computing Machinery* **18**(6):341–343 (1975).

Hirschberg, D. S., An information theoretic lower bound for the longest common subsequence problem, *Information Processing Letters* **7**:40–41 (1978).

Hörnsten, Å. Summary of discussion of the measurements and identification of varves, pages 215–217, in Olsson, Ingrid, U. (ed.), *Radiocarbon variations and absolute chronology*, New York: Wiley Interscience, and Stockholm: Almqvist and Wiksell (1970).

Hunt, J. W., and McIlroy, M. D., An algorithm for differential file comparison, *Bell Laboratories Computing Science Technical Report* #41, (1976).

Itakura, F., Minimum prediction residual principle applied to speech recognition, *IEEE Transactions on Acoustics, Speech, and Signal Processing*, ASSP-**23**:67–72 (1975).

Johnson, S. C., Language-development tools on the UNIX System, *Computer* **13**(8):16–21 (1980).

Julesz, B., *Foundations of Cyclopean Perception*, Chicago: University of Chicago Press, (1971).

Julesz, Bela, Global stereopsis: Cooperative phenomena in stereoscopic depth perception, pages 215–256, in Held, R. et al. (eds.) *Handbook of Sensory Physiology,* **8**: *Perception,* Berlin: Springer-Verlag (1978).

Kanehisa, M. I., and Goad, W. B., Pattern recognition in nucleic acid sequences. II. An efficient method for finding locally stable secondary structures, *Nucleic Acids Research* **10**(1):265–277 (1982).

Kernighan, B. W., Lesk, M. E., and Ossanna, J. F., Document preparation, *Bell System Technical Journal* **57**:2115–2135 (1978).

Kruskal, J. B., and Wish, M. *Multidimensional scaling.* Beverly Hills: Sage University Papers 07–011, and London: Sage Publications (1978).

Kunze, M., and Thierrin, G., Maximal common subsequences of pairs of strings, *Congressus Numerantium* **34**:299–311 (1982).

Larson, R. E., and Casti, J. L., *Principles of Dynamic Programming.* New York: Marcell Dekker, (1978).

Levenshtein, V. I., Binary codes capable of correcting deletions, insertions, and reversals. *Cybernetics and Control Theory* **10**(8):707–710, (1966); Russian *Doklady Akademii Nauk SSR* **163**(4):845–848 (1965a).

Levenshtein, V. I., Binary codes capable of correcting spurious insertions and deletions of ones. (Original in Russian.) *Problems of Information Transmission* **1**(1):8–17, (1965b); Russian *Problemy Peredachi Informatsii* **1**(1):12–25 (1965b).

Levenshtein, V. I., A method of constructing quasilinear codes providing synchronization in the presence of errors. (Original in Russian.) *Problems of Information Transmission* **7**(3):215–222 (1971).

Lowrance, R., and Wagner, R. A., An extension of the string-to-string correction problem, *Journal of the Association for Computing Machinery* **22**:177–183 (1975).

MacWilliams, F. J., and Sloane, N. J. A., *Theory of error-correcting codes.* New York: North-Holland (1977).

Mills, D. L., A new algorithm to determine the Levenshtein distance between two strings (1978). Distributed at *Conference on Sequence Comparison,* Université de Montréal, April 1979.

Morgan, H. L., Spelling correction in systems programs, *Communications of the Association for Computing Machinery* **13**(2):90–94 (1970).

Myers, C. S., and Rabiner, L. R., A dynamic time warping algorithm for connected word recognition, *IEEE Transactions for Acoustics, Speech, and Signal Processing* ASSP-29:284–297 (1981a).

Myers, C. S., and Rabiner, L. R., Connected digit recognition using a level building DTW algorithm, *IEEE Transactions on Acoustics, Speech, and Signal Processing* ASSP-29:351–363 (1981b).

Myers, C. S., Rabiner, L. R., and Rosenberg, A. E., Performance tradeoffs in dynamic time-warping algorithms for isolated word recognition, *IEEE Transactions on Acoustics, Speech, and Signal Processing,* ASSP-28:622–635 (1980).

Needleman, Saul B., and Wunsch, C. D., A general method applicable to the search for similarities in the amino-acid sequence of two proteins. *Journal of Molecular Biology* **48**:443–453 (1970).

Norman, J. M., *Elementary dynamic programming,* New York: Crane, Russak and Co., (1975).

Nussinov, R., Tinoco, I., Jr., and Jacobson, A. B., Small changes in free energy assignments for unpaired bases do not affect predicted secondary structures in single stranded RNA, *Nucleic Acids Research* **10**(1):341–349 (1982a).

Nussinov, R., Tinoco, I., Jr., and Jacobson, A., Secondary structure model for the complete simian virus-40 late precursor RNA, *Nucleic Acids Research* **10**(1):351–363 (1982b).

Rabiner, L. R., Rosenberg, A. E., and Levinson, S. E., Considerations in dynamic time-warping for discrete word recognition, *IEEE Transactions on Acoustics, Speech, and Signal Processing,* ASSP-**26**:575–582 (1978).

Rabiner, L. R. and Schmidt, C. E., Application of dynamic time-warping to connected digit recognition, *IEEE Transactions on Acoustics, Speech, and Signal Processing,* ASSP-**28**:337–388 (1980).

Reichert, T. A., Cohen, D. N., and Wong, A. K. C., An application of information theory to genetic mutations and the matching of polypeptide sequences, *Journal of Theoretical Biology* **42**:245–261 (1973).

Reiner, E., and Kubica, G. P., Predictive value of pyrolysis-gas liquid chromatography in the differentiation of mycobacteria, *American Review of Respiratory Disease* **99**:42–49 (1969).

Reiner, E., and Bayer, F. L., Botulism: a pyrolysis gas-liquid chromatographic study, *Journal of Chromatographic Science* **16**(12):623–629 (1978).

Reiner, E., *et al.,* Characterization of normal human cells by pyrolysis gas chromatography mass spectrometry, *Biomedical Mass Spectrometry* **6**(11):491–498 (1979).

Sakoe, H., and Chiba, S., A similarity evaluation of speech patterns by dynamic programming (Japanese) *Institute of Electronic Communications Engineering of Japan,* July 1970:136 (1970).

Sakoe, H., and Chiba, S., A dynamic-programming approach to continuous speech recognition. *1971 Proceedings of the International Congress of Acoustics, Budapest, Hungary,* Paper 20 C 13 (1971).

Sakoe, H., and Chiba, S., Dynamic-programming algorithm optimization for spoken-word recognition, *IEEE Transactions for Acoustics, Speech, and Signal Processing* ASSP-**26**:43–49 (1978).

Sankoff, D., Matching sequences under deletion–insertion constraints. *Proceedings of the National Academy of Sciences of the U.S.A.,* **69**:4–6 (1972).

Sankoff, D., Minimal mutation trees of sequences, *SIAM Journal on Applied Mathematics* **78**:35–42 (1975).

Sankoff, D., Cedergren, R. J., and Lapalme, G., Frequency of insertion–deletion, transversion, and transition in the evolution of 5S ribosomal RNA, *Journal of Molecular Evolution* **7**:133–149 (1976).

Sankoff, D., Morel, C., and Cedergren, R. J., Evolution of 5S RNA and the nonrandomness of base replacement, *Nature New Biology* **245**:232–234 (1973).

Sellers, F. F., Bit loss and gain correction code, *IRE Transactions on Information Theory* **8**:35–38 (1962).

Sellers, P., An algorithm for the distance between two finite sequences, *Journal of Combinational Theory* **16**(2):253–258 (1974a).

Sellers, P., On the theory and computation of evolutionary distances, *SIAM Journal on Applied Mathematics* **26**(4):787–793 (1974b).

Sellers, P. H., The theory and computation of evolutionary distances: Pattern recognition, *Journal of Algorithms* **1**:359–373 (1980).

Smith, T. F., and Waterman, M. S., New stratigraphic correlation techniques, *Journal of Geology* **88**:451–457 (1980).

Smith, T. F., and Waterman, M. S., Identification of common molecular subsequences, *Journal of Molecular Biology* **147**:195–197 (1981).

Sneath, P. H. A., and Sokal, R. R., *Numerical taxonomy,* San Francisco: Freeman, (1973).

Tanaka, E., and Kasai, T., Synchronization and substitution error-correcting codes for the Levenshtein metric, *IEEE Transactions on Information Theory* IT-**22**:156–162 (1976).

Tauber, H., The Scandinavian varve chronology and C14 dating, pages 173–196, in Olsson, Ingrid U. (ed.), *Radiocarbon variations and absolute chronology,* New York: Wiley Interscience, and Stockholm: Almqvist and Wiksell, (1970).

Ullman, J. D., Near-optimal, single-synchronization-error-correcting code, *IEEE Transactions on Information Theory* **12**:418–424 (1966).

Ullman, J. D., On the capabilities of codes to correct synchronization errors, *IEEE Transactions on Information Theory* IT-**13**:95–105 (1967).

Velichko, V. M., and Zagoruyko, N. G., Automatic recognition of 200 words. *International Journal of Man–Machine Studies* **2**:223–234 (1970).

Vintsyuk, T. K., Speech discrimination by dynamic programming, *Cybernetics* **4**(1):52–57, Russian *Kibernetika* **4**(1):81–88 (1968).

Wagner, R. A., and Fischer, M. J., The string-to-string correction problem. *Journal of the Association for Computing Machinery* **21**(1):168–173 (1974).

Wallace, D., Clustering, pages 47–53, in Kruskal, W. H., and Tanur, J. M. (eds.), *International Encyclopedia of Statistics,* New York: The Free Press and Macmillan Publishing Co., (1978).

White, G. M., and Neely, R. B., Speech-recognition experiments with linear prediction, bandpass filtering, and dynamic programming, *IEEE Transactions on Acoustics, Speech, and Signal Processing,* ASSP-**24**:183–188 (1976).

White, G. M., Dynamic programming, the Viterbi algorithm, and low-cost speech recognition, pp. 413–417 in *Proceedings of the 1978 IEEE International Conference on Acoustics, Speech, and Signal Processing,* (1978).

Wilbur, W. J. and Lipman, D. J. (1983) Rapid similarity searches of nucleic acid and protein data banks, *Proceedings of the National Academy of Science of the USA,* **80**:726–730.

Wong, C. K., and Chandra, A. K., Bounds for the string-editing problem, *Journal of the Association for Computing Machinery* **23**:13–16 (1976).

Yasuhara, M., and Oka, M., Signature-verification experiment based on nonlinear time alignment: a feasibility study, *IEEE Transactions on Systems, Man, and Cybernetics* SMC-**7**:212–216 (1977).

PART ONE

MACROMOLECULAR SEQUENCES

1. INTRODUCTION

The interest of molecular biologists in methods of sequence comparison, first stimulated by the determination of amino acid sequences for a great many proteins in the 1960s, is now undergoing a sharp increase due to the recent flood of nucleotide sequence determinations for nucleic acids. (Readers who wish some explanation of biochemical terms may refer to the section entitled Biological Background, below.) The main use of sequence comparison in molecular biology is to detect and characterize the homology (i.e., correspondence) between two or more related sequences.

The dynamic-programming method for sequence comparison was introduced into molecular biology before 1970, and has many advantages over the other methods currently in use. However, despite these advantages, which are described below, and despite the wide use of the dynamic-programming method in the field of speech processing (see Part II), its use in molecular biology is still rather restricted. Among 12 papers in a recent issue of *Nucleic Acids Research* (Volume 10, No. 1) that present methods or programs for finding homologous (i.e., highly similar) regions of two sequences, only four use the dynamic-programming method, namely, Orcutt *et al.* (1982), Osterburg *et al.* (1982), Goad and Kanehisa (1982), and Kanehisa (1982). We hope this volume will encourage greater use of this method.

Part I of this volume contains two chapters that exemplify advanced uses of sequence comparison in molecular biology, and several chapters from other parts of the volume are also relevant to the field. Chapter 2, by Erickson and Sellers, deals with the question of how to find the consecutive string (or strings) in a longer sequence **a** with "best possible agreement" to a shorter sequence **b**. (In speech research, the analogous question is called "word-spotting.") It presents algorithms to solve two different versions of this question, and describes three diverse biological applications. In one version, best agreement means that the string has the smallest possible distance to **b**. In the other, best agreement is intended "locally" and means that no substring or superstring has smaller distance to **b**. The first application is the detection of a common pattern in two repeating sequences known as "satellite DNA." The second application is based on the discovery of a remarkable situation, where the sequences of two proteins are related by a cyclical permutation. The third application involves recognition of a likely candidate for the sequence that serves as particular "translocation signal."

Chapter 3, by Sankoff, Kruskal, Mainville, and Cedergren, is about methods of predicting, for an RNA molecule having a known nucleotide sequence, its secondary structure—the way different parts of the molecule will bond to each other. A vast number of different secondary structures can be imagined for a given RNA molecule: the secondary structure that actually occurs in nature is generally the one that optimizes the "free energy." Methods for predicting this structure involve comparison of different parts of the molecule with each other, using dynamic-programming algorithms.

Other parts of the volume also contain topics of direct interest to molecular biologists, though they are presented in a more general setting and they have other uses. Chapter 9 presents a method for comparing many sequences simultaneously when they are related by a known evolutionary tree. This method is helpful in making inferences about sequences in ancestral organisms and other aspects of evolutionary history. Chapter 10 contains several methods of interest in molecular biology, including a method for finding a partial homology (i.e., a homology between part of one sequence and part of another), methods for finding homologies subject to a variety of constraints, and an application to 5S ribosomal RNA. Part V contains material that is helpful when we are trying to decide whether or not the similarity between two sequences is statistically significant.

ADVANTAGES OF THE DYNAMIC PROGRAMMING METHOD

The dynamic-programming method is preferable to other methods because it produces better results (i.e., homologies that conform more closely to what biologists are likely to consider appropriate and realistic), and because it displays more clearly the basis for evaluating the quality of homologies. We describe several properties of the method to illustrate these advantages. Very short sequences are used as illustrations, because longer ones are not practicable here, but they correctly illustrate what happens in comparisons of genuine interest.

During this discussion, we shall represent the alternative methods by one that is reasonably characteristic of the others, namely, the routine for finding homologies constituting *one portion* of the deservedly popular system of Korn, Queen, and Wegman (1977). We refer to this as KQW, but we wish to emphasize that this abbreviation and our comments refer only to the one routine, since their system contains many excellent features.

1. Separation of evaluation from algorithm. The dynamic-programming method clearly separates the matter of *evaluating* the quality of possible homologies from the algorithm used to *find* the good ones. In some other methods, the evaluation is implicit in the algorithm, so that the question of what constitutes a good homology (or the best one) is inextricably intertwined with the question of how to do the calculations. The separation of these two questions leads to several advantages. It permits the user to understand the biologically important aspect of the method, namely, the evaluation system, without studying the algorithm. It even permits the user to change the evaluation system, where appropriate, without changing the algorithm. Finally, the greater clarity of concept leads to better results.

For example, in the comparison of AACAAA and AAAAA, KQW will recognize the homology $\begin{smallmatrix} AACAAA \\ AA\text{-}AAA \end{smallmatrix}$, but if we reverse both sequences and compare AAACAA with AAAAA, it will *not* recognize the reverse homology $\begin{smallmatrix} AAACAA \\ AAA\text{-}AA \end{smallmatrix}$. There is no biological basis for this kind of asymmetry, and the dynamic-programming method does not suffer from it.

Of course, KQW contains a variety of parameters that can be adjusted to correct the anomaly for these particular sequences, but this fact does not remove the disadvantage for several reasons. First, no setting of the parameters will eliminate similar deficiencies for all comparisons of genuine interest. Second, in such comparisons the deficiency may not be noticed and the need for corrective action may be overlooked. Third, learning to use the parameters skillfully requires considerable experience.

2. Use of a simple evaluation score to select homologies. In dynamic programming any possible homology is evaluated by a simple score that is based either on a penalty for each substitution, deletion, and insertion required to make the sequences match, and/or on a merit point for each pair of matched elements in the two sequences. Several other methods either fail to have any explicit evaluation score, or else use several different scores without a clear understanding of when and why they agree or disagree among themselves.

This score provides a well-motivated criterion for choosing among homologies. Many other methods do not clearly define their basis for selecting one homology over another, or fail to link up their selection of homologies with an evaluation score. As a consequence, poorer homologies may be recognized and better ones neglected. For example, in the comparison of AACCGU with AACGU, the dynamic-programming method will produce the two obviously desirable homologies $\begin{smallmatrix} AACCGU \\ AA\text{-}CGU \end{smallmatrix}$ and $\begin{smallmatrix} AACCGU \\ AAC\text{-}GU \end{smallmatrix}$. KQW will instead produce two homologous regions, $\begin{smallmatrix} AAC \\ AAC \end{smallmatrix}$ and $\begin{smallmatrix} CGU \\ CGU \end{smallmatrix}$, which are incompatible, a much less desirable result. Of course, the deficiency for these particular sequences can be corrected by adjusting the KQW parameters but, as above, this does not remove the disadvantage.

It is worth noting that the evaluation score is available and appropriate for other uses, e.g., to help answer the question of whether or not two sequences are homologous. For this purpose, the score of the best tentative homology may be compared with scores that would arise by chance.

3. Global optimality. The algorithm of dynamic programming is guaranteed to find the best homology (or several best homologies) out of all possible homologies between the sequences, i.e., to find the homology or homologies with the best evaluation score. Many methods do not explicitly link their algorithm to an evaluation score, or do not guarantee to find the homologies with the best score.

4. Stable parameters and soft limits. All methods incorporate parameters that can be adjusted to alter the results. For the ordinary dynamic-programming method, these parameters are the penalty values assigned to various deletions, insertions, and substitutions, and their meaning is simple: The less likely a particular type of evolutionary change is considered to be, the larger the penalty value assigned to it. While these parameters are available to adjust the method to truly different kinds of applications, it is not necessary to keep adjusting them every time new sequences are compared. In fact, sensible useful results can be obtained over a wide range of situations without using them at all (i.e., by using the "default" values, which are all equal to 1). In some methods, it is frequently necessary to adjust the parameters for new comparisons, and considerable expertise may be required to make the choice properly.

The penalty parameters impose soft limits of various kinds, e.g., on the number of successive unmatched elements in one sequence: If there are too many of them, the quality of the valuation is reduced, so too many of them will not occur. It is far more realistic to impose this kind of penalty, which increases gradually, than to impose a sharp limit that, for example, may permit three unmatched elements without penalty but absolutely forbid four of them. Other methods generally use sharp limits whose values are parameters, a fact that has much to do with the frequent need, when using such methods, to adjust the values of the parameters. (It is possible in the dynamic-programming method to impose sharp limits when desired, however, as illustrated in Chapter 10.)

5. Versatility and consistency. The dynamic-programming method, based as it is on evaluation and optimization, extends in a natural consistent way to a wide range of situations. For example, as illustrated in Chapter 9, the method extends very gracefully to the simultaneous comparison of several sequences with complete consistency between comparisons of two sequences and comparisons of several. By contrast, while other methods also lend themselves to extension, the extension does not flow naturally out of the basic method and may fail to be consistent with it. For example, KQW was supplemented by a method for the comparison of several sequences in Queen, Wegman, and Korn (1982), but the newer method will generally yield different results from KQW when both are applied to the same pair of sequences. The great versatility of the dynamic-programming method is also illustrated by its ability to handle the very different problem of predicting the secondary structure of RNA, as illustrated in Chapter 3, Kanehisa and Goad (1982), Nussinov and Jacobson (1980), and Auron *et al.* (1982).

BIOLOGICAL BACKGROUND

In molecular biology, sequence comparison is applied to the nucleic acids and to proteins. The nucleic acids are of two types, DNA and RNA. DNA is the primary medium of genetic information for all organisms (with some very primitive exceptions). The DNA molecules within a single cell contain essentially complete instructions for the development and differentiation of the entire organism, as well as instructions for the production and consumption of the materials for the life of the cell and organism. Some RNA molecules serve a variety of direct biochemical functions, while others (known as messenger RNA) serve as intermediaries in the formation of protein.

Each DNA molecule is a sequence of four types of (deoxyribo)nucleotides, A, G, C, and T, which can be millions of units long or more. Each RNA molecule is a sequence of four types of (ribo)nucleotides, A, G, C, and U, up to thousands of units long. Each nucleotide is characterized by the base it contains: adenine(A), guanine(G), cytosine(C), thymine(T) for DNA only, or uracil(U) for RNA only (T and U play corresponding roles).

There are thousands of different proteins in a single organism, which serve in a wide variety of structural and functional roles. For example, enzymes are proteins, and practically every biochemical process requires one or several of its own characteristic enzymes to catalyze its action. Each protein molecule is a sequence of twenty types of amino acids, up to hundreds of units long, or occasionally more. The names of the amino acids may be found in Fig. 1 (the structure of Fig. 1 will be explained later). Each amino acid contains an NH_2 group and a COOH group, and adjacent amino acids are joined by a peptide bond between these groups. Chains of peptide-linked amino acids, such as those in protein, are referred to as polypeptides. The terminal amino acid that has a free NH_2 group is called the NH_2-terminus or "beginning" of the protein, and the terminal amino acid that has a free COOH-group is called the COOH-terminus of the protein.

Macromolecules that are functionally equivalent and designated by the same name in different species usually have somewhat different sequences, and the difference tends to be greater when the evolutionary distance is greater. Functionally different sequences, such as hemoglobin and myoglobin, have frequently developed by differentiation from a common ancestor sequence, and corresponding portions of related sequences often have similar biochemical activity. Thus comparison of sequences is very important in understanding the evolution, structure, and function of many molecules. Sequences derived from a

	U	C	A	G
U	UUU UUC Phenyl-alanine UUA UUG Leucine	UCU UCC UCA UCG Serine	UAU UAC Tyrosine UAA UAG TERMINATE	UGU UGC Cysteine UGA TERMINATE UGG Tryptophan
C	CUU CUC CUA CUG Leucine	CCU CCC CCA CCG Proline	CAU CAC Histidine CAA CAG Glutamine	CGU CGC CGA CGG Arginine
A	AUU AUC AUA Isoleucine AUG Methionine	ACU ACC ACA ACG Threonine	AAU AAC Asparagine AAA AAG Lysine	AGU AGC Serine AGA AGG Arginine
G	GUU GUC GUA GUG Valine	GCU GCC GCA GCG Alanine	GAU GAC Aspartic acid GAA GAG Glutamic acid	GGU GGC GGA GGG Glycine

Figure 1. Codons and their amino acids. The codons are shown in their mRNA form.

common ancestor are called *homologous,* and a homology is a matching showing which elements in one are presumed to correspond to which elements in the other.

Basic to the function and structure of the nucleic acids is a natural capacity of the nucleotides to bond in pairs, as first described by Watson and Crick:

> G bonds with C,
> A bonds with T or U.

This pairing is referred to as *complementarity.* Nucleotides are loosely referred to in terms of the bases they contain, and two bonded nucleotides are generally referred to as a *base pair.* DNA and RNA make use of this pairing in several ways.

DNA ordinarily occurs not as a single sequence, but as two complementary sequences (running in opposite directions), joined together along their entire length by Watson–Crick pairing, for example,

$$
\overrightarrow{A}-\overrightarrow{G}-\overrightarrow{C}-\overrightarrow{T}-\overrightarrow{T}-\overrightarrow{T}-\overrightarrow{C}-\overrightarrow{T}-\overrightarrow{A}-\overrightarrow{A}- \cdots
$$
$$
\bullet \quad \bullet \quad \bullet \quad \bullet \quad \bullet \quad \bullet \quad \bullet \quad \bullet \quad \bullet \quad \bullet
$$
$$
\overleftarrow{T}-\overleftarrow{C}-\overleftarrow{G}-\overleftarrow{A}-\overleftarrow{A}-\overleftarrow{A}-\overleftarrow{G}-\overleftarrow{A}-\overleftarrow{T}-\overleftarrow{T}- \cdots
$$

Such double-stranded DNA, which arranges itself into the double helix discovered by Watson and Crick, can also be thought of as a sequence of base pairs. For convenience, double-stranded DNA can be adequately described by just one of the complementary sequences, since complementarity permits either strand to be inferred from the other. Prior to cell reproduction, the two strands of the double helix separate and each serves as a template for the formation of a complementary sequence through *replication.* Thus the two daughter cells can both receive faithful copies of the DNA from the parent cell. An important source of evolutionary change is the rare event in which DNA undergoes change, through imperfect replication or other means.

Under normal conditions, RNA molecules do not occur as complementary pairs. Instead, parts of a single RNA molecule bond to each other through complementarity, to define its *secondary structure,* e.g.,

$$
\overrightarrow{A}-\overrightarrow{A}-\overrightarrow{A}-\overrightarrow{A}-\overrightarrow{A}-\overrightarrow{A}-\overrightarrow{G}-\overrightarrow{G}-\overrightarrow{G}
$$

The secondary structure, which may be very complicated, plays a vital role in the shape and function of the molecule.

The way in which the DNA molecules guide the formation of the substances needed for the reproduction, development, and life of the organism is summarized here:

$$
\text{DNA} \xrightarrow[\text{Transcription}]{} \left\{ \begin{array}{l} \text{Untranslated RNA} \\ \text{Messenger RNA} \xrightarrow[\text{Translation}]{} \text{Protein} \end{array} \right.
$$

Replication

Both untranslated and messenger RNA are formed from DNA by *transcription.* During transcription, the double-stranded DNA separates in a limited region, and an appropriate portion of one strand (often one or several genes) serves as a template for the formation of a complementary strand of RNA.

In the case of messenger RNA, the strand attaches to a complex structure called a *ribosome,* which is the site for *translation* in which a protein molecule is formed. During translation, the strand is read as a succession of triplets, called *codons.* As shown in Fig. 1, each of the 64 possible codons corresponds to an amino acid, except for three that serve to terminate the translation process. This correspondence, called the genetic code, is mediated by several dozen types of *transfer RNA.* Each type can carry one specific amino acid and contains a codon-complementary sequence that permits it to recognize a codon for this amino acid by pairing. It is interesting to note that the ribosome is composed of several dozen protein molecules and three RNA molecules, one of which is discussed in Chapter 10.

REFERENCES

Auron, P. E., Rindone, W. P., Vary, C. P. H., Celentano, J. J., and Vournakis, J. N., Computer-aided prediction of RNA secondary structures, *Nucleic Acids Research* **10**(1):403–419 (1982).

Goad, W. B., and Kanehisa, M. I., Pattern recognition in nucleic acid sequences. I. A general method for finding homologies and symmetries, *Nucleic Acids Research* **10**(1):247–263 (1982).

Kanehisa, M. I., and Goad, W. B., Pattern recognition in nucleic acid sequences. II. An efficient method for finding locally stable secondary structures, *Nucleic Acids Research* **10**(1):265–277 (1982).

Kanehisa, M. I., Los Alamos sequence package for nucleic acids and proteins. *Nucleic Acids Research* **10**(1):183–196 (1982).

Korn, L. J., Queen, C. L., and Wegman, M. N., Computer analysis of nucleic acid regulatory sequences, *Proceedings of the National Academy of Sciences* **74**(10):4401–4405 (1977).

Nussinov, R., and Jacobson, A., Fast algorithm for predicting the secondary structure of single-stranded RNA, *Proceedings of the National Academy of Sciences* **77**:6309–6313 (1980).

Orcutt, B. C., George, D. G., Frederickson, J. A., and Dayhoff, M. O., Nucleic acid sequence database computer system, *Nucleic Acids Research* **10**(1):157–174 (1982).

Osterburg, G., Glatting, K. H., and Sommer, R., Computer programs for the analysis and the management of DNA sequences, *Nucleic Acids Research* **10**(1):207–216 (1982).

Queen, C., Wegman, M. N., and Korn, L. J., Improvements to a program for DNA analysis: a procedure to find homologies among many sequences, *Nucleic Acids Research* **10**(1):449–456 (1982).

CHAPTER TWO

RECOGNITION OF PATTERNS IN GENETIC SEQUENCES

Bruce W. Erickson and Peter H. Sellers

1. EVOLUTIONARY DISTANCE AND METRIC ALIGNMENTS

A *genetic sequence* is a linear array of information recorded in a genetic macromolecule, such as a DNA or RNA segment or a protein. For the purposes of this chapter, it is an abstract sequence whose terms come from a given alphabet. The alphabet contains four letters in the case of DNA or RNA, and twenty letters in the case of a protein. If **a** and **b** are such sequences, we say that **a** *contains pattern* **b** if there is an interval in **a** that is either identical to **b** or recognizably close to **b**. The recognition of pattern **b** in sequence **a** depends on our ability to measure the closeness of two genetic sequences. Let us consider what it means for the entire sequence **a** to be close to sequence **b** in the genetic context. In other words, we need a suitable measure of the distance between **a** and **b**, which will be denoted by $d(\mathbf{a}, \mathbf{b})$.

Let us assume that the genetic sequences **a** and **b** have evolved from a common ancestral sequence and that the value of $d(\mathbf{a}, \mathbf{b})$ is a measure of the dissimilarity of **a** and **b** produced by a finite number of evolutionary *steps*. Taken together, these steps form a *path* between **a** and **b** and each step in the path contributes an increment to the dissimilarity of **a** and **b**. Accordingly, the value of $d(\mathbf{a}, \mathbf{b})$ is the sum of the incremental values assigned to the hypothetical evolutionary steps in a path between **a** and **b**. We assume that **a** and **b** are *homologous* in the biological sense, which means that they are joined by a historical path of evolutionary steps, but we do not assume that this historical path is known or that it is the most suitable path to use in evaluating $d(\mathbf{a}, \mathbf{b})$.

Let us assume that certain types of evolutionary steps are possible among a class of genetic sequences, that a positive cost can be assigned to each step, and that there is a sufficient variety of possible evolutionary steps that any two genetic sequences **a** and **b** can be joined by at least one path of evolutionary steps. Then we can define the *length* of a path of steps as the total cost of the steps, and the *evolutionary distance* $d(\mathbf{a}, \mathbf{b})$ as the length of the shortest path of evolutionary steps between **a** and **b**.

A given class of genetic sequences, together with the above function d of two variables, constitutes a *metric space* in the mathematical sense. In other words, for any genetic sequences **a**, **b**, and **c**, this function satisfies the following formal properties:

1. (*Positive definiteness*) $d(\mathbf{a}, \mathbf{b})$ is a positive real number for $\mathbf{a} \neq \mathbf{b}$ and is zero for $\mathbf{a} = \mathbf{b}$.
2. (*Symmetry*) $d(\mathbf{a}, \mathbf{b}) = d(\mathbf{b}, \mathbf{a})$.
3. (*Triangle inequality*) $d(\mathbf{a}, \mathbf{c}) \leq d(\mathbf{a}, \mathbf{b}) + d(\mathbf{b}, \mathbf{c})$.

Note that evolutionary distance is a metric regardless of what kinds of evolutionary steps are regarded as possible. The idea that genetic sequences constitute a metric space does not depend on the specific assumptions made below about the nature of an evolutionary step.

Let us assume that an evolutionary step is the *mutation, deletion,* or *insertion* of a single term in a genetic sequence and that each such step is assigned a positive cost. If they were all assigned a cost of one, the length of a path would be simply the number of steps in the path. A more general approach is to assign costs that vary with the terms of the sequences affected by the evolutionary step. For every letter in the alphabet, there is a deletion cost and an insertion cost. For every ordered pair of letters, there is a mutation cost, although only half of these costs need to be assigned to define $d(\mathbf{a}, \mathbf{b})$ because d is a symmetric function. This concept of an evolutionary distance based on mutations, deletions, and insertions of single terms was introduced by S. M. Ulam (1972).*

If the genetic sequences under consideration are nucleotide sequences, their terms come from an alphabet of four letters representing the four types of nucleotides. If they are protein sequences, however, their terms come from an alphabet of twenty letters corresponding to the twenty types of genetically coded amino acids. In either case, the fixed alphabet of ℓ letters requires the assignment of exactly $\frac{1}{2}\ell \, (\ell - 1)$ kinds of mutations and ℓ kinds of deletions for a total of $\frac{1}{2}\ell \, (\ell + 1)$ kinds of evolutionary steps. Insertions are not a separate kind of evolutionary step because, if **a** is formed by inserting one term in **b**, then **b** is formed by deleting one term from **a**. In either case $d(\mathbf{a}, \mathbf{b})$ has the same

---

*Ulam's metric is a special case of the distance function described elsewhere in this book as the *Levenshtein distance,* which in general is not a metric.

minimum total cost, namely the length of the shortest path between **a** and **b**. Before calculating an evolutionary distance, a table of $\frac{1}{2}\ell$ $(\ell - 1)$ *mutation costs* and a list of ℓ *deletion costs* must be assigned. In this terminology the *evolutionary distance* $d(\mathbf{a}, \mathbf{b})$ between any two genetic sequences **a** and **b** can be defined as the minimum total cost of the mutations and deletions required to convert **a** and **b** into a common sequence.

An algorithm for calculating this distance was introduced independently in Sellers (1974a) and Wagner and Fischer (1974), and a description of this algorithm was given in Sellers (1974b) from the same point of view as outlined above. It is a dynamic-programming algorithm that takes on the order of $m \cdot n$ steps to determine the evolutionary distance between two sequences of lengths m and n. It also displays each path of evolutionary steps that has as its length the minimum total cost required by the definition of evolutionary distance. In general, more than one such path exists.

The most obvious way to display each minimum-cost path of evolutionary steps between sequences **a** and **b** is to list the terms of **a** on one line and the terms of **b** on a second line below **a**, where the terms are spaced so that every mutation is represented by two terms in vertical alignment and every deletion is represented by a term of one sequence in vertical alignment with a dash (–) in the other sequence. Every remaining term of **a** is identical to the term of **b** below it. A dash in a sequence of the display is called a *null* and represents the deletion of a term from that sequence or the insertion of the term with which it is aligned in the other sequence. This term-by-term display of a path of mutations and deletions, whose total cost equals the minimum total cost $d(\mathbf{a}, \mathbf{b})$, is an optimal alignment of **a** and **b** called a *metric alignment.* The concepts of evolutionary distance and metric alignment may be characterized more formally as follows:

Definition 1. Let **a** be a sequence of the form $a_1 a_2 \ldots a_m$ and **b** a sequence of the form $b_1 b_2 \ldots b_n$, whose terms belong to a given metric space

$$\{ -, a_1, a_2, \cdot \cdot \cdot, b_1, b_2, \cdot \cdot \cdot \}$$

which includes the null (–) as one of its elements and whose distance function is denoted by d. (*Note:* The given values $d(a_i, b_j)$, $d(a_i, -)$, etc., are the assigned mutation and deletion costs described above.)

Let $\bar{\mathbf{a}}$ denote the set of all sequences of the form $\bar{a}_1 \bar{a}_2 \ldots \bar{a}_{m+n}$ formed by inserting n nulls into **a**, and let $\bar{\mathbf{b}}$ denote the set of all sequences of the form $\bar{b}_1 \bar{b}_2 \ldots \bar{b}_{m+n}$ formed by inserting m nulls into **b**.

Then the *evolutionary distance* $d(\mathbf{a}, \mathbf{b})$ is defined by

$$d(\mathbf{a}, \mathbf{b}) = \min \left\{ \sum_{i=1}^{m+n} d(\bar{a}_i, \bar{b}_i) \right\},$$

where the minimum is taken over all pairs of sequences

$$\bar{a}_1 \bar{a}_2 \ldots \bar{a}_{m+n} \quad \text{and} \quad \bar{b}_1 \bar{b}_2 \ldots \bar{b}_{m+n}$$

in $\bar{\mathbf{a}}$ and $\bar{\mathbf{b}}$, respectively. Any two such sequences in the format

$$\bar{a}_1 \bar{a}_2 \ldots \bar{a}_{m+n}$$
$$\bar{b}_1 \bar{b}_2 \ldots \bar{b}_{m+n}$$

that achieve the minimum in the above expression for $d(\mathbf{a}, \mathbf{b})$ constitute a *metric alignment* of **a** and **b**.

This definition rephrases the preceding discussion in order to achieve a simple and conventional mathematical notation. When a genetic sequence **a** is expanded by inserting nulls, it represents the same genetic sequence; but the formal distinction between **a** and the expanded forms of **a** must be maintained to frame a mathematical definition. Each sequence formed by inserting nulls into **a** is *equivalent* (but not equal) to **a**. In order to construct a metric alignment of two sequences **a** and **b**, they must be expanded to the same length. If they have lengths m and n, then each needs to be expanded at most to length $m + n$. Therefore, it is sufficient from a mathematical viewpoint to represent **a** by any member of the *equivalence class* $\bar{\mathbf{a}}$ of sequences obtained by expanding **a** to length $m + n$, and likewise for **b**. This means that the metric alignment of Definition 1 may contain pairs of nulls in vertical alignment. As with sequences, a metric alignment may be represented by any member of the equivalence class formed by inserting vertical pairs of nulls into it. In Sellers (1974b) this mathematical device is carried further by expanding all sequences to infinite length, which results in even simpler mathematical language.

The purpose of displaying a metric alignment is to reveal the similarity of two sequences by placing similar terms in vertical alignment in such a way that the total cost

$$\sum_{i=1}^{m+n} d(\bar{a}_i, \bar{b}_i)$$

between vertically aligned terms is an absolute minimum. The alignment is *metric*, because its total cost is the evolutionary distance d, a metric function.

As an example of Definition 1, consider the sequences abc and ac. The equivalence classes $\overline{\text{abc}}$ and $\overline{\text{ac}}$ are listed in Table 1. Since each class contains ten sequences, there are 100 ways of selecting a pair of expanded sequences by taking one from each class. Each pair has a total cost associated with it, which is the sum of the mutation and deletion costs belonging to its corresponding pairs

Table 1
Sets of equivalent sequences $\overline{abc}$
and $\overline{ac}$ formed by expanding the
sequences abc and ac, respectively.

$\overline{abc}$	$\overline{ac}$
− − a b c	− − − a c
− a − b c	− − a − c
− a b − c	− − a c −
− a b c −	− a − − c
a − − b c	− a − c −
a − b − c	− a c − −
a − b c −	a − − − c
a b − − c	a − − c −
a b − c −	a − c − −
a b c − −	a c − − −

of terms. For instance, the total cost associated with the pair of expanded
sequences (− − abc, − − a–c) is

$$d(-, -) + d(-, -) + d(a, a) + d(b, -) + d(c, c).$$

By the metric property of positive definiteness, every term in this sum equals
zero except $d(b, -)$, which is the cost of deleting b. According to the formula in
Definition 1, $d(abc, ac)$ is the minimum of the 100 total costs associated with
the 100 pairs of expanded sequences. If all mutation and deletion costs have the
value one, then the pair of expanded sequences shown above would have a total
cost equal to the minimum total cost, and a metric alignment of abc and ac
would be represented as

$$- - \text{abc}$$
$$- - \text{a–c}$$

or more simply as

$$\text{abc}$$
$$\text{a–c} .$$

Although the metric alignment of this example is unique (except for equivalent
representations), metric alignments are generally not unique.

The dynamic-programming algorithm for generating the evolutionary distance and the corresponding metric alignments has been implemented at The Rockefeller University as a computer algorithm called SS. The two S's indicate that both of the genetic sequences are *specified* before the algorithm is executed. Algorithm SS is written in the C programming language and currently runs under the UNIX operating system. When run on a Digital Equipment Corporation PDP 11/70 or VAX 11/780 minicomputer, algorithm SS takes only a few seconds to find the evolutionary distance and all metric alignments of two 100-residue sequences. This algorithm is useful for detecting the possible evolutionary relatedness of specific pairs of nucleic acid or protein sequences (Erickson *et al.,* 1979).

2. TWO PATTERN-RECOGNITION THEOREMS

This section describes two pattern-recognition theorems introduced in Sellers (1979). Proofs of these theorems are given in Sellers (1980). Theorem 1 is an algorithm for finding every interval **i** in **a** that most resembles a given pattern **b**, and Theorem 2 is a generalization of it. These dynamic-programming algorithms are similar to the algorithm for computing $d(\mathbf{a}, \mathbf{b})$ described in Sellers (1974a, b) and Wagner and Fischer (1974).

Consider the finite sequences

$$\mathbf{a} = a_1 a_2 \ldots a_m$$

and

$$\mathbf{b} = b_1 b_2 \ldots b_n,$$

where **b** is a specified sequence, or *pattern,* being sought in **a**. In other words, sequence **a** is considered to be much longer than **b** and to contain one or more *unspecified* intervals **i** that resemble *specified* sequence **b**. The role of these algorithms is to identify these intervals **i**, to calculate their evolutionary distance from pattern **b**, and to determine their metric alignments with pattern **b**.

Let the notation **i** ⊂ **a** (read "**i** in **a**") mean that **i** is an interval in **a** of the form

$$\mathbf{i} = a_p a_{p+1} \ldots a_q,$$

where $1 \leq p \leq q \leq m$.

Definition 2. Let d be the evolutionary distance of Definition 1. Then an interval **i** ⊂ **a** *most resembles* **b** *globally* if and only if

$$d(\mathbf{i}, \mathbf{b}) \leq d(\mathbf{j}, \mathbf{b})$$

for all $\mathbf{j} \subset \mathbf{a}$.

Theorem 1. Given the sequences

$$\mathbf{a} = a_1 a_2 \ldots a_m$$

and

$$\mathbf{b} = b_1 b_2 \ldots b_n,$$

the following procedure determines every interval

$$(a_p a_{p+1} \ldots a_q) = \mathbf{i} \subset \mathbf{a}$$

that most resembles $\mathbf{b}$ globally:

i) Construct an $(m + 1) \times (n + 1)$ matrix $(e(i, j))$ by induction. The meaning of $e(i, j)$ is the minimum total cost of the mutations, deletions, and insertions required to convert the initial segment $b_1 b_2 \ldots b_j$ of $\mathbf{b}$ into any interval $\mathbf{i} \subset \mathbf{a}$ of any length that terminates at a_i. The initial values of the matrix elements in the first column and first row are set as

$$e(i, 0) = 0$$

for $i = 0, 1, \ldots, m$ and

$$e(0, j) = \sum_{h=1}^{j} d(-, b_h)$$

for $j = 1, 2, \ldots, n$. (These initial conditions allow interval $\mathbf{i}$ to begin at any term a_p and end at any later term a_q without any penalty for ignoring the terms before a_p and after a_q.) The values of the remaining matrix elements are determined inductively by making $e(i, j)$ equal to the smallest of three values:

$$[e(i - 1, j) + d(a_i, -)],$$
$$[e(i - 1, j - 1) + d(a_i, b_j)],$$
$$[e(i, j - 1) + d(-, b_j)].$$

Depending on which of the three values is smallest, we say that $e(i - 1, j)$ or $e(i - 1, j - 1)$ or $e(i, j - 1)$ is *connected* to $e(i, j)$ by this inductive step.

Each such connection is displayed by a line segment in the graph P_f of Fig. 1. If two or three of the values are all equally small, then there are two or three connections to $e(i, j)$.

ii) In the last column of the matrix, find each position q having the lowest value of that column. In other words, find each q such that

$$e(q, n) = \min_h e(h, n)$$

for all h. Then each desired interval has one of the associated a_q as its last term.

iii) For each q found in (ii), find a position p in the first column of the matrix such that $e(p - 1, 0)$ is connected to $e(q, n)$ by the sequence of inductive steps described in (i). Then $a_p a_{p+1} \ldots a_q$ is an interval $i \subset a$ which most resembles pattern b globally. All other such intervals may be found in the same way.

This theorem has the shortcoming that it cannot identify intervals i in a that most resemble b locally but not globally. Consider the situation of sequence a containing two widely separated intervals i_1 and i_2 that both strongly resemble b. For simplicity, let us assume that no other interval in a resembles b as well as either i_1 or i_2 do. Thus the distances $d(i_1, b)$ and $d(i_2, b)$ are both relatively small. But if

$$d(i_1, b) < d(i_2, b),$$

only i_1 would be found by Theorem 1 to resemble b. Thus we shall define a new type of resemblance between $i \subset a$ and b and state a more general theorem for finding intervals such as i_2 that closely resemble pattern b.

Definition 3. Let d be the evolutionary distance of Definition 1. Then an interval $i \subset a$ *most resembles* b *locally* if and only if

$$d(i, b) \leq d(h, b)$$

and

$$d(i, b) \leq d(j, b)$$

for all h and j, where $h \subset i \subset j \subset a$.

Theorem 2. Given the sequences

$$a = a_1 a_2 \ldots a_m$$

and

$$\mathbf{b} = b_1 b_2 \ldots b_n,$$

the following procedure determines every interval

$$a_p a_{p+1} \ldots a_q = \mathbf{i} \subset \mathbf{a}$$

that most resembles **b** locally:

i) The *forward distance matrix*

$$D_f = (e(i, j))$$

is constructed by induction as in Theorem 1, starting with the first terms of **a** and **b** and finishing with the last terms.

ii) The *reverse distance matrix*

$$D_r = (f(k, j))$$

is also constructed by induction but starting with the last terms of **a** and **b** and ending with the first. The meaning of $f(k, j)$ is the minimum total cost of the mutations, deletions, and insertions required to convert the terminal segment $b_k b_{k+1} \ldots b_n$ into any interval $\mathbf{i} \subset \mathbf{a}$ of any length that starts at a_k. The initial values of the matrix elements in the last column and last row are set as

$$f(i, n + 1) = 0$$

for $i = 1, 2, \ldots, m + 1$ and

$$f(m + 1, j) = \sum_{h=0}^{n-j} d(-, b_{n-h})$$

for $j = 1, 2, \ldots, n$. The remaining values are determined by making $f(i, j)$ equal to the smallest of the three values:

$$[f(i + 1, j) + d(a_i, -)],$$
$$[f(i + 1, j + 1) + d(a_i, b_j)],$$
$$[f(i, j + 1) + d(-, b_j)].$$

iii) Find a sequence of matrix positions with $(p - 1, 0)$ at one end and (q, n) at the other end such that every pair of successive positions in this sequence is connected by an inductive step in both the D_f matrix and the D_r

matrix. Then $a_p a_{p+1} \ldots a_q$ is an interval $\mathbf{i} \subset \mathbf{a}$ that most resembles pattern **b** locally. All other such intervals may be found in the same way.

3. RECOGNITION OF PATTERN s IN SEQUENCE l

Consider the problem of finding the intervals **i** in the longer sequence

$$\mathbf{l} = \text{abcacac}$$

that most resemble the shorter pattern

$$\mathbf{s} = \text{bcab}.$$

Let us use the evolutionary distance based on the costs $d(x, y) = 0$ for $x = y$ and $d(x, y) = 1$ for $x \neq y$, where x and y are from the set $\{-, \text{a}, \text{b}, \text{c}\}$. First, the distance matrices D_f and D_r are calculated, as shown in Fig. 1. Second, *forward path graph* P_f is constructed from forward distance matrix D_f by connecting each pair of matrix positions connected by an inductive step with a line segment and by replacing each matrix position with a dot. *Reverse path graph* P_r is constructed from reverse distance matrix D_r similarly. These path graphs are also shown in Fig. 1.

A *transverse path* is a succession of line segments connecting the first column and the last column of a path graph. Since each line segment corresponds to an inductive step in a distance matrix, a transverse path in graph P_f starts in the first column and can move to the right, diagonally, or down on each step, but it cannot move up or to the left. In contrast, a transverse path in graph P_r starts in the last column and can move to the left, diagonally, or up on each step, but it cannot move down or to the right. Horizontal lines in a path graph correspond to the deletion of terms from pattern **s**, diagonal lines represent the identity or mutation or two terms, and vertical lines correspond to the deletion of terms from sequence **l**.

Each transverse path defines a metric alignment of pattern **s** with a unique interval in **l**. For instance, the transverse path that ends at the fourth position of the last column of graph P_f defines the metric alignment of **s** with the interval bc in **l** as follows:

$$\mathbf{s} = \text{b} \quad \text{c} \quad \text{a} \quad \text{b}$$
$$\mathbf{i}_0 = \text{b} \quad \text{c} \quad - \quad -$$

The corresponding number 2 in the fourth position of the last column of matrix D_f is the evolutionary distance for that metric alignment (transverse path), so that

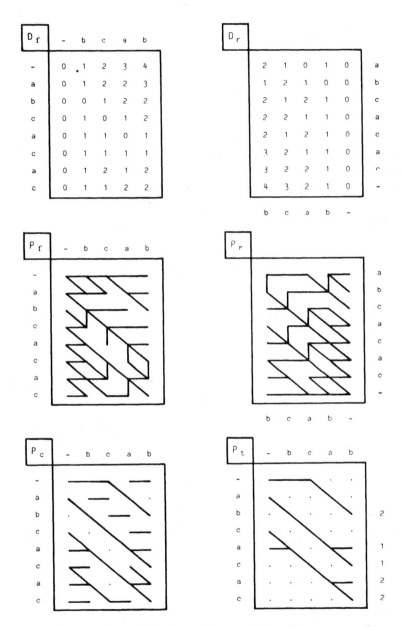

Figure 1. Matrices and graphs for determining each interval in sequence abcacac that most resembles pattern bcab globally or locally. D_f, forward distance matrix; D_r, reverse distance matrix; P_f, forward path graph; P_r, reverse path graph; P_c, common path graph; P_t, transverse path graph.

$$d(\text{bcab, bc}) = 2.$$

Since 2 is not the smallest number in the last column of D_f, interval bc does not most resemble bcab globally. In addition, since this transverse path does not appear in both path graphs P_f and P_r, bc does not even most resemble bcab locally.

There are four paths ending at the last two positions in the last column of graph P_f, however, that are also present in graph P_r. These paths are more readily seen by constructing two new graphs, P_c and P_t. *Common path graph P_c* is the intersection of the path graphs P_f and P_r and consists of all line segments (inductive steps) common to both. The *transverse path graph P_t* contains only those line segments in graph P_c that are part of a transverse path between the first and last columns. Graphs P_c and P_t are also shown in Fig. 1. For convenience, the numbers shown to the right of the positions in the last column of graph P_t are the evolutionary distances for each transverse path, which were taken from the corresponding positions of the last column of the forward distance matrix D_f.

Graph P_t clearly illustrates that there are seven transverse paths common to graphs P_f and P_r. The single transverse path at the top of the graph corresponds to the interval ab in I, which is at an evolutionary distance of 2 from pattern s and is metrically aligned with s as follows:

$$
\begin{array}{cccc}
s = b & c & a & b \\
i_1 = - & - & a & b \\
\end{array}
$$
$$1 + 1 + 0 + 0 = 2 = d(\text{bcab, ab})$$

The four overlapping transverse paths in the lower part of graph P_t correspond to four intervals in I that resemble pattern s. Note that *intersecting* transverse paths always specify intervals that are the *same* distance from the pattern. These four intervals in I (acac, aca, cac, ca) are each at an evolutionary distance of 2 from pattern s and are metrically aligned with s as follows:

$$
\begin{array}{cccc}
s = b & c & a & b \\
i_2 = a & c & a & c \\
\end{array}
$$
$$1 + 0 + 0 + 1 = 2 = d(\text{bcab, acac})$$

$$
\begin{array}{cccc}
s = b & c & a & b \\
i_3 = a & c & a & - \\
\end{array}
$$
$$1 + 0 + 0 + 1 = 2 = d(\text{bcab, aca})$$

$$s = b \quad c \quad a \quad b$$

$$i_4 = - \quad c \quad a \quad c$$

$$1 + 0 + 0 + 1 = 2 = d(\text{bcab, cac})$$

$$s = b \quad c \quad a \quad b$$

$$i_5 = - \quad c \quad a \quad -$$

$$1 + 0 + 0 + 1 = 2 = d(\text{bcab, ca})$$

Each of the five intervals in I mentioned above most resembles s *locally*, which means that for each interval no other interval that contains it or is contained within it is closer to s. But since the evolutionary distance of these transverse paths (metric alignments) is not the lowest distance in the last column of D_f, these five intervals do not most resemble pattern s globally.

The two paths in the center of graph P_t correspond to two intervals in I (bcac, bca) that are both at an evolutionary distance of 1 from s. They are metrically aligned with pattern s as follows:

$$s = b \quad c \quad a \quad b$$

$$i_6 = b \quad c \quad a \quad c$$

$$0 + 0 + 0 + 1 = 1 = d(\text{bcab, bcac})$$

$$s = b \quad c \quad a \quad b$$

$$i_7 = b \quad c \quad a \quad -$$

$$0 + 0 + 0 + 1 = 1 = d(\text{bcab, bca})$$

Each of these intervals most resembles s *globally*, which means that no other interval in I is closer to s.

In practice, Theorems 1 and 2 have been implemented as a computer algorithm called SU. The letters S and U indicate that pattern **b** is *specified* but the intervals **i** in **a** are *unspecified* before the algorithm is executed. Algorithm SU calculates the distance matrices D_f and D_r and the path graphs P_f, P_r, and P_c. Then it generates the transverse path graph P_t by discarding all line segments that are not part of a transverse path between the first and last columns of P_c. Finally, for each interval **i** in I that most resembles pattern s either globally or locally, algorithm SU can obtain the evolutionary distance $d(s, i)$ from matrix D_f and generate the corresponding metric alignments from path graph P_t. A flow chart of these operations is shown in Fig. 2.

Algorithm SU has been used successfully to align metrically the 45-residue sequence of a segment of an antibody from a phylogenetically primitive shark

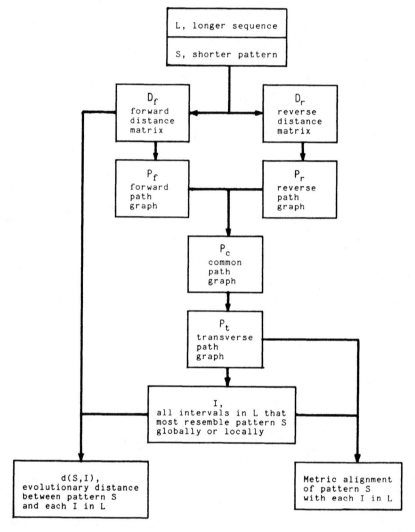

Figure 2. Flow chart of sequence information produced by computer algorithm SU.

with unspecified intervals of human antibody sequences containing from 446 to 568 residues (Litman *et al.*, 1982). The following sections describe three other examples of the metric analysis of genetic sequences using Theorem 2 and algorithm SU.

4. RECOGNITION OF A PATTERN IN TWO DNA SEQUENCES

Metric analysis of two nucleotide sequences is simpler than analysis of two protein sequences. The four-letter alphabets for DNA (A, C, G, T) and RNA (A, C, G, U) are much smaller than the 20-letter alphabet of the genetically coded amino acids present in proteins. The DNA letter T and the RNA letter U are equivalent because T is transcribed as U. Table 2 shows the metric set of deletion and mutation costs we currently use for RNA and DNA sequences. The units are called base changes or bc. The deletion costs have arbitrarily been set equal to 2 bc so that deletion of a base costs twice as much as its mutation.

A biologically interesting example of the metric analysis of two nucleotide sequences is the comparison of two repeated sequences from satellite DNA of the fruitfly, *Drosophila melanogaster*. Satellite DNA molecules are relatively small pieces of double-stranded DNA located in the chromosome but different from transcribed DNA. They usually contain a DNA sequence a few hundred base pairs (bp) in length that is repeated about 20–40 times in head-to-tail fashion. Their biological function is not presently known. Brutlag and his collaborators have determined the nucleotide sequences of two such satellite DNA molecules. One contains a repeated sequence 359–bp in length (Hsieh and Brutlag, 1979), and the other contains a 254-bp repeated sequence (Carlson and Brutlag, 1979), as shown in Fig. 3. For simplicity, the sequence of only one strand of each double-stranded nucleotide is shown. The other strand is implied because T of one strand always pairs with A of the other and similarly for G with C. Each repeated sequence was isolated by cleavage with the restriction endonuclease *Hae*III, a highly specific enzyme that happens to cut

Table 2
Mutation and deletion costs for nucleic acids*

	A	C	G	T	U	X†	–
A	0	1	1	1	1	1	2
C	1	0	1	1	1	1	2
G	1	1	0	1	1	1	2
T	1	1	1	0	0	1	2
U	1	1	1	0	0	1	2
X	1	1	1	1	1	1	2
–	2	2	2	2	2	2	0

*Units are base changes.
†The unknown base X is assumed to be different from another unknown base when they are paired in a metric alignment. The RNA base U and the DNA base T represent the same genetic information.

(a) The 359-base pair repeat

```
CCACATTTTGCAAATTTTGATGACCCCCCTCCTTACAAAAAATGCGAAAATTGATCCAAA
AATTAATTTCCCTAAATCCTTCAAAAAGTAATAGGGATCGTTAGCACTGGTAATTAGCTG
CTCAAAACAGATATTCGTACATCTATGTGACCATTTTTAGCCAAGTTATAACGAAAATTT
CGTTTGTAAATATCCACTTTTTTGCAGAGTCTGTTTTTCCAAATTTCGGTCATCAAATAA
TCATTTATTTTGCCACAACATAAAAAATAATTGTCTGAATATGGAATGTCATATCTCACT
GAGCTCGTAATAAAATTTCCAATCAAACTGTGTTCAAAAATGGAAATTAAATTTTTTGG
```

(b) The 254-base pair repeat

```
CAXATTTGCAAATTTAATGAACCCCCCTTCAAAAAATGCGAAAATTAACGCAAAAATTGA
TTTCCCTAAATCCTTCAAAAAGTAAATAACAACTTTTTGGCAAAATCTGATTCCCTAATT
TCGGTCATTAAATAATCAGTTTTTTTTGCCACAACTTTAAAAATAATTGTCTGAATATGGA
ATGTCATACCTCGCXXAGCTXGTAATTAAATTTCCAATGAAACTGTGTTCAACAATGAAA
ATTACATTTTTCGG
```

Figure 3. Nucleotide sequences for two repeated segments of the 1.688 g/cm^3 satellite DNA from the fruitfly *Drosophila melanogaster*.

both strands of the double-stranded satellite DNA at only one position per repeated sequence. The resulting double-stranded DNA fragment was numbered from one end of one strand to the other end.

The use of Theorem 2 and algorithm SU in the metric analysis of these sequences requires that a pattern be specified before the algorithm is executed. In the present case, a trial pattern was chosen through aligning the two DNA sequences visually by sliding them back and forth until a reasonable fit was seen. This process suggested that the first 85 base pairs of the 254-bp sequence could be aligned well with a comparable region from the beginning of the 359-bp sequence. In addition, the remainder of the 254-bp sequence could be aligned well with the part of the 359-bp sequence following position 190. These tentative alignments were found to be the best alignments by using Theorem 2 and algorithm SU. Thus interval (2–91) of the 359-bp sequence most resembled the pattern 254-bp–(1–85) globally, and interval (190–359) of the 359-bp sequence most resembled the pattern 254-bp–(86–254) globally, as shown in Fig. 4.

This diagram suggested that the two repeated sequences have in common not two separate intervals but one large interval. In order to test this proposal, a doubled sequence was constructed by joining the end of the 359-bp sequence to the beginning of another copy of the same sequence. Then another sequence was constructed by joining the end of interval (86–254) to the beginning of interval (1–85) from the 254-bp sequence. Metric analysis using algorithm SU showed that interval (190–359/1–91) from the doubled 359-bp sequence most resembled the pattern 254-bp–(86–254/1–85) globally, as shown in Fig. 5.

This metric-analysis result suggests that the numbering convention that starts after the *Hae*III cleavage site may not define the natural repeated sequence. A more natural scheme for the 254-bp sequence would start at nucleotide 86 of one copy and end at nucleotide 85 of the next copy. Similarly, a more natural numbering scheme for the 359-bp repeated sequence would start at nucleotide 92 of one copy and end at nucleotide 91 of the next copy. Alternatively, numbering would start at nucleotide 190 of one copy and end at nucleotide 189 of the next. In either case, it is clear that interval (92–189) of the 359-bp sequence is absent from the 254-bp sequence.

The evolutionary distance between the pattern, 254-bp–(86–254/1–85), and the found interval, 359-bp–(190–359/1–91), is 45 bc, which consists of 31 base changes (mutations) and 7 base deletions (scored as 2 bc each). The metric alignments that correspond to this distance are indicated in Fig. 5, which uses the *standard format* illustrated in Table 3. The first of the six lines of the standard format shows the position numbering for the first sequence, here the pattern from the 254-bp sequence. The second line shows the valence of the alignment, as explained below. The third line shows the first sequence. The fourth line indicates the *similarity* of the two sequences by echoing all terms identical in both sequences and by leaving positions blank that have either different terms in the two sequences or a null in one of the sequences. The fifth line shows the second sequence, here the interval found in the 359-bp sequence, and the sixth line shows its position numbering.

Algorithm SU located a single interval in the sequence of two adjacent copies of the 359-bp sequence as globally most resembling the pattern, but the sequences of the pattern and this interval can be metrically aligned in precisely 1600 different ways. Only one of the 1600 metric alignments is explicitly shown in Fig. 5. The relatedness of these many alignments is indicated by the property of *valence,* as shown in the second line. A dash (–) is shown above each alignment position that is conserved in all metric alignments. These uniquely

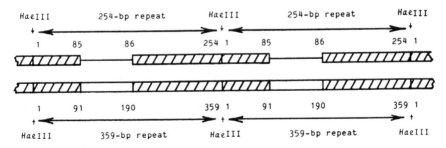

Figure 4. Schematic alignment of two repeated sequences of satellite DNA from the fruitfly. The sites cleaved by the restriction endonuclease *Hae*III are shown by arrows. The similar regions of the 254-bp sequence and the 359-bp sequence are hatched.

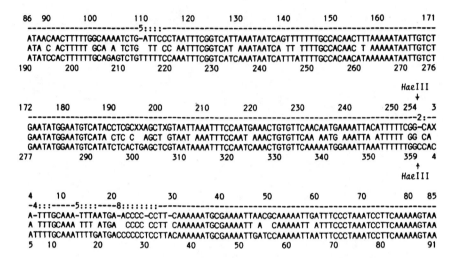

Figure 5. Metric alignment of two repeated sequences of satellite DNA from the fruitfly. See Table 3 for explanation of the format. Interval (86-254/1-85) of the repeated 254-bp sequence is aligned with interval (190-359/1-91) of the repeated 359-bp sequence. The evolutionary distance of 45 bc consists of 223 identities, 31 base changes, and 7 base deletions (costing 2 bc each).

aligned positions are termed *univalent*. The present valence line contains 236 dashes located as follows: 207 over identical positions, 28 over mutated positions, and one over a position containing a null. The remaining 25 alignment positions have either a numeral or a colon (:) in the valence line. Since they cannot be uniquely aligned in all metric alignments, these positions are called *multivalent*. The alignment in Fig. 5 contains five short multivalent regions. For example, the notation "5: : : :" near the beginning of the valence line indicates that for this region of five alignment positions (from the first numeral through the last colon), the four bases ATTC of the 254-bp sequence can be aligned with the five bases TTTTT of the 359-bp sequence in *five* different ways that preserve the *local evolutionary distance* of four base changes. The valence of this multivalent region is thus five. Similarly, near the end of the valence line the numeral "8" is followed by eight colons, which denotes that this region of nine alignment positions can be aligned in *eight* different ways that still preserve the local evolutionary distance of five base changes. The valence of this alignment region is thus eight.

The *total alignment valence,* which is defined as the product of the valences of all alignment regions, is equal to the total number of different metric alignments in the metric alignment set. Since each univalent position is an alignment region with valence of one, this product reduces to the product of the valences of the multivalent alignment regions. In the present example, the total alignment valence is

$$5 \cdot 2 \cdot 4 \cdot 5 \cdot 8 = 1600.$$

Thus the metric alignment set contains 1600 different metric alignments having the evolutionary distance of 45 base changes between the DNA sequences. This result is due to the presence of six nulls (the absence of six bases) in the 254-bp pattern relative to the 359-bp interval that can be aligned metrically at 25 alignment positions. Thus 90 percent of the alignment positions are common to all 1600 metric alignments.

The single metric alignment shown in the standard format places the nulls as early as possible in the first sequence and as late as possible in the second sequence. The *local metric alignments* for a given multivalent alignment region can often be deduced by inspection. For example, the first multivalent region of Fig. 5 shows one local metric alignment as

–ATTC

TTTTT.

The other four local metric alignments can be generated in this case by moving the null down the sequence one position at a time to give

A–TTC

TTTTT

Table 3
Brief explanation of the six-line standard format for describing the metric alignment of two genetic sequences

Line	Information	Example*		
1	POSITION NUMBERS for first sequence	1 5 9		
2	VALENCE of alignment regions	– – 2 : – – 4 : : : –		
3	FIRST SEQUENCE and inserted nulls	CG – AGA – TTA C		
4	SIMILARITY of alignment positions	CG AGA TT C		
5	SECOND SEQUENCE and inserted nulls	CGAAGATTTG C		
6	POSITION NUMBERS for second sequence	1 5 11		

*If interpreted as two DNA sequences, this alignment has an evolutionary distance of 5 bc, by using the mutation and deletion costs shown in Table 2. If interpreted as two amino acid sequences, it has a genetic distance of 7 mbc, by using the metric set of genetic distances between amino acids shown in Table 5. In either case, the standard format shows one of eight metric alignments. See text for explanation of the format lines.

AT–TC

TTTTT

ATT–C

TTTTT

ATTC–

TTTTT.

The alternative local metric alignments for the next three multivalent alignment regions are generated similarly. The eight local metric alignments for the last multivalent region are less obvious but can be generated from the transverse path graph.

The close resemblance between these two repeated segments of satellite DNA has suggested that they evolved from a common ancestral DNA sequence (Carlson and Brutlag, 1979). Metric analysis has provided not just one best alignment but all 1600 different metric alignments of these DNA sequences.

5. A METRIC SET OF GENETIC DISTANCES BETWEEN AMINO ACIDS

Metric analysis of the amino acid sequences of proteins is more involved than analysis of the nucleotide sequences of DNA or RNA. The 20-letter alphabet of the genetically coded amino acids requires a significantly larger metric set of mutation and deletion costs. Amino acid sequences are generated from DNA sequences by way of RNA sequences through the translation of specific base triplets, or codons, as specific amino acids according to the genetic code. Thus the cost of mutation between two amino acids is most appropriately correlated with the number of base changes needed to interconvert those amino acids. For metric analysis of amino acid sequences, we introduce a metric function called genetic distance that operates on the 20 codon sets corresponding to the 20 genetically coded amino acids.

> *Definition 4.* The *genetic distance* between two genetically coded amino acids is defined as the minimum number of base changes ("minimum base changes" or mbc) needed to convert any codon from the set coding for one amino acid into any codon from the set coding for the other amino acid.

The metric property of positive definiteness requires that the genetic distance between an amino acid and itself be zero. Thus we define the genetic distance between the codon set for an amino acid and itself to be 0 mbc. Mutations between codons from the same codon set are called *silent mutations* because the the encoded amino acid does not change. Silent mutations are ignored in assigning genetic distances. For example, the genetic distance $d(V, V)$ between the valine codon set and itself is 0 mbc even though one valine residue may be coded as GUG and another as GUC. Thus

$$d(V, V) = 0 \text{ mbc}$$

where

$$\text{valine} = V = \{\underline{GUG}, GUA, GUC, GUU\}$$

$$\uparrow \downarrow \qquad\qquad 0 \text{ mbc}$$

$$\text{valine} = V = \{GUG, GUA, \underline{GUC}, GUU\}$$

and codon sets are enclosed in braces.

The metric property of positive definiteness also requires that the genetic distance between two codon sets coding for different amino acids be greater than 0 mbc. For example, the genetic distance $d(M, K)$ between the codon sets for methionine and lysine is 1 mbc because a minimum of one base change is needed to interconvert the methionine codon AUG and the lysine codon AAG. Thus

$$d(M, K) = 1 \text{ mbc}$$

where

$$\text{methionine} = M = \{\underline{AUG}\}$$

$$\uparrow \downarrow \qquad\qquad 1 \text{ mbc}$$

$$\text{lysine} = K = \{\underline{AAG}, AAA\}$$

In another example, the genetic distance $d(M, A)$ between the codons sets for methionine and alanine is 2 mbc because a third codon set (either threonine or valine) must intervene. Thus

$$d(M, A) = d(M, L) + d(L, A) = 1 \text{ mbc} + 1 \text{ mbc} = 2 \text{ mbc}$$

where

$$\text{methionine} = M = \{\underline{AUG}\}$$

$$\uparrow \downarrow \qquad\qquad 1 \text{ mbc}$$

$$\text{valine} = V = \{\underline{GUG}, GUA, GUC, GUU\}$$

$$\uparrow \downarrow \qquad\qquad 1 \text{ mbc}$$

$$\text{alanine} = A = \{\underline{GCG}, GCA, GCC, GCU\}$$

In 13 cases, the genetic distance between two codon sets is 2 mbc because a silent mutation can be invoked in the intervening codon set. For example, the methionine codon set {AUG} shares no bases at the same codon position with either member of the aspartic acid codon set {GAC, GAU}. Although three base changes must in fact occur to interconvert the methionine codon AUG and the aspartic acid codon GAC, the genetic distance $d(M, D)$ between the codon sets for methionine and aspartic acid is only 2 mbc because both intervening codon sets can be the same codon set, such as valine. Thus

$$d(M, D) = d(M, V) + d(V, V) + d(V, D)$$
$$= 1 \text{ mbc} + 0 \text{ mbc} + 1 \text{mbc} = 2 \text{ mbc}$$

where

$$\text{methionine} = M = \{\underline{AUG}\}$$

$$\uparrow \downarrow \qquad\qquad 1 \text{ mbc}$$

$$\text{valine} = V = \{\underline{GUG}, GUA, GUC, GUU\}$$

$$\uparrow \downarrow \qquad\qquad 0 \text{ mbc}$$

$$\text{valine} = V = \{GUG, GUA, \underline{GUC}, GUU\}$$

$$\uparrow \downarrow \qquad\qquad 1 \text{ mbc}$$

$$\text{aspartic acid} = D = \{\underline{GAC}, GAU\}.$$

All 13 cases involving a silent mutation in the intervening codon set are listed in Table 4. In only one case is the genetic distance between two codon sets 3 mbc. The genetic distance $d(M, Y)$ between the codon sets for methionine and tyrosine cannot involve a silent mutation in a single intervening codon set. For example, the three base changes needed to interconvert the AUG codon for methionine and the UAC codon for tyrosine can be achieved only by invoking two different intervening codon sets. Thus

$$d(M, Y) = d(M, I) + d(I, F) + d(F, Y)$$
$$= 1 \text{ mbc} + 1 \text{ mbc} + 1 \text{ mbc} = 3 \text{ mbc}$$

where

Table 4
Thirteen genetic distances involving a silent mutation*

Codon set $\xrightarrow{\text{1 mbc}}$	Intervening codon set (Silent mutation)	$\xrightarrow{\text{1 mbc}}$ Codon set
K{AAA, AAG}	I{AUA, <u>AUC</u>, AUU}	F{<u>UUC</u>, UUU}
K{AAA, AAG}	R{AGA, AGG, CGA, <u>CGC</u>, CGG, CGU}	C{<u>UGC</u>, UGU}
E{<u>GAA</u>, GAG}	V{GUA, <u>GUC</u>, GUG, GUU}	F{<u>UUC</u>, UUU}
E{<u>GAA</u>, GAG}	G{<u>GGA</u>, <u>GGC</u>, GGG, GGU}	C{<u>UGC</u>, UGU}
D{<u>GAC</u>, GAU}	V{GUA, <u>GUC</u>, <u>GUG</u>, GUU}	M{AUG}
D{<u>GAC</u>, GAU}	G{GGA, <u>GGC</u>, <u>GGG</u>, GGU}	W{UGG}
N{<u>AAC</u>, AAU}	S{<u>AGC</u>, AGU, UCA, UCC, <u>UCG</u>, UCU}	W{UGG}
I{AUA, <u>AUC</u>, AUU}	S{<u>AGC</u>, AGU, UCA, UCC, <u>UCG</u>, UCU}	W{UGG}
Q{<u>CAA</u>, CAG}	L{<u>CUA</u>, CUC, CUG, CUU, <u>UUA</u>, UUG}	F{<u>UUC</u>, UUU}
Q{<u>CAA</u>, CAG}	R{AGA, AGG, <u>CGA</u>, <u>CGC</u>, CGG, CGU}	C{<u>UGC</u>, UGU}
C{<u>UGC</u>, UGU}	R{CGA, <u>CGC</u>, CGG, CGU, AGA, <u>AGG</u>}	M{AUG}
H{<u>CAC</u>, CAU}	L{<u>CUA</u>, <u>CUC</u>, CUG, CUU, UUA, UUG}	M{AUG}
H{<u>CAC</u>, CAU}	L{CUA, <u>CUC</u>, CUG, CUU, UUA, <u>UUG</u>}	W{UGG}

*The silent mutation within the intervening codon set involves the underlined codons.

methionine = M = {A<u>UG</u>}

↑ ↓　　　　1 mbc

isoleucine = I = {<u>AUC</u>, AUA, AUU}

↑ ↓　　　　1 mbc

phenylalanine = F = {<u>UUC</u>, UUU}

↑ ↓　　　　1 mbc

tyrosine = Y = {U<u>AC</u>, UAU}.

A complete set of genetic distances between the codon sets for all 20 genetically coded amino acids is shown in Table 5. On each side of the long diagonal, the 13 distances of 2 mbc involving a silent mutation are underlined and the single distance of 3 mbc is circled. In addition to these mutation costs, a set of deletion costs needs to be specified. We propose that it should cost at least as much to delete a base as to change it. Deletion of an amino acid, which is equivalent to deletion of one codon or three bases, should thus cost a

Table 5
Mutation and deletion costs for amino acids*

	A	S	G	L	K	V	T	P	E	D	N	I	Q	R	F	Y	C	H	M	W	Z	B	X	–
Ala = A	0	1	1	2	2	1	1	1	1	1	2	2	2	2	2	2	2	2	2	2	2	2	2	3
Ser = S	1	0	1	1	2	2	1	1	2	2	1	1	2	1	1	1	1	2	2	1	2	2	2	3
Gly = G	1	1	0	2	2	1	2	2	1	1	2	2	2	1	2	2	1	2	2	1	2	2	2	3
Leu = L	2	1	2	0	2	1	2	1	2	2	2	1	1	1	1	2	2	1	1	1	2	2	2	3
Lys = K	2	2	2	2	0	2	1	2	1	2	1	1	1	1	2	2	2	1	2	1	2	2	2	3
Val = V	1	2	1	1	2	0	2	2	1	1	2	1	2	2	1	2	2	1	2	2	2	2	2	3
Thr = T	1	1	2	2	1	2	0	1	2	2	1	1	2	1	2	2	2	2	1	2	2	2	2	3
Pro = P	1	1	2	1	2	2	1	0	2	2	2	2	1	1	2	2	2	1	2	2	2	2	2	3
Glu = E	1	2	1	2	1	1	2	2	0	1	2	2	1	2	2	2	2	2	2	1	2	2	2	3
Asp = D	1	2	1	2	2	1	2	2	1	0	1	2	2	2	2	1	2	1	2	2	2	1	2	3
Asn = N	2	1	2	2	1	2	1	2	2	1	0	1	2	2	2	1	2	1	2	2	2	1	2	3
Ile = I	2	1	2	1	1	1	1	2	2	2	1	0	2	1	1	2	2	2	1	2	2	2	2	3
Gln = Q	2	2	2	1	1	2	2	1	1	2	2	2	0	1	2	2	2	1	2	2	1	2	2	3
Arg = R	2	1	1	1	1	2	1	1	2	2	2	1	1	0	2	1	1	1	1	1	2	2	2	3
Phe = F	2	1	2	1	2	1	2	2	2	2	2	1	2	2	0	1	1	2	2	2	2	2	2	3
Tyr = Y	2	1	2	2	2	2	2	2	2	1	1	2	2	1	1	0	1	1	③	2	2	1	2	3
Cys = C	2	1	1	2	2	2	2	2	2	2	2	2	2	1	1	1	0	2	2	1	2	2	2	3
His = H	2	2	2	1	1	1	2	1	2	1	1	2	1	1	2	1	2	0	2	2	2	1	2	3
Met = M	2	2	2	1	2	2	1	2	2	2	2	1	2	1	2	③	2	2	0	2	2	2	2	3
Trp = W	2	1	1	1	1	2	2	2	1	2	2	2	2	1	2	2	1	2	2	0	2	2	2	3
Glx = Z	2	2	2	2	1	2	2	2	1	2	2	2	1	2	2	2	2	2	2	2	1	2	2	3
Asx = B	2	2	2	2	2	2	2	2	2	1	1	2	2	2	2	1	2	1	2	2	2	1	2	3
??? = X	2	2	2	2	2	2	2	2	2	2	2	2	2	2	2	2	2	2	2	2	2	2	2	3
null = –	3	3	3	3	3	3	3	3	3	3	3	3	3	3	3	3	3	3	3	3	3	3	3	0

*Units are minimum base changes (mbc). Mutation costs involving a silent mutation are underlined and mutations having a cost of 3 mbc are circled. When complete amino acid sequence data are not available, the following conservative assumptions are made: costs for Glx are the larger of the costs for Glu and Gln, costs for Asx are the larger of the costs for Asp and Asn, and mutation costs involving an unknown residue are 2 mbc.

minimum of three base changes. Thus we define the genetic distance $d(J, -)$ between an amino acid J and the null element "–" to be 3 mbc, as shown in the last row and column of Table 5. For completeness, the metric property of positive definiteness requires that the distance $d(-, -)$ between two null elements be 0 mbc. This set of 441 genetic distances satisfies the metric properties of positive definiteness, symmetry, and triangle inequality. It constitutes a metric set of 400 mutation costs and 41 deletion costs that is useful in calculating the dissimilarity of two amino acid sequences.

Definition 5. The *genetic distance* between two amino acid sequences is defined as the evolutionary distance between them calculated by using as the mutation costs the 400 genetic distances of 0–3 mbc between the 20 types of genetically coded amino acids and as the deletion costs the genetic distances $d(J, -) = 3$ mbc and $d(-, -) = 0$ mbc for all amino acids J and the null element "–".

When one or both amino acid sequences have not been completely determined, it is often desirable to approximate the genetic distance between them. The difficulty of unambiguously assigning a residue as either aspartic acid (Asp, D) or its amide derivative asparagine (Asn, N) occasionally leads to the ambiguous assignment of the residue as "either D or N," which is symbolized by "Asx" or "B." We conservatively define the *approximate genetic distance* $d(J, B)$ between any amino acid residue J and the ambiguous residue B as the larger of the two genetic distances $d(J, D)$ and $d(J, N)$. Analogously, problems in assigning a residue as either glutamic acid (Glu, E) or its amide derivative glutamine (Gln, Q) can result in its assignment as "either E or Q," which is symbolized by "Glx" or "Z." We similarly define the approximate genetic distance $d(J, Z)$ between any amino acid residue J and the ambiguous residue Z as the larger of the two genetic distances $d(J, E)$ and $d(J, Q)$. Finally, we define the approximate genetic distance $d(J, X)$ between any amino acid residue J and an unassigned residue X as 2 mbc, which is as large as or larger than all genetic distance except $d(M, Y)$ between methionine and tyrosine.

The 155 approximate genetic distances involving B, Z, or X are also shown in Table 5. Strictly speaking, the total set of 576 distances shown in Table 5 is not metric because

$$d(B, B) = 1 \text{ mbc}, \qquad d(Z, Z) = 1 \text{ mbc}, \qquad \text{and} \qquad d(X, X) = 2 \text{ mbc}.$$

This violates the metric principle of positive definiteness. In practice, however, the ability to calculate an approximate genetic distance between two slightly ambiguous amino acid sequences by using these values is still very useful.

6. RECOGNITION OF A CYCLICALLY PERMUTED PATTERN IN TWO PROTEIN SEQUENCES

An interesting biological example of the use of metric analysis for comparing two proteins is the relationship between the amino acid sequences of two plant proteins, concanavalin A (Con A) and favin. Con A, which is isolated from the jack bean (*Canavalia ensiformis*), and favin, which comes from the fava bean (*Vicia faba*), share the ability to bind the same type of sugar molecule. Favin contains two amino acid chains, an alpha chain of 51 amino acid residues (Hemperly *et al.*, 1979) and a beta chain of 185 residues (Cunningham *et al.*, 1979) for a combined length of 236 residues. Con A is a single chain of 237 amino acid residues (Cunningham *et al.*, 1975). In 1979, we collaborated with

Cunningham, Edelman, and their colleagues (Cunningham *et al.,* 1979) to confirm the significance of their observation that the amino acid sequences of Con A and favin are related by cyclic permutation. They had noted that the beginning of Con A resembles the end of favin beta chain, and that the end of Con A resembles the beginning of favin beta chain.

These unprecedented relationships were examined through determining the genetic distances between a pattern from one sequence and all intervals of the other sequence that most resemble that pattern globally, by using the SU algorithm. Initially, intervals of favin were specified as the pattern and unspecified intervals of Con A were sought. As shown schematically in Fig. 6, Con A–(70–120) most resembles favin alpha chain globally, Con A–(169–237) most resembles favin beta chain–(41–111) globally, and Con A–(1–69) most resembles favin beta chain–(117–185) globally.

These results were consistent with favin being initially synthesized as a single-chain precursor protein, but it was not known if the order of the chains was alpha–beta or beta–alpha. Thus an artificial amino acid sequence called favin beta/alpha/beta was created by joining the end of favin beta chain to the beginning of favin alpha chain and the end of the latter to the beginning of a second copy of favin beta chain. This 421-residue sequence effectively encompassed both orderings of the favin chains. Then intervals of Con A were specified as the pattern and unspecified intervals of favin beta/alpha/beta were sought. Metric analysis using algorithm SU showed that (a) favin beta chain–(117–185) most resembles Con A–(1–69) globally, (b) favin alpha chain–(2–51) most resembles Con A–(70–120) globally, (c) favin beta chain–(1–40) most resembles Con A–(123–162) globally, and (d) favin beta chain–(41–111) most resembles Con A–(169–237) globally. These four pairs of intervals are aligned schematically in Fig. 6; their genetic distances and metric alignments are shown in Fig. 7.

The amino acid sequences of concanavalin A and favin are thus related by cyclic permutation. Successive intervals of Con A most resemble globally the favin intervals beta chain–(117–185), alpha chain–(2–51), beta chain–(1–40), and beta chain–(41–111). Two short intervals of each protein sequence [Con A–(121–122), Con A–(163–168), favin alpha chain–(1), and favin beta chain–(112–116)] have no counterpart in the other sequence.

The observed genetic distance between two amino acid sequences might be due only to similarity of amino acid *composition* (the number of each type of amino acid) or might in part be due to similarity of amino acid *order*. The importance of amino acid order can be established by showing that the observed genetic distance is significantly lower than the genetic distance expected for a pair of sequences having the same amino acid compositions as the original sequences but having their amino acid residues in a random order. For example, the importance of amino acid order in producing metric-analysis result (c) of Fig. 7 was examined by the following procedure.

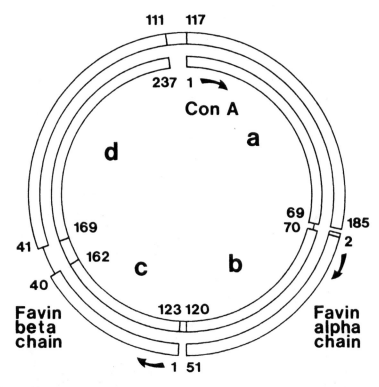

Figure 6. Schematic diagram showing the cyclic permutation of the amino acid sequences of concanavalin A and favin.

A Monte Carlo method was used to permute the sequence Con A–(123–162) in order to generate a random sample of 3350 sequences from the set of all sequences having the same amino acid composition as Con A–(123–162). A random sample of 3350 sequences from the set of all sequences having the same amino acid composition as favin beta chain–(1–40) was also generated. Then 3350 genetic distances were calculated between 3350 pairs of these permuted sequences. These distances ranged from 46 mbc to 65 mbc, with a mean of about 56 mbc. By using the one-sided Student's *t*-test at the 99% confidence level, it can be calculated that less than 0.1% of the time will the genetic distance between *any* pair of such permuted sequences have a value less than 46 mbc. Since the genetic distance between the original sequences is only 39 mbc, the probability that this distance is due to *only* the similarity of their amino acid *compositions* is much less than 0.1%. Thus the relatively low value of the genetic distance between Con A–(123–162) and favin beta chain–(1–40) is due at least in part to the similarity of the *order* of their amino acids. This Monte Carlo method was used with each of the other three metric-analysis results

Figure 7. Four results of metric analysis of intervals from concanavalin A and favin. The four pairs of intervals shown are aligned in the standard format (see Table 3). The similarity line between the sequences echoes identities and marks 1-mbc mutations with a plus ("+"). The genetic distance d is broken down into the number of 0-mbc identities, 1-mbc mutations, 2-mbc mutations, and 3-mbc deletions (nulls). In each case, sequence similarity is due at least in part to similarity of amino acid order (see text).

shown in Fig. 7. In each case, the probability that the result is due to amino acid composition alone is less than 0.1%. Thus, all four results shown in Fig. 7 are due at least in part to similarity of amino acid order.

The observation that the amino acid sequences of concanavalin A and favin are related by cyclic permutation suggests that the genes for these plant proteins probably evolved from a common ancestor. As previously discussed (Cunningham *et al.*, 1979), an unusual genetic event must be invoked to account for the present cyclic permutation of their amino acid sequences. The metric analysis results presented here confirm that the sequence of favin alpha chain is best aligned with the middle of the Con A sequence and that favin beta chain is best aligned by pairing the first 60 percent of its sequence with the end of Con A and the last 40 percent with the beginning of Con A. The metric-analysis results initially reported in 1979 (Cunningham *et al.*) and the more detailed results presented here represent improvements in the initial visual alignment of favin alpha chain with concanavalin A (Hemperly *et al.*, 1979). They also provide a more quantitative assessment of the sequence resemblance of these plant proteins.

7. RECOGNITION OF A POTENTIAL TRANSLOCATION SIGNAL IN A PROTEIN SEQUENCE

Another example of the use of Theorem 2 and the SU algorithm to locate a biologically relevant pattern within a protein sequence involves the potential translocation signal in ovalbumin, the major protein in chicken egg white. Trypsin and other secreted proteins differ from hemoglobin and other cytosolic proteins by their ability to enter the endoplasmic reticulum. This intracellular compartment is separated by a membrane from the cytosol, which is where amino acids are assembled into proteins.

Almost without exception, each secretory protein is synthesized in a slightly larger form called a presecretory protein, which bears a hydrophobic NH_2-terminal segment containing 15–30 amino acid residues not found in the secreted form of the protein. According to the signal hypothesis (Blobel and Dobberstein, 1975), this pre-segment contains the translocation signal that starts the movement of the presecretory protein through the endoplasmic reticulum membrane. Translocation generally begins soon after a small portion (roughly 100 residues) of the NH_2-terminus of the presecretory protein has been synthesized. Once the pre-segment has passed through the membrane, it is cut from the protein by signal peptidase, an enzyme bound to the inside of the endoplasmic reticulum membrane. In living cells, removal of the short-lived pre-segment normally occurs before synthesis of the presecretory protein is complete.

A surprising exception to this general scheme was reported by Palmiter *et al.* in early 1978. The initial methionine residue of the presecretory form of albumin is removed and replaced by an acetyl group in the secretory form. Ovalbumin is exceptional because it lacks a short-lived hydrophobic segment at its NH_2-terminus. Among several explanations for this unusual result, these workers suggested that an NH_2-terminal segment of ovalbumin functions as a translocation signal but is not removed, or that a hydrophobic translocation signal is located not at the NH_2-terminus but elsewhere within the protein. In any case, they concluded that at least two distinct mechanisms for protein secretion probably operate in the chicken oviduct, one that functions for ovomucoid and lysozyme, which have short-lived NH_2-terminal segments containing a translocation signal, and another that functions for ovalbumin, which lacks such a segment.

Table 6
Partial amino acid sequences at the NH_2-terminus of three proteins secreted from chicken oviduct

Protein segment	Amino acid sequence*
Preovomucoid-(1–29)	M A M A G V F V L F S F V L C G F L P D A A F G A E V D C
Prelysozyme-(1–24)	M R S L L I L V L C F L P L A A L G K V F G R C
Ovalbumin-(0–24)	M G S I G A A S M E F C F D V F K E L K V H H A N

*Initial cleavage site is shown by the arrow. Translocation signal is shown by the bracket.

Late in 1978, Lingappa *et al.* (1978) reported that partially synthesized ovalbumin crosses the microsomal membrane, an isolated form of the endoplasmic reticulum membrane. In addition, partially synthesized ovalbumin competes* with bovine preprolactin, a presecretory protein, for translocation across the membrane. These authors proposed that ovalbumin does contain a translocation signal at its NH$_2$-terminus but that it is not removed by the signal peptidase. Early in 1979, Lukas and Erickson prepared two synthetic peptides corresponding to the first 19 residues of chicken ovalbumin. When assayed by Lingappa and Blobel, neither synthetic ovalbumin–(1–19) nor synthetic N^α-acetyl-ovalbumin–(1–19) inhibited the translocation of bovine preprolactin across the microsomal membrane. Thus ovalbumin does not contain a translocation signal at its NH$_2$-terminus.

Later in 1979, Lingappa *et al.* (1979) described the isolation of a tryptic fragment† of chicken ovalbumin consisting of residues 229 to 276. This 48-residue fragment, called OV$_s$, did inhibit the translocation of preprolactin across the microsomal membrane. This result suggests that the translocation signal of ovalbumin is located within the interval (229–276).

In April 1979, shortly after the sequence of OV$_s$ was established, we independently used algorithm SU to determine which interval of the entire ovalbumin sequence globally most resembles the translocation signals of two normal presecretory proteins from chicken oviduct, prelysozyme (Palmiter *et al.*, 1977) and preovomucoid (Thibodeau *et al.*, 1978). Both of these preproteins bear a short-lived hydrophobic NH$_2$-terminal segment containing a translocation signal, as shown in Table 6. These pre-segments are removed by signal peptidase through selective cleavage after a glycine residue. In contrast, the NH$_2$-terminal methionine residue is removed but no larger fragment is lost during synthesis of ovalbumin.

For the purpose of metric analysis, the translocation signals of preovomucoid and prelysozyme were chosen to be preovomucoid–(9–24) and prelysozyme–(4–18) on the basis of their sequence similarity and their occurrence immediately before the signal peptidase cleavage site. As shown in Fig. 8, the genetic distance between these translocation signals is quite low. Ten of the 16 alignment positions contain the same residue in both sequences.

Each of these translocation signals was used as a pattern to search the 385-residue sequence of ovalbumin (McReynolds *et al.,* 1978) for similar sequences using algorithm SU. For example, part of the transverse path graph for alignment of ovalbumin with the prelysozyme pattern is shown in Fig. 9. The bent diagonal path between positions 239 and 251 indicates the alignment of

---

Editors' footnote: Two substances are said to compete for some process if both participate in it and an increase in the concentration of either one reduces the rate at which the other participates in the process.

†*Editors' footnote:* Tryptic fragments are the cleavage fragments produced by the enzyme trypsin.

```
                           9      15  19    24
                          ----------------
Preovomucoid-(9-24)       LFSFVLCGFLPDAAFG
                          L+++VLC FLP AA+G
Prelysozyme-(4-18)        LLILVLC-FLPLAALG
                          4      10  13    18
```

Figure 8. Metric alignment of the translocation signals of two presecretory proteins from the chicken oviduct. The genetic distance of 9 mbc is due to 10 identities, four 1-mbc mutations ("+"), one 2-mbc mutation, and one 3-mbc deletion.

ovalbumin–(239–251) with prelysozyme–(4–18). This ovalbumin interval is one of two intervals that most resemble the prelysozyme translocation signal pattern globally. Also shown is a path between positions 245 and 259, representing the interval (245–259), which most resembles the prelysozyme pattern locally.

This metric analysis located six ovalbumin intervals that most resemble the translocation signal patterns of *both* preovomucoid and prelysozyme locally and that are a combined genetic distance from these patterns of 34 mbc or less. Searching with two patterns was more discriminating than using only one. For example, ovalbumin–(242–256) is only 16 mbc from the prelysozyme pattern (Fig. 9) but it does not most resemble the preovomucoid pattern either globally or locally. The metric alignments of these six intervals with each of the translocation signal patterns are shown in Fig. 10 in an abbreviated format. For each of these six pairs of alignments, the sum of the evolutionary distances from the two translocation signal patterns are shown in Table 7.

The ovalbumin interval (239–251) is the only interval that most resembles *both* translocation signal patterns globally, being 16 mbc from the preovo-

Table 7
Genetic distances of six ovalbumin intervals from two translocation-signal patterns.

Ovalbumin interval, distance (mbc)		
Preovomucoid-(9–24)	Prelysozyme-(4–18)	Sum of distances (mbc)
(29–44), 19	(30–44), 15	34
(52–66), 18	(52–66), 14	32
(96–110), 16	(96–110), 18	34
(236–251), 16	(239–251), 14	30
(244–259), 18	(245–259), 16	34
(307–322), 16	(307–321), 17	33

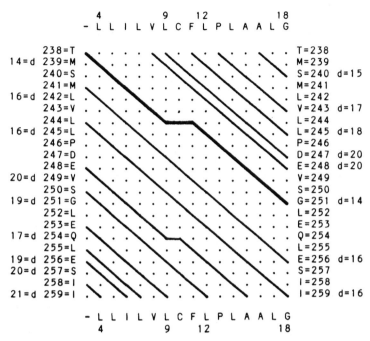

Figure 9. Part of the transverse path graph for metric alignment of intervals from chicken ovalbumin with chicken prelysozyme-(4-18). The 6 percent of the graph shown here corresponds to the ovalbumin region (238-259). The darker path represents the ovalbumin interval (239-251), which most resembles the prelysozyme signal globally.

mucoid pattern and 14 mbc from the prelysozyme pattern. As such, it is the single ovalbumin interval that is the lowest combined distance from both patterns. Thus ovalbumin–(239–251) is the interval most likely to be genetically related to the translocation signals of preovomucoid and prelysozyme at the DNA level and likely to be the interval that functions as the translocation signal of ovalbumin.

The translocation signals of preovomucoid and prelysozyme are aligned with this potential translocation signal of ovalbumin in Fig. 11. The alignment of the first two sequences is the same as the metric alignment shown in Fig. 8, and the alignment of the prelysozyme signal and ovalbumin–(239–251) is the same as the metric alignment shown in Fig. 10. But the alignment between the preovomucoid signal and ovalbumin–(239–251) is different from the metric alignment shown in Fig. 10 because three nulls have been inserted into the ovalbumin interval to maintain the other two alignments. This combined alignment shows identical residues for all three sequences in five of the alignment positions. This result is similar to an alignment proposed by Lingappa

```
preovomucoid-(9-24)     LFSFVLCGFLPDAAFG
                        +++++ + L+ + +G    d=19
ovalbumin-(29-44)       YCPIAIMSALAMVYLG

prelysozyme-(4-18)      LLILVLCFLPLAALG
                        +I +++ L+++ LG     d=15
ovalbumin-(30-44)       CPIAIMSALAMVYLG


preovomucoid-(9-24)     LFSFVLCGFLPDAAFG
                        +++ V++ F   ++FG   d=18
ovalbumin-(52-66)       QINKVVR-FDKLPGFG

prelysozyme-(4-18)      LLILVLCFLPLAALG
                        +++ V++F  L+++G    d=14
ovalbumin-(52-66)       QINKVVRFDKLPGFG


preovomucoid-(9-24)     LFSFVLCGFLPDAAFG
                        ++SF L ++++++A+ +  d=16
ovalbumin-(96-110)      VYSFSL-ASRLYAEER

prelysozyme-(4-18)      LLILVLCFLPLAALG
                        + ++ L +++ A+ +    d=18
ovalbumin-(96-110)      VYSFSLASRLYAEER


preovomucoid-(9-24)     LFSFVLCGFLPDAAFG
                        + +  + ++LPD+++G   d=16
ovalbumin-(236-251)     SGTMSMLVLLPDEVSG

prelysozyme-(4-18)      LLILVLCFLPLAALG
                        +++LVL  LP +++G    d=14
ovalbumin-(239-251)     MSMLVL--LPDEVSG


preovomucoid-(9-24)     LFSFVLCGFLPDAAFG
                        L++ +++G+ + +++    d=18
ovalbumin-(244-259)     LLPDEVSGLEQLESII

prelysozyme-(4-18)      LLILVLCFLPLAALG
                        L+   V+++ +L+++    d=16
ovalbumin-(245-259)     LPDEVSGLEQLESII


preovomucoid-(9-24)     LFSFVLCGFLPDAAFG
                        ++S  L+G+++++++    d=16
ovalbumin-(307-322)     SSSANLSGISSAESLK

prelysozyme-(4-18)      LLILVLCFLPLAALG
                        +++  L+ +++A++     d=17
ovalbumin-(307-321)     SSSANLSGISSAESL
```

Figure 10. Metric alignment for the six ovalbumin intervals that most resemble two translocation signal patterns. The middle line echoes identities, marks 1-mbc mutations with a plus ("+"), and indicates the genetic distance between the interval and the pattern.

```
Preovomucoid-(9-24)      LFSFVLCGFLPDAAFG

Prelysozyme-(4-18)       LLILVLC-FLPLAALG

Ovalbumin-(239-251)      MSMLVL---LPDEVSG

Common residues              VL    LP    G
```

Figure 11. Combined alignment of the functional translocation signals of two pre-secretory proteins from chicken oviduct and the potential translocation signal from chicken ovalbumin.

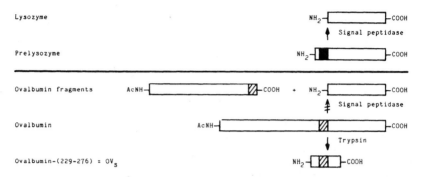

Figure 12. Different cleavage patterns for prelysozyme and ovalbumin. The translocation signal of prelysozyme is shown in black and the potential translocation signal of ovalbumin is hatched.

et al. (1979) after visual comparison of OV_s with the NH_2-terminal sequences of preovomucoid and prelysozyme. But metric analysis has shown that ovalbumin–(239–251) most resembles the translocation signals of these presecretory proteins not just within the 48-residue sequence of OV_s but within the entire 385-residue sequence of ovalbumin.

A diagram of the relative positions of the translocation signals of prelysozyme and ovalbumin within their respective proteins is shown in Fig. 12. Both of these signals begin about 150 residues from the COOH-terminus and are about 15 residues in length. When prelysozyme is cleaved by the signal peptidase to generate lysozyme, the translocation signal is lost. In contrast, signal peptidase does not cleave ovalbumin at the COOH-terminal side of the potential translocation signal, so the signal remains within the secreted protein. This translocation signal evidently remains functionally competent to interact with the microsomal membrane even after tryptic digestion has trimmed away most of ovalbumin to leave OV_s.

The similarity of ovalbumin–(239–251) to two functional translocation signal patterns may be biologically significant because this sequence is contained within the sequence of the known translocation inhibitor OV_s (residues 229–276). Of the five other ovalbumin intervals listed in Table 7, only ovalbumin–(245–259) is located in OV_s. But interval (245–259) is 4 mbc farther than (239–251) from the translocation signal patterns. In addition, Fig. 10 shows that (245–259) aligns metrically with only five identical residues in these patterns but (239–251) aligns with ten.

ACKNOWLEDGMENTS

We thank Mark A. Fulk and especially Janet M. Sekulski for computer implementation of the algorithms SS and SU. This work was supported by

Public Health Service research grants AI 15301, AI 18362, CA 28571, and RR 07065 and by funds from The Rockefeller University.

REFERENCES

Blobel, G., and Dobberstein, B., Transfer of proteins across membranes. I. Presence of proteolytically processed and unprocessed nascent immunoglobulin light chains on membrane-bound ribosomes of murine myeloma. *J. Cell Biol.* **67**, 835–851 (1975).

Carlson, M., and Brutlag, D., Different regions of a complex satellite DNA vary in size and sequence of the repeating unit. *J. Mol. Biol.* **135**, 483–500 (1979).

Cunningham, B.A., Hemperly, J.J., Hopp, T.P., and Edelman, G.M., Favin versus concanavalin A: Circularly permuted amino acid sequences. *Proc. Natl. Acad. Sci. USA* **76**, 3218–3222 (1979).

Cunninhgam, B.A., Wang, J.L., Waxdal, M.J., and Edelman, G.M., The covalent and three-dimensional structure of concanavalin A. II. Amino acid sequence of cyanogen bromide fragment F_3. *J. Biol. Chem.* **250**, 1503–1512 (1975).

Erickson, B.W., Fulk, M.A., and Sellers, P.H., Metric analysis of nucleic acid and protein sequences. *Abstr., 11th Int. Congr. Biochem.,* 29 (1979).

Hemperly, J.J., Hopp, T.P., Becker, J.W., and Cunningham, B.A., The chemical characterization of favin, a lectin isolated from *Vicia faba. J. Biol. Chem.* **254**, 6803–6810 (1979).

Hsieh, T.S., and Brutlag, D., Sequence and sequence variation within the 1.688 g/cm^3 satellite DNA of *Drosophila melanogaster. J. Mol. Biol.* **135**, 465–481 (1979).

Lingappa, V.R., Lingappa, J.R., and Blobel, G., Chicken ovalbumin contains an internal signal sequence. *Nature* **281**, 117–121 (1979).

Lingappa, V.R., Shields, D., Woo, S.L.C., and Blobel, G., Nascent chicken oval-bumin contains the functional equivalent of a signal sequence. *J. Cell Biol.* **79**, 567–572 (1978).

Litman, G.W., Erickson, B.W., Lederman, L., and Mäkelä, O., Antibody response in *Heterodontus. Mol. Cell. Biochem.* **45**, 49–57 (1982).

McReynolds, L., O'Malley, B.W., Nisbet, A.D., Fothergill, J.E., Givol, D., Fields, S., Robertson, N., and Brownlee, G.G., Sequence of chicken ovalbumin mRNA. *Nature* **273**, 723–728 (1978).

Palmiter, R.D., Gagnon, J., Ericsson, L.H., and Walsh, K.A., Precursor of egg white lysozyme. *J. Biol. Chem.* **252**, 6386–6393 (1977).

Palmiter, R.D., Gagnon, J., and Walsh, K.A., Ovalbumin: A secreted protein without a transient hydrophobic leader sequence. *Biochemistry* **75**, 94–98 (1978).

Sellers, P.H., An algorithm for the distance between two finite sequences. *J. Combinator. Theor.* **A16**, 253–258 (1974a).

Sellers, P.H., On the theory and computation of evolutionary distances. *SIAM J. Appl. Math.* **26**, 787–793 (1974b).

Sellers, P.H., Pattern recognition in genetic sequences. *Proc. Natl. Acad. Sci. USA* **76**, 3041 (1979).

Sellers, P.H., The theory and computation of evolutionary distances; pattern recognition. *J. Algorithms* **1**, 359–373 (1980).

Thibodeau, S.M., Lee, D.C., and Palmiter, R.D., Precursor of egg white ovomucoid. Amino acid sequence of an NH$_2$-terminal extension. *J. Biol. Chem.* **253**, 9018–9023 (1978).

Ulam, S.M., Some combinatorial problems studied experimentally on computing machines. In *Applications of Number Theory to Numerical Analysis,* S.K. Zaremba (Ed.), pp. 1–3. New York: Academic Press (1972).

Wagner, R.A., and Fischer, M.J., The string-to-string correction problem. *J. Assoc. Comput. Mach.* **21**, 168–173 (1974).

CHAPTER THREE

FAST ALGORITHMS TO DETERMINE RNA SECONDARY STRUCTURES CONTAINING MULTIPLE LOOPS

David Sankoff, Joseph B. Kruskal, Sylvie Mainville, and
Robert J. Cedergren

1. INTRODUCTION

A molecule of ribonucleic acid (RNA) is made up of a long chain of subunits—ribonucleotides—linked together. Each ribonucleotide contains one of four possible bases, abbreviated A, C, G, and U, and it is the sequence of bases in the chain that distinguishes one type of RNA from another. This base sequence is called the primary structure of the RNA molecule.

Under natural conditions, a ribonucleotide chain will twist and bend, and the bases will form bonds with one another in a complicated pattern, so that the molecule forms a coiled and looped conformation. Both the conformation and the pattern of bonding are called the secondary structure, but this paper deals only with the pattern of bonding. The base-to-base interactions that form the RNA secondary structure are of two kinds, namely, hydrogen bonding between an A and a U, and hydrogen bonding between a C and a G, as first described by Watson and Crick. Figure 1 depicts the secondary structure of a transfer RNA molecule. (Other macromolecules, such as proteins, also have secondary structures involving coiling and bending, but their biochemical basis and geometric character are different from that of RNA.)

Given the primary structure of an RNA, one may imagine a vast number of different secondary structures based on pairing A's with U's and pairing C's with G's in different ways. According to the laws of chemical thermodynamics, however, generally only one such secondary structure will be stable, namely, the secondary structure that optimizes the free energy. A method for deducing the stable secondary structure directly from the primary structure would be very useful, since empirical results, from x-ray diffraction of crystalline RNA and from various biochemical techniques, are difficult and costly to obtain and can often be interpreted in several ways. Such a method requires two parts: (1) a method to specify the free energy of any proposed secondary structure; (2) a reasonably fast algorithm to search among the many conceivable secondary structures to find the optimal one.

Part (2) is the primary focus of this chapter. We present three dynamic-programming algorithms with different properties, which somewhat resemble the algorithms discussed elsewhere in this volume. Though the first widely used program in this field (Pipas and McMahon, 1975) was of a different type, the dynamic-programming approach has since become prevalent. We may cite Sankoff (1976), Nussinov et al. (1978, 1980), Waterman (1978), Waterman and Smith (1978), and Zuker and Stiegler (1981). The most dramatic results obtained within this framework are those of Nussinov et al. (1982). Other important references include Salser (1977), Studnicka et al. (1978), Dumas and Ninio (1982), and Auron et al. (1982).

The algorithms presented in Secs. 6 through 9 improve on all previous work in that they are based on a completely rigorous analysis of the problem of "multiple loops" in secondary structure, whereas earlier algorithms ignore these, consider them as having no effect on free energy, or incorrectly assume they may be treated in the same way as other components of secondary structure. At the same time our methods in Secs. 8 and 9 achieve the same efficient operation reported in the recent literature (Nussinov and Jacobson, 1980; Zuker and Steigler, 1981), requiring only one "pass" of the algorithm and time that is proportional only to n^3 or n^4, respectively, where n is the length of the RNA sequence.

Such algorithms, however, depend on the method used to carry out Part (1) of the problem. Our method for this is essentially that of Tinoco, Uhlenbeck, and others (Tinoco et al., 1973), but includes careful attention to multiple loops. Our mathematical formalization and extension is presented in Secs. 2, 3, and 4.

The two faster algorithms are practical for n in the hundreds. For longer RNA molecules, where n is in the low thousands, however, the algorithms, though possibly feasible, are probably too expensive to be of interest. (See Nussinov et al., 1982, for description of a three-month computer run for the case $n = 1790$.) In Sec. 9 we sketch a heuristic technique that can be used to extend algorithms like ours and may make it practical to handle such long molecules.

2. CONSTRAINTS ON SECONDARY STRUCTURE

Let the molecule be the sequence $\mathbf{a} = a_1 \ldots a_n$, where each a_i is either A, G, C, or U. The sequence $\mathbf{a}$ is called the primary structure. We use $a_i \cdot a_j$ with $i < j$ to indicate that a_i and a_j form a pair, or are paired to each other in the Watson–Crick manner. A secondary structure S for $\mathbf{a}$ is a set of such pairs. The set S is required to satisfy the following constraints:

  a) Watson–Crick: If S contains $a_i \cdot a_j$, then a_i and a_j are either A and U, or U and A, or C and G, or G and C.
  b) No overlap of pairs: If S contains $a_i \cdot a_k$, then it cannot contain $a_i \cdot a_j$ (with $j \neq k$) or $a_j \cdot a_k$ (with $j \neq i$).
  c) No knots: If $h < i < j < k$, then S cannot contain both $a_h \cdot a_j$ and $a_i \cdot a_k$.
  d) No sharp turns: If S contains $a_i \cdot a_j$, then $|j - i| \geq 4$.

It follows from (b) that each a_i occurs either in exactly one pair or in no pairs, and a_i is described as paired or unpaired accordingly.

Because the constraints all restrict the possible shape of secondary structures, we may loosely refer to (a)–(d) as geometrical constraints, though they derive from a variety of stereochemical, mechanical, and thermodynamic considerations. Constraint (d) is local, while (b) and (c) constrain the pairing in a global way. Constraint (a) is the only one that makes reference to the actual values of the a_i in the alphabet {A, C, G, U}.

Constraints (b) and (c) may occasionally be violated in nature, but the resulting "triplets," "knots," etc., are considered features of a higher level of structure—tertiary structure—and their exclusion here reflects the prevailing hypothesis that it is meaningful to analyze secondary structures independently of such tertiary interactions, at least as a first step. Constraint (a) may also be violated occasionally by the presence of non-Watson–Crick pairs, primarily G · U, but there is no loss in generality in ignoring these pairs, as we shall show in Sec. 4.

Because the sequence $\mathbf{a}$ stays constant as we consider many different secondary structures on it, it is possible and generally convenient to refer to i and j, for example, rather than a_i and a_j. Thus we may refer to a pair $i \cdot j$ rather than $a_i \cdot a_j$, and similarly for other constructs that appear later on.

3. LOOPS

Secondary structures may have many different and complicated shapes: Figure 1a depicts a relatively simple example. However, as we shall describe in detail below, constraints (b) and (c) limit the possible shapes in a very significant way. In fact, any secondary structure S can be described in a unique and natural way as made up of substructures called *loops, stacked pairs,* and *external single-stranded regions.* The *cycles* of S, written $\mathbf{s}_1, \ldots, \mathbf{s}_t$, refer to both its loops and

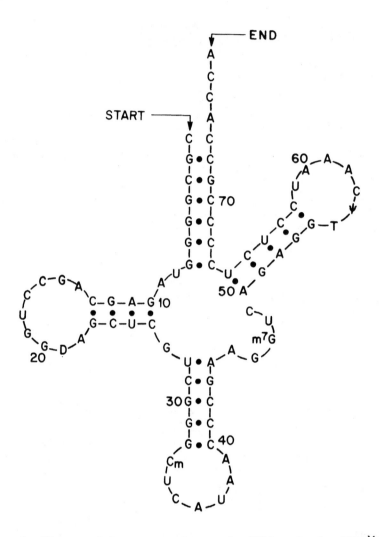

Figure 1a. The secondary structure of a transfer RNA molecule, tRNA$_F^{Met}$ from *Anacystis nidulans*. Letters other than A, G, C, U indicate chemical modifications of these four basic units. (Adapted from Ecarot-Charrier and Cedergren (1976).)

its stacked pairs. The Tinoco–Uhlenbeck approach to specifying the free energy $E(S)$ of any proposed secondary structure S rests on the hypothesis that the free energy is a sum of terms associated with the cycles,

$$E(S) = \sum_r e(s_r),$$

where $e(s_r)$ depends only on the nature of the cycle s_r.

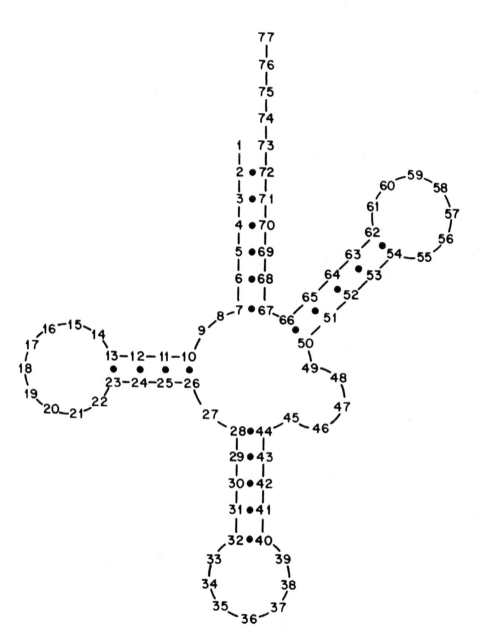

Figure 1b. Numbering of nucleotides for same tRNA molecule.

Following this approach, we distinguish the following types of substructures, as illustrated in Fig. 2 and described informally here. (A more formal characterization is given later.) If $i \cdot j$ is a pair and $i < r < j$, we say that $i \cdot j$ *surrounds* r. Similarly, $i \cdot j$ surrounds a pair $p \cdot q$ if it surrounds both p and q; due to constraint (c) forbidding knots, if $i \cdot j$ surrounds either p or q, it must surround both p and q.

1. If S contains $i \cdot j$ but none of the surrounded elements $i + 1, \ldots, j - 1$ are paired, the loop thus formed is called a *hairpin*.

2. If S contains $i \cdot j$, $(i + 1) \cdot (j - 1), \ldots, (i + h) \cdot (j - h)$, each of these pairs (except the last) is said to *stack* on the following pair. Two consecutive pairs may be referred to as a *stacked pair* or as a *stacked-pair cycle*.

3. If $i + 1 < p < q < j - 1$ and S contains $i \cdot j$ and $p \cdot q$, but the elements between i and p are unpaired and the elements between q and j are unpaired, then the two unpaired regions are said to constitute an *interior loop*.

4. If S contains $i \cdot j$ and $(i + 1) \cdot q$, and there are some unpaired elements between q and j, these unpaired elements form a *bulge*. Symmetrically, a bulge also occurs if S contains $i \cdot j$, $p \cdot (j - 1)$ and some unpaired elements between i and p.

5. If S contains $i \cdot j$ and $i \cdot j$ surrounds two or more pairs $p \cdot q$, $r \cdot s, \ldots$ which do not surround one another, then a *multiple loop* is formed. Figure 1 shows a multiple loop with three such pairs at the center of the RNA structure.

6. If r is unpaired and there is no pair in S surrounding r then we say that r is in an *(external) single-stranded region*.

Note that any cycle containing unpaired base(s) is a loop, and that the only kind of cycle that is not a loop is a stacked-pair cycle.

We now present a formal description of how the secondary structure S may be analyzed. We start with some notation and terminology. The *string* from i to j is written $[i, j]$ and means the sequence $i, i + 1, \ldots, j - 1, j$. A string is called *proper* (with respect to S) if, for every paired element in the string, its partner is also in the string. In Fig. 1b, the strings $[10, 26]$ and $[10, 27]$ are proper with respect to the structure shown, while $[10, 28]$ is not. The string $[i, j]$ is said to be *closed* (with respect to S) if i is paired to j. In Fig. 1b, $[10, 26]$ is closed but $[10, 27]$ is not. Due to constraint (c) forbidding knots, every closed string is also proper.

Suppose $[i, j]$ is a proper string. A pair $p \cdot q$ or an unpaired element r in the string is said to be *accessible in* $[i, j]$ if it is not surrounded by any pair in $[i, j]$ except possibly for $i \cdot j$ itself (if i and j are paired). If i and j are paired, we can also say that $p \cdot q$ or r is *accessible from* $i \cdot j$. In Fig. 1b, $28 \cdot 44$ and 45 are accessible in $[9, 61]$ and from $7 \cdot 67$, but not from $6 \cdot 68$. We always list the accessible pairs in a proper string $p_1 \cdot q_1, p_2 \cdot q_2, \ldots$ in natural order, i.e., so that $p_1 < q_1 < p_2 < q_2 < \ldots$.

Any pair $i \cdot j$ defines a *cycle* s *from i to j:* s consists of the *closing pair* $i \cdot j$ together with any pairs $p_1 \cdot q_1, p_2 \cdot q_2, \ldots$ accessible from $i \cdot j$ and any unpaired elements accessible from $i \cdot j$. If s contains k pairs (including the closing pair), it is said to be a *k-cycle* or to have *order k*. In Fig. 1b, the cycle

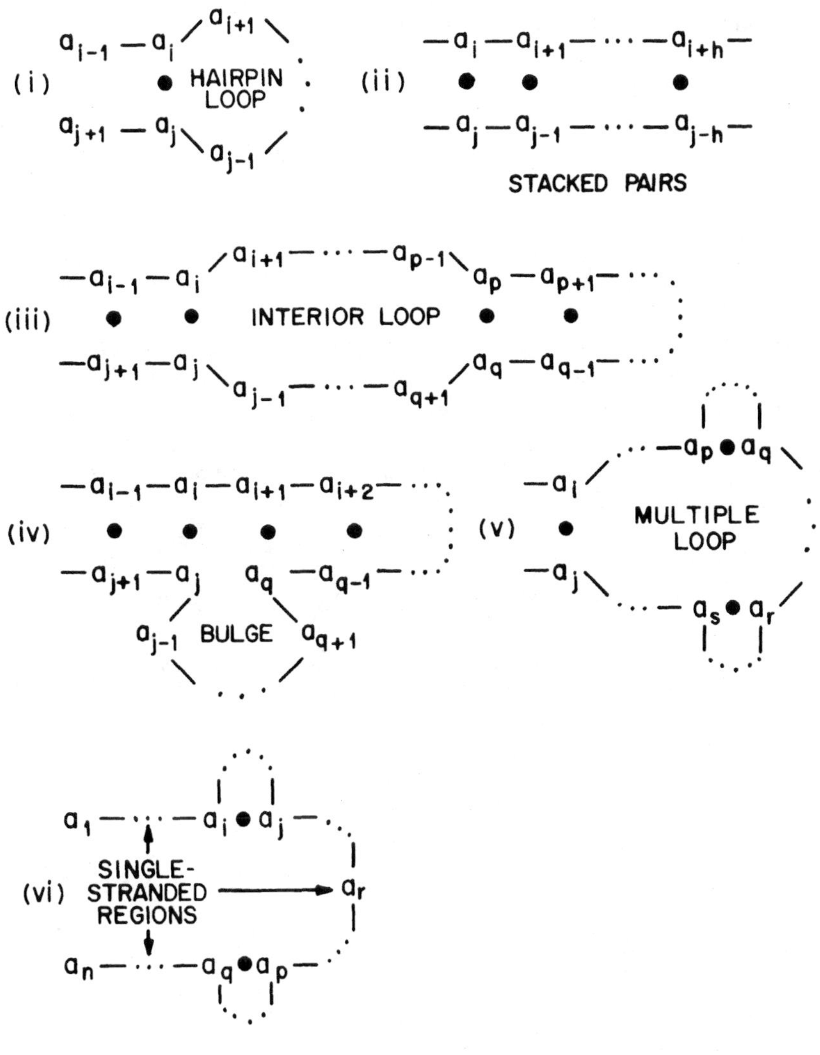

Figure 2. The six substructures.

from 13 to 23 is a 1-cycle and a hairpin loop: its only pair is the closing pair, $13 \cdot 23$, and its unpaired elements are $14, \ldots, 22$. The cycle from 7 to 67 is a 4-cycle and a multiple loop: its four pairs are $7 \cdot 67$, $10 \cdot 26$, $28 \cdot 44$, and $50 \cdot 66$, and its unpaired elements are $8, 9, 27$, and $45, \ldots, 49$. The cycle from 6 to 68 is a 2-cycle and a stacked pair: it contains pairs $6 \cdot 68$ and $7 \cdot 67$ but no unpaired elements.

Every pair is contained in one or two loops, but never in more. Specifically, each pair is contained in the loop it defines, and in most cases a pair is also accessible in one other loop. Because a cycle may have many accessible pairs but only one closing pair, accessibility has a nonsymmetric character: Many pairs may be accessible from a given pair $i \cdot j$, but a given pair $i \cdot j$ is accessible from at most one other pair.

It is sometimes conceptually helpful to represent the secondary structure by a "tree diagram" like that in Fig. 3, which shows the structure of Fig. 1. Each pair is represented by a node, and two nodes are connected by an edge if one is accessible from the other (in either order). It is possible to arrange the nodes so that "accessibility from" is pointed upwards: if $p \cdot q$ is accessible from $i \cdot j$, then the edge from the $p \cdot q$ node to the $i \cdot j$ node runs generally upward. Of course, the edges going generally downward from the $i \cdot j$ node lead to the $p_h \cdot q_h$ nodes, where $p_1 \cdot q_1, p_2 \cdot q_2, \ldots$ indicate the accessible pairs in the cycle from i to j. At the top of the diagram is the node for the pair $1 \cdot n$, if this pair exists; otherwise the highest node(s) represent the pair(s) accessible in $[1, n]$. If there is only one highest node, the diagram is a tree; if there are several of them, the diagram consists of several trees. This diagram is helpful in understanding the optimality principle, which underlies the dynamic-programming method introduced in Sec. 6, and also in enumerating the different possible secondary structures for a sequence of given length.

To analyze a secondary structure S, we will need to consider structures on strings that are less than the whole sequence $[1, n]$. A secondary structure on $[i, j]$ will be indicated by S_{ij}, so that S could also be designated S_{1n}. A secondary structure S is said to *induce* a structure on any proper string $[i, j]$: the induced structure is defined to contain all pairs in S whose elements lie in $[i, j]$.

A cycle of order k from i to j can be classified as follows, where p and q (if used) satisfy $i + 1 < p$, $p < q$, and $q < j - 1$.

   i. If $k = 1$, the 1-cycle is a hairpin.
   ii. If $k = 2$, and the accessible pair is $(i + 1) \cdot (j - 1)$, then the 2-cycle is a stacked pair (cycle).
   iii. If $k = 2$ and the accessible pair is $p \cdot q$, then the 2-cycle is an interior loop.
   iv. If $k = 2$ and the accessible pair is either $(i + 1) \cdot q$ or $p \cdot (j - 1)$, then the 2-cycle is a bulge.
   v. If $k \geq 3$, then the k-cycle is a multiple loop.

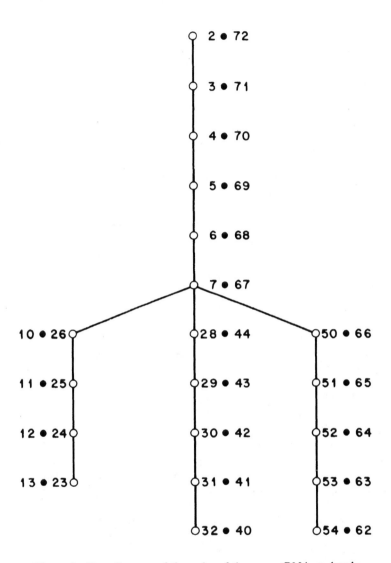

Figure 3. Tree diagram of the pairs of the same tRNA molecule.

4. THE THERMODYNAMICS OF A STRUCTURE

An RNA chain will tend to exist in a conformation characterized by the minimum free energy E. The exact calculation (from basic physical chemical principles) of E for an arbitrary proposed secondary structure S of an RNA chain is not feasible; problems of definition and computation are unsurmounted even for far smaller molecules. Experimentally, it is also impossible to determine the free energy of most conceivable structures since they do not occur; a structure can have stable existence only if it is the "true" secondary structure having minimum value for E or, rarely, if it is one of a limited number of metastable structures for which E is almost the minimum.

It is possible, however, to estimate the free energy of a proposed structure S with some degree of confidence by combining experiment with the hypothesis of the Tinoco–Uhlenbeck theory, namely, that the free energy can be expressed as a sum of terms where each term is associated with one of the cycles s_r,

$$E(S) = e(s_1) + e(s_2) + \ldots + e(s_t).$$

Experimental evidence is used to determine $e(s)$. Our first algorithm permits $e(s)$ to be an arbitrary function. The other algorithms achieve better computation time by assuming $e(s)$ to have a specific plausible form.

Let us consider what is known about $e(s)$, and the experimental evidence available to determine it. The most fundamental fact is that $e(s)$ cannot yet be determined with great accuracy, but only to a rough approximation. The uncertainty is large enough to have a significant effect on the nature of the optimizing structure in many cases. With a sufficiently fast algorithm to determine the optimizing structure—we hope the faster algorithms introduced in this chapter may play this role—it will be possible to experiment with alternative values, as Ninio (unpublished) does, using molecules about which some experimental information on secondary structure is available, and from these results refine the values.

Another fundamental fact is that in general $e(s)$ is negative only when s is a stacked pair, i.e., it is only stacked pairs that contribute directly to the stability of the molecule. The formation of hairpins and other loops makes a positive contribution to E and thus tends to reduce the stability of the structure. In practice, all stable secondary structures contain chains of at least four or five stacked pairs, and the vast majority of cycles in actual molecules are stacked pairs, as illustrated in Fig. 1.

To determine the value of $e(s)$ where s is a stacked pair, a chemical solution containing quantities of two complementary molecules **a** and **b** of the same length can be studied. For example, suppose **a** consists of n ribonucleotides containing G and **b** consists of n ribonucleotides containing C. Under suitable experimental conditions, each **a** will tend to bond with a **b** molecule, all along its length,

$$\text{G–G–G– } \ldots \text{ –G–G}$$
$$\cdot \quad \cdot \quad \cdot \qquad \cdot \quad \cdot$$
$$\text{C–C–C– } \ldots \text{ C–C}$$

thus providing $n - 1$ stacked-pair cycles

$$s = \begin{matrix} G{-}G \\ \bullet \quad \bullet \\ C{-}C \end{matrix},$$

i.e., 2-cycles with closing pair G · C, accessible pair G · C, and no accessible unpaired elements. According to the Tinoco–Uhlenbeck hypothesis, the change in free energy from the case n to the case $n + 1$ is $e(s)$, which is the quantity we want to measure. This change can be measured experimentally, being proportional to the difference in "melting point" of the molecules (i.e., highest temperature at which the bonded-chain molecules are stable). By experiments like these, it has been found, for example, that the free-energy contribution for the stacked pair depicted above is two or three times as much as the free energy contribution for

$$\begin{matrix} U{-}U \\ \bullet \quad \bullet \\ A{-}A \end{matrix}.$$

As another example, to determine $e(s^m)$ for a hairpin loop s^m containing m unpaired A ribonucleotides, the experimenter might use molecules of the form

$$\overbrace{G{-}G{-}\ldots{-}G{-}G}^{r}\underbrace{{-}A{-}A{-}\ldots{-}A{-}A}_{m}\overbrace{{-}C{-}C{-}\ldots{-}C{-}C}^{r}$$

for some fixed r that is large enough, say 6. At temperatures below the characteristic "melting point" for this structure, the r G's will bond with the r C's in the obvious way to form a hairpin loop. Changes in the melting point as m increases are proportional to changes in $e(s^m)$. From such experiments it is discovered that as m increases from its smallest possible value of 3, $e(s^m)$ decreases for a while. When m gets large enough, however, $e(s^m)$ starts to increase.

This type of experiment can also be done for interior loops and bulges. A summary of known results for stacked pairs and the different types of loops can be found in Tinoco *et al.* (1973). We note that external single-stranded regions, substructure type 6 of Section 3, contain no cycles, and hence neither enhance nor reduce the stability of the structure as a whole. (Their contribution to E is implicitly taken to be zero.)

Certain generalizations emerge from the experimental results (in addition to the fundamental facts stated above):

a) If $E(S)$ is positive, S cannot be stable, since E can always be reduced to 0 by leaving the entire sequence single-stranded (i.e., using no pairs at all). It is quite possible for a given RNA sequence that no S will have a negative value for $E(S)$.

b) For any pair $i \cdot j$ in a stable structure S, if the cycle from i to j is a loop (i.e., anything other than a stacked pair), then S must also contain $(i - 1) \cdot (j + 1)$. Otherwise $E(S') < E(S)$, where S' is the structure derived from S by unpairing i and j. Indeed, if $e(s)$ is the energy contribution of a loop s from i to j, then there must be a sequence of stacked pairs $i \cdot j$, $(i - 1) \cdot (j + 1), \ldots, (i - h) \cdot (j + h)$ with

$$e(s) \leqq - [e(\mathbf{t}_1) + \ldots + e(\mathbf{t}_h)],$$

where $\mathbf{t}_p$ is the stacked pair closed by $(i - p) \cdot (j + p)$. That is, the positive contribution to E for forming and maintaining the loop must not exceed the negative contribution by the stacked pairs that "close" the loop. Otherwise the entire structure S is unstable. (Note that this "local stability constraint" is based on experimental results, and has implications for the nature of $e(s)$. Mathematically, it is possible to construct a hypothetical function $e(s)$ under which the optimal secondary structure would not satisfy this property.)

c) It follows that the shortest string that can have a thermodynamically stable secondary structure (i.e., a structure with E negative) has length 7, and consists of a minimum-length hairpin loop inside a stacked pair, as shown in Fig. 4.

d) It further follows that a thermodynamically stable multiple loop from i to j can have at most $(j - i - 2)/7$ accessible pairs, i.e., in a thermodynamically stable k-cycle from i to j, we must have $k - 1 \leqq (j - i - 2)/7$. Roughly, $k \leqq (j - i)/7$.

e) If s is an interior loop of the form

$$i—(i + 1)—(i + 2)$$
$$\bullet \qquad \bullet$$
$$j—(j - 1)—(j - 2)$$

and if the two unpaired bases $i + 1$ and $j - 1$ are G and U (or U and G), respectively, then $e(s) = 0$. This reflects a stereochemically well understood phenomenon that apparent $G \cdot U$ pairs occasionally occur interspersed among Watson–Crick pairs when many pairs of bases are stacked upon each other. G and U do not bond with each other as Watson–Crick pairs do, but they can fit together without causing a structural deformation requiring energy to maintain. It would thus be possible to permit $G \cdot U$ pairing as an exception to constraint (a) of Sec. 2, where the stacking of such pairs on each other or on other

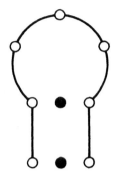

Figure 4. Smallest stable loop structure.

Watson–Crick pairs contributes no free energy. For computational purposes, however, it is more efficient not to allow G · U pairs, and to consider them rather as forming part of zero-cost interior loops.

5. SEQUENCE COMPARISON AND SELF-COMPARISON

We return now to the question of how to find the secondary structure S (for a given RNA sequence) which optimizes $E(S)$. Out of the myriad possibilities imaginable that satisfy the constraints (a)–(d) of Section 2, how can we find which structure makes $E(S)$ smallest, without explicitly constructing and evaluating every possibility?

To motivate the approach we shall take, note that the chains of stacked pairs in secondary structures resemble the traces or alignments of sequence-comparison theory (see Chapter 1), despite the following important differences.

1. In sequence comparison, generally two (and sometimes more) sequences are aligned and compared. Here we are aligning or "comparing" part of one sequence with another part of the same sequence.

2. In sequence comparison, one objective is to align as many identical terms in the two sequences as possible. Here we are aligning A's with U's and G's with C's.

3. In sequence comparison, the best alignment is the one that minimizes a length function based on the cost of aligning non-identical terms (as is sometimes unavoidable), and the cost of introducing gaps between two adjacent terms (in order to shift part of one sequence so that it is aligned with a more similar part of the other). Here we are minimizing a more

complicated free-energy function based on negative contributions from stacked Watson–Crick pairs and positive contributions from the loops containing unpaired bases.

These differences, while substantive, do not preclude a strong resemblance between the two problem areas. Indeed, this resemblance suggests that the dynamic-programming approach of sequence-comparison theory may help us in the search for the optimal secondary structure.

6. A RECURSION

In this section we present a recursion which makes this search feasible for arbitrary functions $e(s)$. Implemented directly, as in the next section, the computation time grows exponentially with n, and is too slow to be really practicable for even moderately long sequences whose secondary structures contain multiple loops. In Secs. 8 and 9 we propose and discuss various restrictions on the form of $e(s)$ which enable us to devise faster algorithms, for which the computation time is proportional to n^3 or n^4.

Suppose S is a secondary structure and suppose that $[i, j]$ is proper. Consider the secondary structure S_{ij} on $[i, j]$ induced by S (i.e., S_{ij} consists of all pairs from S whose elements belong to $[i, j]$). Suppose S'_{ij} is any other secondary structure on $[i, j]$ whatsoever. If we substitute S'_{ij} for S_{ij}, then the result is a valid secondary structure, i.e.,

$$S^* = (S - S_{ij}) \cup S'_{ij}$$

is a valid secondary structure. It is obvious from Fig. 3 that this is correct, since we are merely removing the partial tree(s) below certain node(s) and replacing them by other tree(s). Mathematical proof requires only that properties (a)–(d) in Sec. 2 be verified for S^*, which is an easy exercise.

From the decomposition of free energy into terms associated with the cycles,

$$E(S) = \sum e(s),$$

it is obvious that

$$E(S^*) = E(S) - E(S_{ij}) + E(S'_{ij}).$$

Therefore, if S is optimal on $[1, n]$, then S_{ij} is optimal on $[i, j]$ (since if we could improve S_{ij}, we could incorporate this improvement in S and hence improve it also). These facts are a vital mathematical basis for the algorithms that follow.

We will need the following elementary recurrence over the secondary structure. Suppose $[i, j]$ is proper, and let $p_h \cdot q_h$ for $h = 1$ to $k - 1$ be its accessible pairs. Let S_{ij} and S_{p_h, q_h} be induced by S. Then from the decomposition of E into terms associated with the cycles, it is easy to see that

$$E(S_{ij}) = \begin{cases} \sum_{h=1}^{k-1} E(S_{p_h, q_h}) & \text{if } [i, j] \text{ is not closed,} \\ e(s) + \sum_{h=1}^{k-1} E(S_{p_h, q_h}) & \text{if } [i, j] \text{ is closed,} \end{cases}$$

where s is the loop from i to j.

Now define $F(i, j)$ to be the optimum value of E in $[i, j]$, i.e.,

$$F(i,j) = \min_{S_{ij}} E(S_{ij})$$

where S_{ij} ranges over all possible secondary structures on $[i, j]$. Then our goal is to find $F(1, n)$ and the secondary structure S^{opt} that yields it.

We shall say that (i, j) is *closable* if it is possible to form a cycle with $i \cdot j$ as closing pair, which means specifically that a_i and a_j are Watson–Crick-pairable and that $j - i \geq 4$. If (i,j) is closable, let $C(i,j)$ be the optimum value of E given that i is paired with j, i.e.,

$$C(i, j) = \min_{\substack{S_{ij} \\ i \cdot j}} E(S_{ij})$$

where S_{ij} ranges over all secondary structures on $[i,j]$ that pair i with j. Note that "C" for "closed" is associated with a closed string, while F is associated with any proper string.

Now it is easy to see, using the elementary recurrence equation above, that if (i, j) is closable

$$C(i, j) = \min_{\substack{k \geq 1 \\ s \text{ is a} \\ k\text{-cycle} \\ \text{from } i \text{ to } j}} \left[e(s) + \sum_{\substack{(p_h, q_h) \\ \text{accessible} \\ \text{pairs of } s}} C(p_h, q_h) \right] .$$

Both here and below, values of C that are undefined may be treated as $+\infty$, so that sums containing them are ignored during minimization. It is also easy to see that

$$F(i, j) = \min \begin{cases} 0 \\ C(i, j) \\ \min_{i \leq h < j} [F(i, h) + F(h + 1, j)] \end{cases}$$

where the first line is selected if $j - i < 4$, the second line is selected if the optimum structure on $[i, j]$ is closed, and the third line is selected otherwise. The basic algorithm below is simply a direct implementation of these two recurrence equations.

The recursion above yields only the free energy $E(S^{opt})$, not S^{opt} itself. In order to obtain S^{opt} it is necessary to define pointers during the recursion, and to use these pointers for "backtracking" following the recursion. The algorithms below also include these operations

Suppose the values of F and C are arranged in a square array, where the (i, j) cell contains both $F(i, j)$ and $C(i, j)$, as illustrated in Fig. 5. We always take $i < j$, so we are interested only in the part of the array above the main diagonal. Since $j - i > q_h - p_h$ in the formulas above, the values in cell (i, j) are determined using the values in cells closer to the main diagonal. Thus it is natural to evaluate the cells in one diagonal after another, starting at the main diagonal and proceeding outward. More specifically, the order we shall use is

$$
\begin{array}{llll}
(1, 1) & (2, 2) & (3, 3), & \cdots & (n, n); \\
(1, 2) & (2, 3) & (3, 4), & \cdots & (n - 1, n); \\
(1, 3), & (2, 4), & (3, 5), & \cdots & (n - 2, n); \\
\cdots \cdots \\
(1, n).
\end{array}
$$

We shall refer to this as the *standard order*. However, in practice we may be able to omit the first few rows of this array.

7. THE BASIC ALGORITHM

The most direct approach to finding $F(1, n)$ and the associated S^{opt} uses two stages. During the first stage, the values of $C(i, j)$ and the associated pointers are calculated for all closable (i, j), by a recursion in the standard order. For each (i,j), it is determined whether (i,j) is closable (which means that $j - i \geq 4$ and a_i is Watson–Crick pairable with a_j. If it is not closable, the step is complete. If it is closable, the first recurrence equation from Sec. 6 is evaluated. In addition, a record is kept of the nature of the optimizing cycle s: its accessible pairs are recorded by means of "triples" in this pointer equation:

$$
\text{pointer}(C, i, j) = \{(C, p_1, q_1), \ldots, (C, p_{k-1}, q_{k-1})\}.
$$

Note that in this equation "C" is simply a flag, not the function $C(i, j)$.

During the second stage, the values of $F(i, j)$ and associated pointers are calculated for all (i, j), by a recursion in the standard order. For each (i, j), the

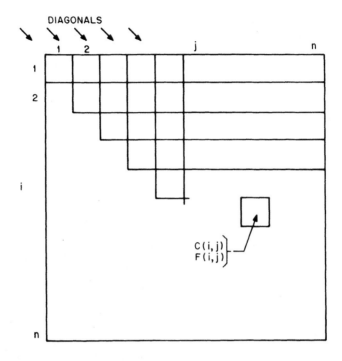

Figure 5. Workspace for the basic algorithm.

second recurrence equation from Sec. 6 is evaluated. In addition, a record is kept of the nature of the optimizing term in that equation by this pointer equation:

$$\text{pointer}(F, i, j) = \left\{ \begin{array}{ll} \phi & \text{or} \\ \{(C, i, j)\} & \text{or} \\ \{(F, i, h), (F, h+1, j)\} & \end{array} \right\}.$$

At the end of this stage, we have the value of $E(S^{\text{opt}}) = F(1, n)$. To obtain S^{opt} itself, we simply "backtrack" starting from the triple $(F, 1, n)$. This means that we find what other triples it points to, and what triples those point to, and so on until the process stops. Each triple (C, i, j) indicates a pair $i \cdot j$ in S^{opt}.

It turns out the computation time for the first stage grows much more quickly as a function of n than computation time for the second stage. (The pointer equations and backtracking have negligible effect on this function of n

and may be ignored.) Consider the recursive step in the first stage. If $j - i = 4$, then only a hairpin loop (with $k = 1$) is possible, so evaluating $C(i, j)$ is trivial. For general (i, j), it is necessary to consider all possible sequences

$$(p_1, q_1), (p_2, q_2), \ldots, (p_{k-1}, q_{k-1})$$

made from closable pairs (p_h, q_h) for which

$$i < p_1 < q_1 < p_2 < q_2 < \ldots < p_{k-1} < q_{k-1} < j.$$

Geometrically, as shown in Fig. 6 these inequalities mean that each (p_h, q_h) is strictly contained within the triangle from (i, j) to the main diagonal, and that the triangles from (p_h, q_h) to the main diagonal do not overlap.

How many such sequences may exist? For any fixed k and for large values of $j - i$, the number of sequences may be proportional to $(j - i)^{2k-2}$, where the constant of proportionality depends on k and the frequency of Watson–Crick pairs (this frequency is roughly $\frac{1}{4}$, but its precise value will depend on the frequencies of the four bases).

One of the generalizations in Sec. 4 is that $k \leq (j - i)/7$. Thus the calculation time for evaluating $C(i, j)$ includes a highest-order term of the form $(j - i)^{2c(j-i)-2}$, where $c = \frac{1}{7}$ and where $j - i$ can be as high as $n - 1$. Suppose r is some fixed fraction such as $\frac{3}{4}$, and we consider all closable pairs (i, j) for which $j - i \geq rn$. The number of such pairs is proportional to n^2, and the calculation time for each pair includes a term greater than $(rn)^{2crn-2}$. Thus the calculation time for the entire matrix is more than proportional to $n^2 \cdot (rn)^{2crn-2} = r^{-2}(rn)^{2crn}$, which increases very rapidly with n.

Now consider the recursive step in the second stage, ignoring the pointer equations. To find $F(i, j)$ requires elementary look-up and calculation time approximately proportional to $j - i$. When we sum over all (i, j), the calculation time for the entire second stage includes a highest-order term proportional to n^3.

There are many ways to increase the efficiency of an algorithm such as this one. Imposing an arbitrary limit on the number of unpaired bases in a loop could reduce computing time, but at the considerable risk of not finding the "true" structure if it happens to contain a longer loop. Careful use of the thermodynamic results (see Sec. 4) to exclude pairs $i \cdot j$ that cannot form part of a stack of two or more pairs, and to exclude loops whose energetic cost cannot be adequately compensated for by stacking on the closing pair, can cut down on unnecessary computation without sacrificing the optimality of the result.

These improvements, however, do not affect the basic cause for the costliness of the algorithm: the complexity of dealing with multiple loops. Even if we assume an upper limit K for the order k of all multiple loops, the time cost of the algorithm is still n^{2K}. The lengths of the sequences being studied, of the order of $n = 100$ or more, mean that anything much worse than

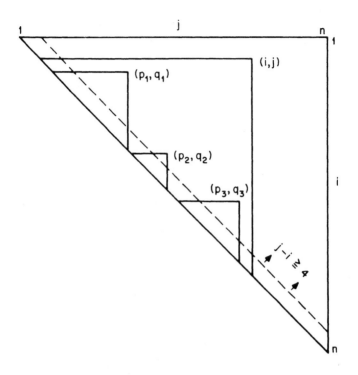

Figure 6. Restrictions on $(p_h,\ q_h)$ in the basic algorithm.

an n^4 algorithm is impractical. But this would mean $2K = 4$, or $k \leq 2$, so that we could not identify even 3-loops in our secondary structures.

8. AN n^3 ALGORITHM*

The large computation time of the basic algorithm results from the interaction of two facts: (i), the algorithm considers all possible multiple loops, including those of high order; (ii), the algorithm will work correctly for an arbitrary function $e(s)$ for the free energy of the cycle s, however complicated or unrealistic the function e may be. There are a great many possible multiple loops, and because no structure is assumed for e it is necessary to consider these loops one by one.

One way to reduce the computation time is to impose some assumption about e. For example, we could simply assume that the free-energy contribution

*Developed by J. B. Kruskal.

of multiple loops is zero—a simple assumption which is implicit in much of the literature (Pipas and McMahon, 1975; Nussinov *et al.,* 1978). However, we shall treat multiple loops less drastically by using an assumption first introduced in Mainville (1981). In this section we use Mainville's assumption more broadly than she did, namely, for *all* cycles. This assumption then permits development of an algorithm using time proportional to n^3. In the next section we limit the Mainville assumption to multiple loops, as she did, increasing the time to n^4.

To describe the Mainville assumption, let **s** be a cycle, let (i, j) be its closing pair, let $p_h \cdot q_h$ be its accessible pairs, and let r run through its accessible unpaired elements. Then we assume in the general case that

$$e(\mathbf{s}) = e_0(i, j) + e_1(\mathbf{s}) + e_2(\mathbf{s})$$

where

$e_0(i, j)$ is an arbitrary function,

$e_1(\mathbf{s}) = \sum_h \bar{e}_1(p_h, q_h)$ where $\bar{e}_1$ is an arbitrary function,

$e_2(\mathbf{s}) = \sum_r \bar{e}_2(r)$ where $\bar{e}_2$ is an arbitrary function.

The *spirit* of this assumption is that these arbitrary functions depend only on the identity of the bases involved (e_0 on $a_i \cdot a_j$, $\bar{e}_1$ on $a_{p_h} \cdot a_{q_h}$, and $\bar{e}_2$ on a_r). Mathematically, however, this is not necessary: e_0 can be an arbitrary function of two indices, and similarly for $\bar{e}_1$ and $\bar{e}_2$. It is easy to imagine useful purposes for this generality.

An important special case of this assumption is formed by choosing three constants c_0, c_1, and c_2 and setting

$$e_0(i, j) = c_0 \quad \text{for all } i \text{ and } j,$$
$$\bar{e}_1(p, q) = c_1 \quad \text{for all } p \text{ and } q,$$
$$\bar{e}_2(r) = c_2 \quad \text{for all } r.$$

In this special case, $e(\mathbf{s})$ simplifies to

$$e(\mathbf{s}) = c_0 + c_1(k - 1) + c_2 u,$$

where k is the order of the cycle and u is the number of unpaired accessible elements in **s**. This is reminiscent of the "linear gap weight" assumption often used in the search for homology between two macromolecules (e.g. Gotoh 1982).

Unfortunately, this assumption permits only a rough approximation to the actual free energy of cycles, particularly those of order 2 and order 1, according to current knowledge based on a combination of experiment and theory. Nevertheless, this assumption can deal with multiple loops far more realistically than simply ignoring their free energy completely—and surprisingly, it permits an algorithm which is at least as efficient as previous algorithms that do ignore their free energy. Furthermore, this algorithm paves the way to other algorithms, such as the one in the next section, which handle cycles of order 2 and 1 realistically.

The algorithm in this section, which is based on the general-case assumption about $e(\mathbf{s})$, makes use of $C(i, j)$ and $F(i, j)$ with exactly the same meanings as in the basic algorithm, and is done in two stages (plus backtracking) like those used there. In fact, the second stage and the backtracking are exactly the same as before. Thus it is only the first stage that is new.

The first stage makes use of $C(i, j)$ and a new variable, $G(i, j)$. We will explain the meaning of G later. The first stage is a simultaneous recursion for $C(i, j)$ and $G(i, j)$ and associated pointers. The recursion examines the pairs in the standard order. For each (i, j), the appropriate recursive step shown below is carried out, in the order shown here. Values of C that are undefined are treated as $+\infty$, so that sums containing them are ignored during minimization.

If $j - i = 0$,

$$G(i, i) = \bar{e}_2(i),$$

pointer $(G, i, i) = \phi.$

If $j - i = 1$ or 2 or 3

$$G(i, j) = G(i, i) + G(i + 1, j),$$

pointer $(G, i, j) = \phi.$

If $j - i \geq 4$,

$$C(i, j) = \begin{cases} G(i + 1, j - 1) + e_0(i, j) & \text{if } (i, j) \text{ is closable,} \\ \text{undefined} & \text{otherwise,} \end{cases}$$

pointer$(C, i, j) = \{(G, i + 1, j - 1)\}$ or $\phi,$

$$G(i, j) = \min \begin{cases} C(i, j) + \bar{e}_1(i, j), \\ \min_{i \leq h < j} [G(i, h) + G(h + 1, j)], \end{cases}$$

pointer$(G, i, j) = \{(C, i, j)\}$ or $\{(G, i, h), (G, h + 1, j)\}.$

In a practical implementation of this algorithm, it would probably save memory space and computer time to modify the algorithm slightly so that $C(i, j)$ does not explicitly appear in either stage. In addition, there may be more efficient if less elegant procedures for handling the pointers.

We now describe the meaning of G with the aid of an associated concept, $\widetilde{E}$. Substitute the formula for $e(s)$ from the Mainville assumption into the recurrence from Sec. 6 for $C(i, j)$,

$$C(i, j) = \min_{k \geq 1} \min_{s} [e_0 (i, j) + e_1(s) + e_2(s) + \sum C(p_h, q_h)],$$

where s ranges over all k-cycles from i to j. Now $e_0(i, j)$ can legitimately be taken outside the two min operators. Conceptually, $G(i + 1, j - 1)$ is the expression that remains after removing $e_0(i, j)$, so

$$C(i, j) = G(i + 1, j - 1) + e_0(i, j).$$

The function $\widetilde{E}$ evaluated at an appropriate argument is the expression inside the two min operators, so $G = \min \min \widetilde{E}$.

To develop the formal definition for $G(i, j)$, we note the unit change in i and j, and consider a cycle s from $i - 1$ to $j + 1$. Any such cycle induces a secondary structure S_{ij} on $[i, j]$. Furthermore, the mapping from cycles to secondary structures is one-to-one (assuming $(i - 1, j + 1)$ is closable). We shall define $\widetilde{E}(S_{ij})$ as the appropriate expression and $G(i, j) = \min \min \widetilde{E}$, whether or not $(i - 1, j + 1)$ is closable. Specifically, if $i \leq j$ and any secondary structure S_{ij} is given, let $p_h \cdot q_h$ run through the accessible pairs and r run through the accessible unpaired elements of S_{ij}. Define

$$\widetilde{E}(S_{ij}) = \sum_h \bar{e}_1(p_h, q_h) + \sum_r \bar{e}_2(r) + \sum_h C(p_h, q_h).$$

Then define

$$G(i, j) = \min_{k \geq 1} \min_{S_{ij}} \widetilde{E}(S_{ij}),$$

where S_{ij} may be any secondary structure having $k - 1$ accessible pairs. The reason G is useful is that the minimization required to calculate it can be accomplished by dividing $[i, j]$ into only two parts instead of into several, owing to the nature of $e_1(s)$ and $e_2(s)$.

To analyze the computation time for the algorithm is quite simple. The first stage and the second stage are extremely similar, and the computation-time analysis given earlier for the second stage applies here to both stages; thus computation time for this algorithm is proportional to n^3.

9. AN n^4 ALGORITHM

Before relaxing the assumption from the previous section about the function $e(s)$, we consider what theory and experiment have to say about this function, so as to explain and justify the particular change made here. Recall that the special case of the assumption is

$$e(s) = c_0 + c_1(k - 1) + c_2u,$$

where the loop has $k - 1$ accessible pairs (hence order k) and u unpaired bases. To predict secondary structure, the cycles s for which it is most important to have good free energy values are the common cycles: fairly short cycles (up to 6 or 8 unpaired bases) of order 1 and 2, particularly including stacked pairs. These are also the cycles for which there is the most experimental evidence, and this evidence shows that $e(s)$ depends on s in a complicated and apparently erratic way in this range. Thus no simple assumption, certainly not that from the previous section, is going to fit $e(s)$ well here.

As the length of the cycle grows, thermodynamic theory suggests that the free energy should contain a term that is proportional to log(length of the cycle), and this term should become dominant so that, for long cycles, free energy is not much affected by just what pairs and bases make up the cycle. For cycles of order 1 and 2, there is some experimental evidence to support the theory. Under the assumption used in the previous section, $e(s)$ can be proportional to length but not to log(length) of s, so the assumption does not permit $e(s)$ to fit both short and long cycles well.

Consequently, the two most important steps in making the form of $e(s)$ more realistic are to accommodate the irregular pattern of values for short cycles of orders 1 and 2 and to accommodate the logarithmic growth of $e(s)$ with length of s. We accomplish the first step fully and the second step partially (i.e., for cycles of order 1 or 2) by the way in which we relax the previous assumption here, namely, by permitting an arbitrary table of values for $e(s)$ for cycles s of order 1 and 2. Specifically, we assume that

$$e(s) = \begin{cases} f_1(i, j) & \text{if } s \text{ has order 1,} \\ f_2(i, j, p_1, q_1) & \text{if } s \text{ has order 2,} \\ \text{as in preceding section,} & \text{otherwise.} \end{cases}$$

The *spirit* of the first two lines is that f_1 and f_2 should depend only on the identity of the paired bases involved and the number of unpaired bases (f_1 on

$a_i \cdot a_j$ and $j - i - 1$, f_2 on $a_i \cdot a_j$, $a_{p_1} \cdot a_{q_1}$, and $j - q_1 + p_1 - i - 2$), but mathematically no such assumption is necessary: f_1 and f_2 may be arbitrary functions.

For multiple loops (order $\geqq 3$), we have retained the assumption from the preceding section. To what extent is this justified? Basically, it is hard to tell. For very short multiple loops, there is little experimental evidence with which to compare our assumption. Fortunately, such loops apparently occur very rarely, so accuracy of $e(\mathbf{s})$ in this range may not be so important. For long multiple loops, the same general theoretical principles that yield the log(length) term described above may well be operative, but there is little or no experimental evidence to bear on the topic, nor on the effect of the $k - 1$ accessible pairs, each of which is attached to a more or less massive portion of the molecule outside the loop. Note that, because the logarithm is concave downwards, while our assumption forces $e(\mathbf{s})$ to increase linearly with length, our assumed values for $e(\mathbf{s})$ are likely to be too large for very long and very short loops but too small for loops of moderate length. Zuker and Stiegler (1981) discuss two versions of their program, one for which they present the algorithm explicitly and assign zero cost to multiple loops, and the other which they do not present but which incorporates a logarithmic cost. Within their mathematical framework, however, which is basically the same as the one presented here, a general logarithmic loop energy does not seem possible for an n^3 or n^4 algorithm without compromising its global optimality. It is the linearity in the Mainville assumption that permits the addition of terms in the loop energy function, and this is not possible with a logarithmic energy function.

Nevertheless, this assumption may provide a far better approximation for the free energy of multiple loops than those used previously. Extrapolating from what we know of loops of order 1 and 2, it seems appropriate to take c_0 to be of about the same magnitude and of opposite sign as the free energy of two or three stacked pairs, and to set c_1 and c_2 to be quite small. For large RNA molecules, where a loop length in the hundreds seems plausible (compare the secondary structures proposed by Stiegler et al., 1981, or Woese et al., 1980 for 16S RNA), great care would be needed and some refinement of the method might turn out to be necessary.

To optimize under the assumption in this section, we need only modify the first-stage recursion from above. In addition to $G(i, j)$ we use a function $G_1(i, j)$ which is the same as G except that S_{ij} ranges over secondary structures which contain *at least one accessible pair*:

$$G_1(i, j) = \min_{k \geqq 2} \min_{S_{ij}} \widetilde{E}(S_{ij})$$

(note the change from 1 to 2), where S_{ij} has $k - 1$ accessible pairs. Now the recursion remains unchanged for $j - i \leq 3$, except that G_1 is left undefined for these cases. For $j - i \geq 4$ we have the following, where undefined values of C and G_1 are treated as $+ \infty$:

$$C(i, j) = \begin{cases} \min \begin{cases} f_1(i, j) \\ \min_{i < p_1 < q_1 < j} [f_2(i, j, p_1, q_1) + C(p_1, q_1)] \\ \min_{i+1 \le h < j-1} [G_1(i + 1, h) + G_1(h + 1, j - 1) + e_0(i, j)] \\ \qquad\qquad\qquad\qquad \text{if } (i, j) \text{ is closable,} \\ \qquad\qquad\qquad\qquad \text{otherwise,} \end{cases} \\ \text{undefined} \end{cases}$$

pointer$(C, i, j) = \phi$ or $\{(C, p_1, q_1)\}$ or $\{(G_1, i + 1, h), (G_1, h + 1, j - 1)\}, \}$,

$$G_1(i, j) = \min \begin{cases} C(i, j) + \bar{e}_1(i, j), \\ \min_{i \le h < j} [G_1(i, h) + G(h + 1, j), G(i, h) + G_1(h + 1, j)], \end{cases}$$

pointer $(G_1, i, j) = \{(C, i, j)\}$ or $\{(G_1, i, h), (G, h + 1, j)\}$
$\qquad\qquad\qquad\qquad$ or $\{(G, i, h), (G, H + 1, j)\}$.

$$G(i, j) = \min \begin{cases} G_1(i, j), \\ \min_{i \le h < j} [G(i, h) + G(h + 1, j)], \end{cases}$$

pointer $(G, i, j) = \{(G_1, i, j)\}$ or $\{(G, i, h), (G, h + 1, j)\}$.

It is the second line in the equation for $C(i, j)$ which makes this algorithm use time proportional to n^4.

The approach used in this section may be extended to permit arbitrary functions f_1, f_2, and f_3 for loops of order 1, 2, and 3, respectively, while retaining linearity for loops of order ≥ 4. With this change, however, computing time would become proportional to n^6 instead of n^4, so that it is questionable whether this further limitation of the linearity assumption would be computationally feasible.

10. A HEURISTIC EXTENSION ALGORITHM
FOR LONGER SEQUENCES

The algorithms in the preceding two sections may be practical for n in the low hundreds, but are probably too expensive for longer RNA molecules, where n

may be in the low thousands. We sketch here a heuristic extension of these algorithms which may be able to handle such molecules. Several of the steps resemble those used by Pipas and McMahon (1975).

First, find possible long strings of stacked pairs, which can be done by a rapid, simple n^2 algorithm. (Pipas and McMahon do this, but set a lower limit of 3 for the length of such strings. In our case, a substantially larger limit is probably appropriate, since the strings play a somewhat different role.) From near the middle of each such string, select one possible pair, and from the resulting pairs select a "suitable" subset and call it a *trial set*. We are going to find the best secondary structure subject to the constraint that the pairs in the trial set are actually paired. We shall repeat this for a variety of trial sets, and select the best result found as the desired structure of the RNA molecule.

Since the trial set will be contained in a secondary structure, it must be itself a secondary structure (though presumably a very unstable one). Thus the first requirement for suitability is that the trial set should satisfy all four constraints (a), (b), (c), and (d) for a secondary structure. Another requirement is that the pieces which it yields (discussed below) should be small enough for the n^3 or n^4 algorithm to handle. Clearly, there is no reason to use a trial set that contains any other trial set, so avoiding this is another requirement for suitability. Key practical questions in applying the method described in this section are whether suitable trial sets exist, how to select them, and how many of them should be used.

Because the trial set is a secondary structure, it divides the sequence into cycles and, possibly, external single-stranded regions. We call each cycle an *inner piece*. If there are any external single-stranded regions, then their union together with the pairs accessible in $[1, n]$ is called the *outer piece* (there can be only one outer piece). It is not hard to see that the secondary structure in each piece may be selected independently of that in any other piece. Thus if we find the best secondary structure in each piece separately, subject to the constraint that possible pairs in the trial set are paired, these partial structures can be joined together to form the optimum secondary structure under the constraint. To find the best structure on each piece, we use a modified version of the n^3 or n^4 algorithm.

Suppose a piece extends from i_0 to j_0, and the accessible pairs in it are (in natural order) $p_h \cdot q_h$ for $h = 1$ to $k - 1$. Then the sequence on which the n^3 or n^4 algorithm is applied for this piece is the following union of strings:

$$[i_0, p_1] \cup [q_1, p_2] \cup \ldots \cup [q_{k-2}, p_{k-1}] \cup [q_{k-1}, j_0].$$

In this discussion, we shall continue to label the rows and the columns by their original labels, so it is necessary to recognize that row i_0 is first row, row j_0 the last one, and that row p_1 is adjacent to row q_1, etc.

Before starting the first-stage recursion, we ensure that p_h is paired with q_h (for each h) by defining

$$C(p_h, q_h) = \text{large negative value,}$$

$$\text{pointer}(C, p_h, q_h) = \phi.$$

During the first-stage recursion, the step that normally is performed whenever $j - i = 1$ or 2 or 3 is further limited by the requirement that i and j both belong to the same string of the union, either $[i_0, p_1]$ or $[q_1, p_2]$ or . . . . The step that is normally performed whenever $j - i \geqq 4$ is further expanded to include any cases in which i and j belong to different strings of the union, with one obvious exception: if $(i, j) = (p_h, q_h)$ for some h, we skip the equation that defines $C(p_h, q_h)$ since this value of C has already been defined.

If the piece is an inner piece, we also need to ensure that i_0 is paired to j_0. To do so, we merely apply the first stage of the algorithm, and select the secondary structure that is associated with $C(i_0, j_0)$; that is, we start the backtracking for this piece with the triple (C, i_0, j_0). Incidentally, using only the first stage of the algorithm for inner pieces yields a substantial saving of computation time.

REFERENCES

Auron, B. E., Rindone, W. P., Vary, C. P. H., Celentano, J. J., and Vournakis, J. N., Computer-aided prediction of RNA secondary structures. *Nucleic Acids Research* **10** (1):403–419, (1982).

Dumas, J. P., and Ninio, J., Efficient algorithms for folding and comparing nucleic acid sequences. *Nucleic Acids Research* **10** (1):197–206, (1982).

Ecarot-Charrier, B., and Cedergren, R. J., The preliminary sequence of $tRNA_F^{Met}$ from *Anacystis nidulans* compared with other initiator tRNAs. *Federation of European Biological Sciences Letters* **63**:287–290, (1976).

Gotoh, O. An improved algorithm for matching biological sequences. *Journal of Molecular Biology* **162**:705–708, (1982).

Mainville, Sylvie, *Comparaisons et auto-comparaisons de chaînes finies.* Ph.D. Thesis, Université de Montréal, (1981).

Nussinov, R., Piecznik, G., Grigg, J. R., and Kleitman, D. J., Algorithms for loop matchings. *SIAM Journal on Applied Mathematics* **35**(1):68–82, (1978).

Nussinov, R., and Jacobson, A., Fast algorithm for predicting the secondary structure of single-stranded RNA. *Proceedings of the National Academy of Sciences* **77**: 6309–6313, (1980).

Nussinov, R., and Tinoco, I., Jr., Secondary structure model for the complete simian virus-40 late precursor mRNA. *Nucleic Acids Research* **10** (1):351–363, (1982).

Pipas, J. M., and McMahon, J. E., Method for predicting RNA secondary structure. *Proceedings of the National Academy of Sciences* **72**(6):2017–2021, (1975).

Salser, W., Globin mRNA sequences: analysis of base pairing and evolutionary implications. *Cold Spring Harbor Symposia on Quantitative Biology* **42**(2):985–1002, (1977).

Sankoff, D., Evolution of secondary structure of 5S ribosomal RNA. Paper presented at *Tenth Numerical Taxonomy Conference*, Lawrence, Kansas, (1976).

Stiegler, P., Carbon, P., Zuker, M., Ebel, M.-P., and Ehresmann, C., Structural organization of the 16S ribosomal RNA from *E. coli.* Topography and secondary structure. *Nucleic Acids Research* **9**:2153–2172, (1981).

Studnicka, G. M., Rahn, G. M., Cummings, I. W. and Salser, W. A., Computer method for predicting the secondary structure of single-stranded RNA. *Nucleic Acids Research* **5**(9):3365–3387, (1978).

Tinoco, I., Borer, P. N., Dengler, B., Levine, M. D., Uhlenbeck, O. C., Crothers, D. M. and Gralla, J., Improved estimation of secondary structure in ribonucleic acids. *Nature New Biology* **246**:40–41, (1973).

Waterman, M. S., Secondary structure of single-stranded nucleic acids. *Studies in Foundations and Combinatorics, Advances in Mathematics, Supplementary Studies* **1**:167–212, (1978).

Waterman, M. S. and Smith, T. F., RNA secondary structure: A complete mathematical analysis. *Mathematical Biosciences* **42**:257–266, (1978).

Woese, C. R., Magrum, L. J., Gupta, R., Siegel, R. B., Stahl, D. L., Kop, B., Crawford, N., Brosius, N. Jr., and Noller, H. F., Secondary structure model for bacterial 16S ribosomal RNA: phylogenetic, enzymatic, and chemical evidence. *Nucleic Acids Research* **8**(10):2275–2293, (1980).

Zuker, M., and Stiegler, P., Optimal computer folding of large RNA sequences using thermodynamics and auxiliary information, *Nucleic Acids Research* **9**:133–148, (1981).

PART TWO

TIME-WARPING, CONTINUOUS FUNCTIONS, AND SPEECH PROCESSING

INTRODUCTION

In some applications of sequence comparison, the basic objects to be compared are not sequences but are continuous functions of a continuous parameter, usually time. The values of the functions are usually multidimensional; i.e., each value consists of several coordinates. The prime application of this kind is the processing of human speech, where the sound waves that make up one utterance are compared with those of another. While the value of a sound wave is simply the pressure at time t, and as such is one-dimensional, sound waves are normally converted into multidimensional form for speech processing: at time t, 10 or 15 coordinates are used to describe the sound wave over an interval of perhaps 30 milliseconds surrounding t.

Suppose the speed of the process underlying the functions is subject to variation, so that a function may be traced out more slowly during one portion and more quickly during another, and suppose these variations differ from one occasion to another. To allow for such variations when comparing functions, it is necessary to distort or "warp" the time axis appropriately, i.e., compressing it at some places and expanding it at others. The process of inferring the necessary compressions and expansions is often called *time-warping*.

In practice, the continuous functions are converted into discrete sequences by some sampling or segmentation process, and comparison of such sequences is also called time-warping. A variety of different methods can be used to do the conversion: The commonest one is to sample the function at regularly spaced times, but other methods are sometimes used, such as dividing the time axis into "natural" segments based on the character of the function and then classifying the segments into categories.

Speech processing makes heavy use of sequence comparison, not only to deal with continuous sound waves, but also at other levels to deal with discrete sequences such as phoneme-like units of sound, syllable-like units, etc. Dynamic programming is a standard method for performing all these sequence comparisons. Chapter 5, by Hunt, Lennig, and Mermelstein, describes and contributes to the use of time-warping and sequence comparison in speech processing at two levels. An utterance, in the form of a continuous multi-dimensional function of time, is converted into a sequence of points in multidimensional space by sampling at regular time intervals. Time-warping is used to match pieces of the given sequence with sequences stored in a dictionary of syllables, and the input converted piece by piece into syllable labels. Now a second level of comparison is used: the sequence of labels is compared with the sequences that correspond to possible phrases from a limited domain of discourse. Thus sequence comparison and a grammar of possible phrases can be used to identify the phrase that was spoken.

Chapter 6, by Bradley and Bradley, is concerned with the comparison of bird songs. A continuous tape recording of a bird song is described as a sequence of notes (and possibly also silences), and the notes are classified into a large number of types. The chapter addresses a variety of practical matters that arise in this framework, such as the method by which the continuous signal is segmented into notes, the definition of distances among notes, the relative effectiveness of simple linear time-warping versus time-warping by dynamic programming, and so forth.

Chapter 4, by Kruskal and Liberman, clarifies the subtle difference between compression and expansion on one hand, which occur in sequences as described above, and deletion and insertion on the other hand, as described in other parts of this volume. It develops symmetric time-warping for continuous functions, and systematically derives concept definitions for sequences by discretizing the corresponding definitions for continuous functions, thereby shedding light on a controversy over "weights" in time-warping. Finally, it also introduces several other topics, such as interpolation time-warping and a method of sequence comparison that uses both compression–expansion and deletion–insertion at the same time.

We mention here some other applications not described fully in this volume. Time-warping can be used in processing handwritten material such as signatures and line drawings. Typically, pen position is recorded as a function of time, with pressure of the pen on the paper or height of the pen above the paper as a third coordinate. Handwriting recognition is addressed by Fujimoto *et al.* (1976) and Burr (1980), verification of signatures by Yasuhara and Oka (1977), and other related topics by Burr (1979, 1981).

Chapter 1 contains several paragraphs concerning the application of time-warping to gas chromatography (GC). Reiner et al. (1979, 1978, 1969) have applied GC to identification of microorganisms and tissue samples, but found that comparison of chromatograms was hindered by nonlinear distortions of the time axis. Initially they compensated for this distortion by sliding one chromatogram relative to the other (over a light box), but found substantial advantages in using time-warping instead.

REFERENCES

Burr, D. J., A technique for comparing curves. Pages 271–277 in *Proceedings of the IEEE Conference on Pattern Recognition and Image Processing, 1979, Chicago.* New York: IEEE (1979).

Burr, D. J., Designing a handwriting reader. Pages 715–722 in *International Conference on Pattern Recognition, 5th, 1980, Miami Beach, Florida.* New York: IEEE (1980).

Burr, D. J., Elastic matching of line drawings. *IEEE Transactions on Pattern Analysis and Machine Intelligence* PAMI-3:708–713 (1981).

Fujimoto, Y. et al., Recognition of handprinted characters by nonlinear elastic matching. Pages 113–119 in *International Joint Conference on Pattern Recognition, 3rd, 1976, Coronado, California.* New York: IEEE (1976).

Reiner, E. and Kubica, G. P., Predictive value of pyrolysis-gas-liquid chromatography in the differentiation of mycobacteria. *American Review of Respiratory Disease* **99**:42–49 (1969).

Reiner E. and Bayer, F. L., Botulism: a pyrolysis-gas-liquid chromatographic study. *Journal of Chromatographic Science* **16**(12):623–629 (1978).

Reiner, E. et al., Characterization of normal human cells by pyrolysis-gas chromatography mass spectrometry. *Biomedical Mass Spectrometry* **6**(11):491–498 (1979).

Yasuhara, M. and Oka, M., Signature-verification experiment based on nonlinear time alignment: a feasibility study. *IEEE Transactions on Systems, Man, and Cybernetics* SMC-7:212–216 (1977).

CHAPTER FOUR

THE SYMMETRIC TIME-WARPING PROBLEM: FROM CONTINUOUS TO DISCRETE

Joseph B. Kruskal and Mark Liberman

1. INTRODUCTION

A trajectory, as illustrated in Fig. 1, means a continuous function of time in multidimensional space, i.e., a time-labelled curve in multidimensional space. Time-warping, as illustrated in Fig. 2, refers to comparison of trajectories, or to comparison of sequences derived from them by time-sampling, when each trajectory is subject not only to alteration by the usual additive random error but also to variations in speed from one portion to another. (In some applications, it is necessary to permit other differences between the trajectories as well, such as deletion and insertion, but we touch on that only lightly.) Such variation in speed appears concretely as compression and expansion with respect to the time axis, and will be referred to as compression–expansion. The chief purpose of time-warping is to deal with such variation. The chief application has been to speech processing, where compression–expansion is of major importance.

Time-warping is used in at least three ways. One is to discover the pattern of compression–expansion that connects two sequences. Another is to measure how different two sequences are in a way that is not sensitive to compression–expansion but is sensitive to other differences. A third use is in forming the weighted "average" of two sequences.

Time-warping of sequences is very similar in form and methodology to the comparison of "naturally discrete" sequences discussed elsewhere in this volume, such as the macromolecules of molecular biology and the character

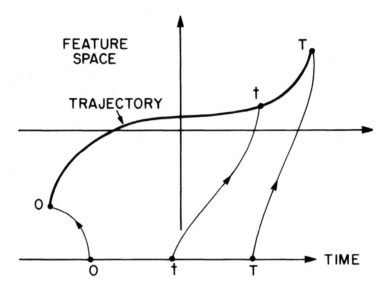

Figure 1a. Trajectory.

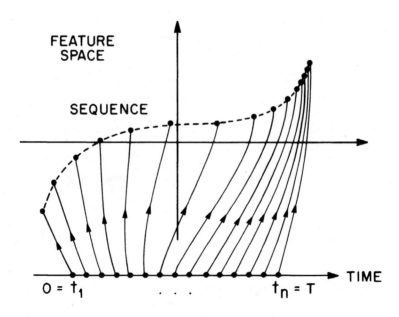

Figure 1b. Sequence derived from trajectory.

strings of computer science, in which the corresponding source of variation is deletion and insertion of units. This similarity is based on the following correspondence:

Compression–Expansion		Deletion–Insertion
Compress 2 units into 1	↔	Delete 1 unit
Expand 1 unit into 2	↔	Insert 1 unit

More generally,

Compress $(k + 1)$ adjacent units into 1	↔	Delete k adjacent units
Expand 1 unit into $(k + 1)$ adjacent units	↔	Insert k adjacent units

It is, however, frequently overlooked that the difference in meaning between compression–expansion and deletion–insertion leads to significantly different definitions of distance between sequences. We shall make the difference very clear below, and illustrate how to use both types of change at the same time when comparing sequences.

In fields where time-warping is used, the basic objects of interest are generally continuous trajectories, so it is natural in concept, though impossible in practice, to compare the trajectories directly. While the conversion of trajectories to sequences by sampling circumvents the practical difficulty, many of the ideas of time-warping can be expressed most naturally in a continuous setting. In this chapter we first develop continuous time-warping, and then systematically "discretize" it, i.e., formulate discrete analogues to all concepts ·and definitions involved. This appears to be the first paper in which continuous time-warping is formulated in a fully symmetric manner, and the first in which the discretization process is systematically examined and a variety of alternative discretizations specified. This approach provides a full justification for some edge weights (such as "$\frac{1}{2}$, 1, $\frac{1}{2}$"), which have been widely used without a fully satisfying rationale.

A method of sequence comparison is symmetric, in the sense used above, if comparing **a** with **b** gives the "same" result as comparing **b** with **a**, that is, the distances are the same and the time-warping of **b** onto **a** is the inverse of the time-warping of **a** onto **b**. Although the methods of sequence comparison in speech recognition are often deliberately asymmetric, treating the "stored template" utterance differently from the utterance to be recognized, our development is almost entirely limited to methods that are symmetric. There are several reasons for this.

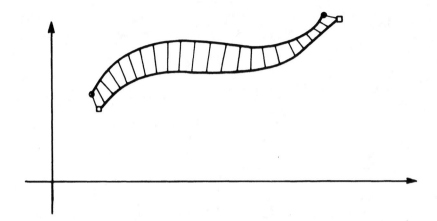

Figure 2a. Intuitive idea of continuous time-warping.

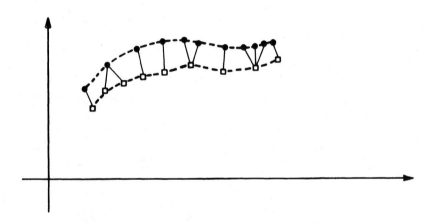

Figure 2b. Intuitive idea of discrete time-warping.

1. One purpose of this paper is to clarify the central difference between the comparison methods of molecular biology and those of speech processing, i.e., the differences between what we now distinguish as deletion–insertion and compression–expansion. The question of symmetry or asymmetry is not important for this purpose, and the symmetric approach is more convenient to work with, and familiar in both molecular biology and speech processing.

2. Another purpose of this paper is a proposed speech application for which symmetric comparison is desired, specifically, the comparison of related utterances so as to study timing variability of normal speech. The data may consist either of x-ray microbeam recordings of articulator motion (as described, e.g., in Fujimura (1981)), or conventional sound wave analyses as used in speech processing.

3. The chief reason for asymmetric comparison in speech recognition lies in the the mild improvement obtained by distinguishing between the stored template and unidentified current utterance. Even in speech recognition, however, there are other uses for comparison in which the desirability of asymmetry is not so clear, e.g., combining of utterances to form an "average" template. Thus, insight into symmetric methods may perhaps be of value even for speech recognition.

In Secs. 2 and 3 we formalize the notion of a time-warping as a "linking" that connects the time scales of the two trajectories or sequences. In the discrete case, the linking concept is similar to the "trace" concept used with deletion–insertion comparisons (see, for example, Chapter 1). In fact, a discrete linking is precisely analogous to a trace, and the differences between linking and trace reflect the differences between compression–expansion and deletion–insertion. In Secs. 4 and 5 we define the length of a linking, and then define distance between two trajectories as the minimum possible length of any linking between them. There is quite a variety of different ways to discretize the concept of length, which lead to mildly different discrete concepts. We explore many of these, including some that have not previously been discussed.

In Sec. 6 we explain the most important difference between compression–expansion and deletion–insertion, namely, the difference between the length of a linking and the length of a trace. Linking length does not use deletion–insertion costs as trace length does, only substitution costs. On the other hand, linking length uses another distinctive element called time-weights, which multiply the substitution costs. In Sec. 7 we explain how compression–expansion and deletion–insertion can be combined into a single potentially useful method, by incorporating both deletion–insertion costs and time-weights in the same comparison.

In Sec. 8, we note that a time-warping between two trajectories may be seriously misleading when the interval at which the trajectories are sampled is large in comparison to the differences between them, and we introduce a new method called *interpolation time-warping* to remedy this difficulty. In Secs. 9 and 10, stimulated by the asymmetric definition of Rabiner and Wilpon (1979, 1980), we give a symmetric definition of a weighted average between two trajectories or two sequences. Averaging is useful in forming a single "typical" sequence that is intended to represent a set of several similar sequences.

We note that when time-warping is applied, numerous related problems need to be dealt with that may not be part of the time-warping itself. These problems include choice of distance function (called w below) in the feature space, local constraints on the time-warping function, and finding where the trajectories begin and end (finding where speech utterances begin and end is surprisingly difficult). The solutions to these problems depend strongly on the domain of application. This paper is devoted to the central time-warping concept itself, and does not deal with problems such as those mentioned.

For information about methodology and the use of time-warping in recognition of isolated words, the reader may consult papers such as Itakura (1975), White (1978), Sakoe and Chiba (1978), Myers, Rabiner, and Rosenberg (1980), Rabiner, Rosenberg and Levinson (1978), and White and Neely (1976). For methodology and the use of time-warping in recognition of connected speech, see Chapter 5 and papers such as Bridle and Brown (1979), Rabiner and Schmidt (1980), and Myers and Rabiner (1981a, 1981b). In addition, a volume of reprints, Dixon and Martin (1979), contains many valuable papers in this field. For applications of time-warping to gas chromatography, see Reiner *et al.* (1979, 1978, 1969). For applications to handwriting recognition and related topics, see Fujimoto *et al.* (1976), Burr (1979, 1980, 1981), and Yasuhara and Oka (1977).

2. TIME-WARPING IN THE CONTINUOUS CASE

In speech processing, gas chromatography, bird song, and other potential applications of sequence comparison, the underlying objects of interest are basically continuous functions $\mathbf{a}(t)$, $\mathbf{b}(t)$, etc., of a continuous variable t, which is often time. Also, the values of the functions lie in a several-dimensional space which we shall call the *feature space*. Thus each object of interest is a continuous *trajectory* or curve through feature space, as shown in Fig. 1(a), in which each point on the curve corresponds to a particular value of the variable t. For practical manipulation, these trajectories are ordinarily converted into sequences by sampling the values of t, as shown in Fig. 1(b). Geometrically, this corresponds to describing the trajectory by a series of points on it.

By way of example, we mention that in speech processing, the dimensionality of the feature space is often in the range from 6 to 15. The ith coordinate of $\mathbf{a}(t)$ might indicate the power present in a speech utterance in the ith frequency band at time t (using a short-time spectral analysis). Alternatively, it might indicate the ith linear predictor coefficient at time t.

Conceptually, time-warping applies most directly to comparisons of continuous trajectories. It has seldom been discussed in this domain, however, because for practical computation it is always used with sequences. We start, however, by discussing time-warping and its uses in the continuous case, for the conceptual guidance this discussion provides in the discrete case.

Two trajectories

$$\mathbf{a}(u), \quad 0 \leqq u \leqq U, \quad \text{and} \quad \mathbf{b}(v), \quad 0 \leqq v \leqq V,$$

are said to be connected by an [approximate] continuous time-warping if they traverse [approximately] the same curve in feature space in the same direction, though at possibly very different rates; for example, $a(u)$ may proceed slowly along an early portion of the curve and quickly along a later portion, while $b(v)$ might do the reverse.

Geometrically, the idea of a time-warping is that each point in one trajectory corresponds to some specific point in the other. One way to visualize this is illustrated in Fig. 2(a), in which corresponding points are connected by line segments. If $a(u)$ corresponds to $b(v)$, we say u is *linked* to v. The correspondence between the trajectories is the central idea of time-warping.

More formally, we say (see Fig. 3(a)) that $a(u)$ and $b(v)$ are *connected by an [approximate] continuous time-warping* $(\mathbf{u}_0, \mathbf{v}_0)$ if $\mathbf{u}_0(t)$ and $\mathbf{v}_0(t)$ are strictly increasing functions defined for $0 \le t \le T$ such that

$$\mathbf{a}(\mathbf{u}_0(t)) = \mathbf{b}(\mathbf{v}_0(t)) \quad [\text{or } \mathbf{a}(\mathbf{u}_0(t)) \cong \mathbf{b}(\mathbf{v}_0(t))].$$

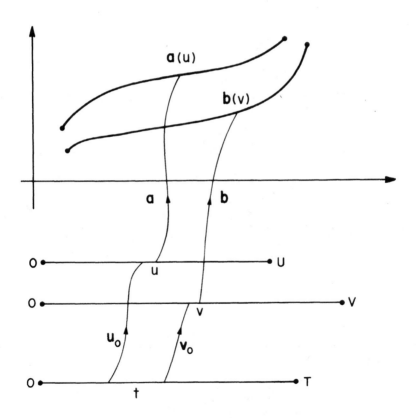

Figure 3a. Continuous time-warping.

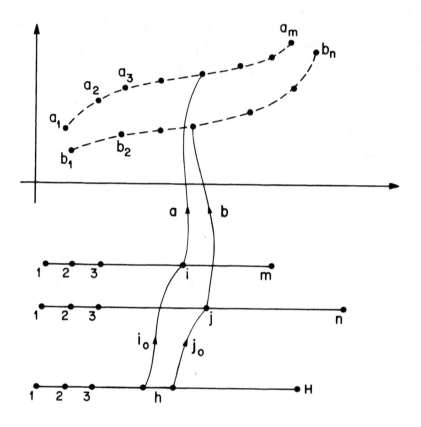

Figure 3b. Discrete time-warping.

Some constraint is generally needed to ensure that the time-warping does not degenerate to some tiny part of the curves involved. For example, the constraint might be that

$$\mathbf{u}_0(0) = 0, \qquad \mathbf{v}_0(0) = 0,$$
$$\mathbf{u}_0(T) = U, \qquad \mathbf{v}_0(T) = V,$$

though a weaker constraint could also be used. The word "approximate" is frequently omitted even when the approximate sense is intended, and the word "continuous" is generally omitted since it is obvious from context.

In this formulation the time-warping correspondence between the two trajectories is mediated by linking the two time-scales u and v. If $u = \mathbf{u}_0(t)$ and $v = \mathbf{v}_0(t)$, we shall say that u and v are linked by $(\mathbf{u}_0, \mathbf{v}_0)$ at t. Thus, points in the trajectories correspond in the time-warping if their times are linked.

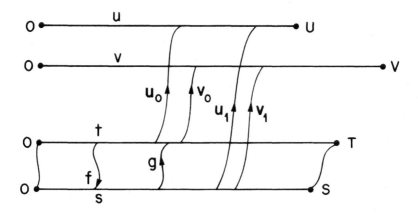

Figure 4. Arbitrariness of time scale.

It is necessary to recognize that the scale on which the parameter t occurs is arbitrary, has no intrinsic meaning, and can be freely distorted. In particular (see Fig. 4), suppose that f is any strictly increasing continuous function for which $f(0) = 0$, and let $s = f(t)$, $S = f(T)$. Let g be the inverse function of f (that is $g(f(t)) = t$, $f(g(s)) = s$), so that $t = g(s)$, $T = g(S)$. Define another time-warping $(\mathbf{u}_1, \mathbf{v}_1)$ by

$$\mathbf{u}_1(s) = \mathbf{u}_0(g(s)), \qquad \mathbf{v}_1(s) = \mathbf{v}_0(g(s)).$$

All we have done is distort the arbitrary scale for t into another arbitrary scale for s. The time-warping $(\mathbf{u}_1, \mathbf{v}_1)$ gives the same correspondence between the two trajectories as $(\mathbf{u}_0, \mathbf{v}_0)$, since if u and v are linked by $(\mathbf{u}_0, \mathbf{v}_0)$ at t, then it is easy to verify that they are linked by $(\mathbf{u}_1, \mathbf{v}_1)$ at $s = f(t)$.

We shall call two time-warpings *equivalent* if they induce the same linking (throughout the entire trajectories). We note without proof that any two equivalent time-warpings must be related in the manner just described. We think of equivalent time-warpings as essentially the same, and differing only in external form, not in any substantive way. This view will have important implications below.

In its various applications, time-warping is used as a method to help overcome the variability among nominally identical trajectories. Conceptually, we can think of a trajectory as composed of two aspects: One is the curve, by which we mean the points swept out; the other is the time pattern, by which we mean the rate at which the curve is followed. The time-warping we construct between two trajectories displays the difference between them in terms of these aspects: $\mathbf{u}_0$ and $\mathbf{v}_0$ compare the time patterns, while the distance is a summary measure of how much the curves differ.

Note that the methods that are used to calculate a time-warping between two trajectories must frequently be applied when the trajectories are, in fact, entirely different and unrelated, e.g., when comparing an observed trajectory with many stored trajectories in order to identify it. Thus, although the basic concept assumes the existence of an approximate time-warping, the methods for calculation must not rely too heavily on this assumption.

Speech processing has generally rested, of course, on the basic assumption that two trajectories of the same word or phrase are connected by an approximate time-warping. While this assumption is reasonable and has been the basis for a great deal of fruitful work, systematic violations are known to occur. For instance, more emphatic pronunciation generally produces not only an increase in duration, but also an "amplification" of the vocal gestures involved. This effect can be seen most clearly in articulatory data, as expansion of some portion of the curve, but formant trajectories also show it plainly. In the filter-bank or linear-prediction feature spaces, such phenomena are equally present, though harder to visualize. Obviously, in such a case the usual time-warping comparison will produce a distance measure that is "too large," because it does not allow for trajectory differences that leave the word or phrase unchanged. (Also, it is observed empirically that when "amplification" of a curve occurs, the usual procedures yield a time-warping that differs quite strongly from our intuitive notion of what it should be.) Such problems are doubtless among the reasons that speech recognition has been such a challenging problem.

3. TIME-WARPING IN THE DISCRETE CASE

To work with the continuous trajectories in practice, one standard approach is to convert them to sequences of points in feature space by sampling (see Fig. 1(b)). To convert $\mathbf{a}(u)$, it is sampled at some suitable set of discrete values $u_1, \ldots, u_m$, and $\mathbf{a}(u)$ is represented by the sequence $(\mathbf{a}(u_1), \ldots, \mathbf{a}(u_m))$. We shall use $a_i = \mathbf{a}(u_i)$ and $\mathbf{a} = a_1 \ldots a_m$. In a similar manner, $\mathbf{b}(v)$ is represented by its values at $v_1, \ldots, v_n$, namely by $\mathbf{b} = b_1 \ldots b_n = (\mathbf{b}(v_1), \ldots, \mathbf{b}(v_n))$.

It is also necessary to convert the time-warping concept from trajectories to sequences. This could be done in more than one way, but we follow the usual definition, which seems very plausible. Following the definition, we justify certain parts of it. As illustrated in Fig. 2(b) and 3(b), two sequences

$$\mathbf{a} = a_1 \ldots a_m \qquad \text{and} \qquad \mathbf{b} = b_1 \ldots b_n$$

with entries in the feature space are said to be *connected by an [approximate] discrete time-warping* $(\mathbf{i}_0, \mathbf{j}_0)$ if $\mathbf{i}_0(h)$ and $\mathbf{j}_0(h)$ are weakly increasing integer functions defined for $1 \leq h \leq H$ satisfying a "continuity constraint" (see below) such that

$$a_{i_0(h)} = b_{j_0(h)} \qquad [\text{or } a_{i_0(h)} \cong b_{j_0(h)}]$$

for all h. Each value of h corresponds to a line in Fig. 2(b) that connects a point in one sequence to a point in the other. To avoid the possibility that the time-warping degenerates to a tiny part of the sequences, we can use the constraint

$$\mathbf{i}_0(1) = 1, \qquad \mathbf{j}_0(1) = 1,$$

$$\mathbf{i}_0(H) = m, \qquad \mathbf{j}_0(H) = n,$$

though a weaker constraint could also be used. The word "approximate" is frequently omitted, even where the approximate sense is intended, and the word "discrete" is generally omitted since it is obvious from context.

The time-warping correspondence between the two sequences is mediated by linking what are in effect discrete time scales, i and j. If $i = \mathbf{i}_0(h)$ and $j = \mathbf{j}_0(h)$, we shall say that i and j are linked by $(\mathbf{i}_0, \mathbf{j}_0)$ at h. The points in the sequence correspond in the time-warping if their times are linked.

Figure 5 illustrates another representation of a discrete time-warping that is particularly important in connection with practical computation. For a given

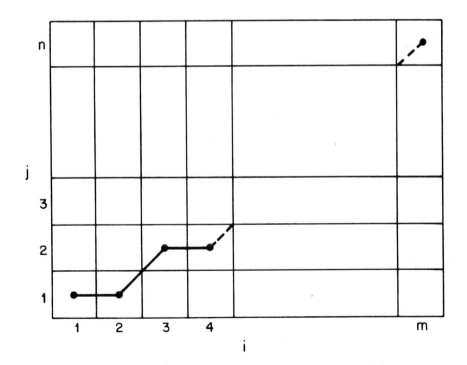

Figure 5. Computational array.

time-warping (i_0, j_0), each h corresponds to one cell in the computational array, namely, to cell $(i_0(h), j_0(h))$. If each such cell is indicated by a dot, and adjacent cells are connected by lines, the entire warping can be visualized as a path through the array.

We describe two different continuity constraints and justify them below. Each description is in terms of the vector or step between adjacent points of the time-warping in Fig. 5, that is, in terms of $(\Delta i_0, \Delta j_0)$, where Δ is defined by $\Delta i_0(h) = i_0(h) - i_0(h - 1)$. The first and most commonly used continuity constraint (see Fig. 6(a)) is

$$(\Delta i_0, \Delta j_0) = \left\{ \begin{array}{ll} (1, 0) & \text{or} \\ (1, 1) & \text{or} \\ (0, 1). \end{array} \right.$$

We remark that the step $(1, 0)$ indicates a time-compression from **a** to **b** that reduces the number of units by one: If there are k adjacent steps of this type, they constitute compression of $k + 1$ units into one. (Readers accustomed to deletion–insertion comparison are reminded that this step does *not* correspond to a deletion. The difference between compression and deletion will be discussed later. In the present notation, a single deletion could be indicated by a step of $(2, 1)$.) Similarly, the vector $(0, 1)$ indicates a time expansion from **a** to **b** that increases the number of units by one.

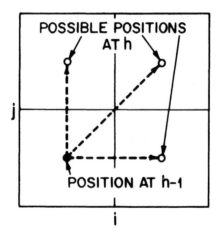

Figure 6a. First continuity constraint.

A second continuity constraint (see Fig. 6(b)), which is used in Chapter 5 by Hunt, Lennig, and Mermelstein and is due to Mermelstein, is

$$(\Delta i_0, \Delta j_0) = \begin{cases} (2, 0) & \text{or} \\ (1, 1) & \text{or} \\ (0, 2). \end{cases}$$

Under this constraint, only cells (i, j) for which $i + j$ is even are used, since the other pairs are skipped over. Also, it is not hard to see that every time-warping uses exactly the same number of pairs (i, j) (that is, same value of H), in contrast to the first constraint. Still other continuity constraints have been used also, but we do not consider them here.

Sometimes weights (most often $\frac{1}{2}, 1, \frac{1}{2}$) are associated with the alternative steps of the first constraint, for use in evaluating the length of a time-warping When we discuss lengths of time-warping later on, weights will arise naturally of their own accord; this fact and the values of these weights are a topic of interest to us.

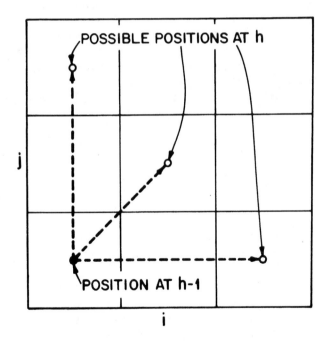

Figure 6b. Second continuity constraint.

Now, however, we simply wish to explain the continuity constraints. In the continuous case, $\mathbf{u}_0$ and $\mathbf{v}_0$ are constrained by monotonicity constraints, and also by continuity conditions that insured that no part of either trajectory can be skipped, i.e., there is no insertion or deletion. The first continuity constraint is exactly what we need to insure monotonicity of $\mathbf{i}_0$ and $\mathbf{j}_0$ and to avoid insertion and deletion. If we are willing to restrict the time-warping to points (i, j) for which $i + j$ is even (i.e., squares of only one color on a checkerboard), then the second continuity constraint is obtained in a similar manner. While the restriction to even values of $i + j$ appears to discard some fine-grain information, it reduces computation time by a factor of two, and its use is favored by Hunt, Lennig, and Mermelstein, partly because of the property that H is the same for all time-warpings. If the sampling rate for converting trajectories into sequences is increased by $\sqrt{2}$, this would appear to balance out the loss of information effectively while restoring the computation time, so the choice of continuity constraint should depend on subtler considerations.

To display a discrete time-warping pictorially, we can use a diagram like the one shown in Fig. 2(b), where each line corresponds to one value of h. If a_i is connected to h consecutive terms $b_j, \ldots, b_{j+h-1}$, this indicates that a region of $\mathbf{a}(u)$ around a_i corresponds in this time-warping to a region of $\mathbf{b}(v)$ around $b_j, \ldots, b_{j+h-1}$. If $u_1, u_2, \ldots$ and $v_1, v_2, \ldots$ are points in time and are regularly spaced using the same interval for the u_i and the v_j, this correspondence indicates that the changes in $\mathbf{a}(u)$ around time u_i occur rapidly and time must be stretched to match the corresponding changes in $\mathbf{b}(v)$ over the interval from v_j to v_{j+h-1}, which occur slowly. Of course, if the multiple connections go the other way, then a similar interpretation holds in reverse.

A diagram somewhat like Fig. 2(b) can be presented more simply:

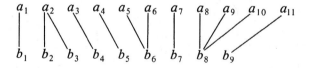

Discrete time-warping diagrams like this are very similar to trace diagrams (see, e.g., Chapter 1). However, such diagrams for symmetric time-warping differ from trace diagrams in two ways. (These remarks do not fully apply to diagrams for asymmetric time-warping, which is frequently used in speech processing.)

1. In a symmetric time-warping diagram, one term of a sequence may be connected to several terms of the other sequence, while in a trace diagram each term can be connected to at most one other term.

2. In a symmetric time-warping diagram, every term is connected to at least one other term, while in a trace diagram not every term need have a connection (i.e., terms that are insertions or deletions).

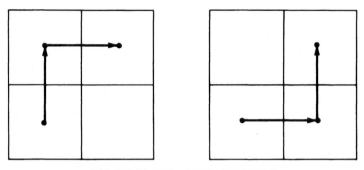

FORBIDDEN TWO-STEP PATTERNS

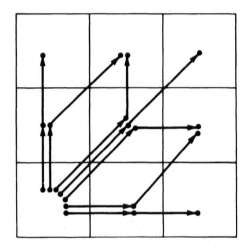

PERMITTED TWO-STEP PATTERNS

Figure 7. An additional constraint.

Additional constraints on the time-warping are common in speech research. Characteristically, they refer to the values of i_0 and j_0 at three or more consecutive values of h. The simplest, least restrictive constraint of this type simply forbids an "N"-shaped configuration in the time-warping diagram like those shown here:

$$\text{Forbidden} \quad \left\{ \begin{array}{cc} a & a \\ \mathbb{N} & \\ b & b \end{array}, \quad \begin{array}{cc} a & a \\ \mathbb{V} & \\ b & b \end{array} \right\};$$

that is, if a term has multiple lines, the terms at the other ends of these lines must not have multiple connections. Other ways of describing this constraint are shown in Fig. 7.

4. DISTANCE AND LENGTH IN THE CONTINUOUS CASE

The basic idea of time-warping is that replications of nominally the same trajectory will trace out approximately the same curve, but with varying time patterns. To measure the extent to which two trajectories $\mathbf{a}(u)$ and $\mathbf{b}(v)$ deviate from having this ideal relationship, we will first define the *length* $\bar{d}(\mathbf{u}_0, \mathbf{v}_0)$ of any given time-warping $(\mathbf{u}_0, \mathbf{v}_0)$ in a suitable way, as the distance between corresponding points in the two trajectories pooled somehow over the entire trajectories. Of course $\bar{d}$ depends on $\mathbf{a}(u)$ and $\mathbf{b}(v)$ as well as on $\mathbf{u}_0$ and $\mathbf{v}_0$, but we omit this dependence from the notation, for simplicity. Once length is defined, then the distance is given by

$$d(\mathbf{a}(u), \mathbf{b}(v)) = \min_{\text{all } (\mathbf{u}_0, \mathbf{v}_0)} \bar{d}(\mathbf{u}_0, \mathbf{v}_0);$$

that is, the distance is the length of the shortest possible time-warping.

To define the length of $\bar{d}(\mathbf{u}_0, \mathbf{v}_0)$, we first need a way to measure how far apart corresponding points $\mathbf{a}(\mathbf{u}_0(t))$ and $\mathbf{b}(\mathbf{v}_0(t))$ are for a fixed t. We shall assume that a distance function $w[a, b]$ suitable for this purpose has been defined on the feature space. This function could be simple Euclidean distance, or weighted Euclidean distance, or a more complicated function in which the distance is sensitive to position in the feature space.

To pool this distance over the whole trajectory, the obvious definition for $\bar{d}(\mathbf{u}_0, \mathbf{v}_0)$ might seem to be

$$\int_0^T w[\mathbf{a}(\mathbf{u}_0(t)), \mathbf{b}(\mathbf{v}_0(t))] \ dt.$$

This definition, however, is incorrect. Recall that we defined time-warpings to be equivalent if they induce the same linking between the trajectories, and that we consider equivalent time-warpings as essentially the same. We want a definition for which equivalent time-warpings have the same length. Using the definition above, however, equivalent time-warpings can result in quite different lengths. To see this, suppose we generate another time-warping equivalent to $(\mathbf{u}_0, \mathbf{v}_0)$, as in Fig. 4, by using $s = f(t)$ with f an increasing continuous function such that $f(0) = 0$ and $f(S) = T$. Writing the integral parallel to the one above but for $(\mathbf{u}_1, \mathbf{v}_1)$ instead, and then transforming it by $s = f(t)$, we have

$$\int_0^S w[\mathbf{a}(\mathbf{u}_1(s)), \mathbf{b}(\mathbf{v}_1(s))] \ ds = \int_0^T w[\mathbf{a}(\mathbf{u}_0(t)), \mathbf{b}(\mathbf{v}_0(t))] f'(t) dt.$$

This is exactly the same as the integral for $(\mathbf{u}_0, \mathbf{v}_0)$ *except for the factor $f'(t)$*, which can be virtually *any positive function.* Obviously, different choices of f' yield different values for the integral, so this definition is not invariant.

The arbitrariness of the integral has a precise geometrical interpretation: The integral runs along the two trajectories at an arbitrary rate that is determined by the arbitrary time scale on which t is measured. If we rush along the trajectories (this will be the case for $(\mathbf{u}_1, \mathbf{v}_1)$ if $g'(s)$ is large, $f'(t)$ small, and S small), then the integral will be small. If we go along the trajectories slowly (corresponding to the reverse situation), the integral will be large. Furthermore, even if the overall rate is the same for two time-warpings, we can still rush along the curves where they are far apart and go slowly where they are close together, in order to get a small value, or use the reverse strategy to get a large value.

Once the problem is stated, there is an obvious solution. The trajectories themselves have natural meaningful time scales, and we should use these time scales to weight each infinitesimal portion of the integral by a weight that corresponds to how long the trajectories linger there. Specifically, suppose u and v are linked by $(\mathbf{u}_0, \mathbf{v}_0)$ at t. The trajectory $\mathbf{a}(u)$ spends time $du = \mathbf{u}_0'(t)\, dt$ in the infinitesimal region around u, and the trajectory $\mathbf{b}(v)$ spends time $dv = \mathbf{v}_0'(t)\, dt$ around v. If we were willing to accept an asymmetric formulation, we could use either $\mathbf{u}_0'(t)$ or $\mathbf{v}_0'(t)$ as the weighting function. Let us disregard the asymmetry for a moment, and consider the use of $\mathbf{u}_0'(t)$. It gives the integral

$$\int_0^T w[\mathbf{a}(\mathbf{u}_0(t)),\ \mathbf{b}(\mathbf{v}_0(t))]\mathbf{u}_0'(t)\quad dt.$$

To test for invariance, we generate $(\mathbf{u}_1, \mathbf{v}_1)$ in the same way as before, and consider its corresponding integral,

$$\int_0^S w[\mathbf{a}(\mathbf{u}_1(s)),\ \mathbf{b}(\mathbf{v}_1(s))]\ \mathbf{u}_1'(s)\quad ds.$$

Consider the new factor $\mathbf{u}_1'(s)$. We have

$$\mathbf{u}_1'(s) = \frac{d}{ds}\,\mathbf{u}_0(g(s)) = \mathbf{u}_0'(g(s))g'(s).$$

Now differentiating $f(g(s)) = s$,

$$f'(g(s)) \cdot g'(s) = 1, \qquad g'(s) = \frac{1}{f'(t)}.$$

Therefore if we transform by $s = f(t)$, the preceding integral equals

$$\int_0^T w[\mathbf{a}(\mathbf{u}_0(t)), \ \mathbf{b}(\mathbf{v}_0(t))] \cdot \frac{\mathbf{u}_0'(t)}{f'(t)} \cdot f'(t) \, dt = \int_0^T w[\cdots] \, \mathbf{u}_0'(t) \, dt.$$

This is the same as the integral corresponding to $(\mathbf{u}_0, \mathbf{v}_0)$, so the new definition of length has the desired invariance property.

Use of $\mathbf{u}_0'(t)$ as the weighting function effectively means that we run along the $\mathbf{a}(u)$ trajectory at the rate set by its time scale, and along the $\mathbf{b}(v)$ trajectory however the correspondence determines. Use of $\mathbf{v}_0'(t)$ reverses the roles of the two trajectories. Either of these gives to the definition of length the invariance property, but neither one treats the two trajectories symmetrically. (This asymmetric approach, incidentally, is used in much speech-recognition work, where the time scale of the unknown utterance trajectory is used to form the distance.)

To give a symmetric formulation that is invariant, we must use a weighting function that combines $\mathbf{u}_0'(t)$ and $\mathbf{v}_0'(t)$ in a symmetric way. The most obvious possibility is the average, $(\mathbf{u}_0'(t) + \mathbf{v}_0'(t))/2$ (or alternatively, the sum). This means that we run along the trajectories at the average of the rates set by their two time scales. Other possibilities that provide both invariance and symmetry include the geometric mean, $(\mathbf{u}_0'(t)\mathbf{v}_0'(t))^{1/2}$, the rth power mean for any r, that is,

$$\left[\frac{\mathbf{u}_0'(t)^r + \mathbf{v}_0'(t)^r}{2} \right]^{1/r},$$

and still more general types of mean value. The case $r = 2$ can be given an arc-length interpretation, and turns out, after manipulation, to be the same as a formula from Myers (1980), which is discussed below.

Generalizing in another direction, a weighted combination of $\mathbf{u}_0'(t)$ and $\mathbf{v}_0'(t)$ with weights U and V (recall that $U = \mathbf{u}_0(T)$, $V = \mathbf{v}_0(T)$), or $1/U$ and $1/V$, or $f(U)$ and $f(V)$ for any function f, is also symmetric and invariant. Using weights $1/U$ and $1/V$ leads to an attractive weighting function $(\mathbf{u}_0'(t)/U + (\mathbf{v}_0'(t)/V)$. Of course, weights could also be incorporated into the generalized means as well.

Lacking a convincing argument for any particular one of these formulations, we choose the ordinary average merely for simplicity. Thus for the remainder of this paper, $d(\mathbf{u}_0, \mathbf{v}_0)$ is formed by minimizing the following length over $(\mathbf{u}_0, \mathbf{v}_0)$:

$$\bar{d}(\mathbf{u}_0, \mathbf{v}_0) \equiv_{\text{def}} \int_0^T w[\mathbf{a}(\mathbf{u}_0(t)), \ \mathbf{b}(\mathbf{v}_0(t))] \frac{\mathbf{u}_0'(t) + \mathbf{v}_0'(t)}{2} \, dt.$$

4.1 Comparison with Other Continuous Formulations

In the many papers that apply time-warping to speech processing, there have been very few discussions of the continuous time-warping problem. In published papers, we note a brief discussion in Velichko and Zagoruyko (1970), a brief mention in Sakoe and Chiba (1971), and a discussion limited largely to the one-dimensional feature space in Levinson (1981). In addition, we note a more extensive discussion in an unpublished paper by Myers (1980).

Sakoe and Chiba (1971) present the following integral (our notation),

$$\int_0^U w[\mathbf{a}(u), \mathbf{b}(\mathbf{f}(u))] \ du.$$

where u is the time parameter for utterance $\mathbf{a}$, and $\mathbf{f}(u)$ (which describes the time-warping) corresponds to $\mathbf{v}_0(\mathbf{u}_0^{-1}(u))$, in our notation. Their integral is essentially the same as our first correct (but asymmetric) integral given above, since the two integrals are connected by an elementary change of variables, $u = \mathbf{u}_0(t)$. They propose minimizing this integral by choice of $\mathbf{f}$. Although their integral is not symmetric in the two utterances, they then state that minimization problems of this type can be "very effectively solved by dynamic-programming technique as follows," and proceed to present a symmetric version of the discrete time-warping problem, but do not indicate how the discrete formulation is derived from the continuous one.

The discussion by Velichko and Zagoruyko (1970) is harder to summarize, because it is less precisely stated. After developing a discrete version of time-warping, they state that "in the continuous approximation, the sum is substituted by the integral"

$$\int_{(\ell)} b(\ell)d\ell,$$

where the integral is taken along a curve in the (u, v)-plane (our notation), ℓ appears to be arc length along the curve, and $b(\ell)$ appears to be a measure of similarity between $\mathbf{a}(u)$ and $\mathbf{b}(v)$ at point ℓ on the curve. Presumably, $b(\ell)$ is intended to be analogous to their discrete measure of similarity ρ^2 defined shortly before. The curve, of course, describes the time-warping, which they refer to as a time normalization. They then argue for introduction (into the integral) of a weighting factor $f(\gamma)$ such as $f(\gamma) = \cos(2\gamma)$, where γ is the angle between the $45°$ line and the tangent to the curve at point ℓ, thus yielding

$$\int_{(\ell)} f(\gamma)b(\ell) \ d\ell.$$

They propose finding the curve that maximizes this integral, subject only to the constraint that the curve denote a monotonic increasing function, and describe this maximization as a variational problem. After this they "reformulate (their) problem for the discrete case," but give no details connecting the continuous and discrete formulations.

The discussion by Levinson (1981) is largely subsumed by that within Appendix I of Myers (1980), which we now discuss. Myers presents the following integral (notation partly changed to ours):

$$\int_0^U w[\mathbf{a}(u), \mathbf{b}(\mathbf{f}(u))] \widetilde{W}(u, \mathbf{f}(u), \dot{\mathbf{f}}(u)) \, du.$$

Note that this is the same as Sakoe and Chiba's integral, except for the introduction of the weighting function $\widetilde{W}$. The use of a weighting factor of this form appears to be largely based on a fact introduced by Myers, namely, that this integral fits within the framework of a much-studied problem in the calculus of variations. Myers introduces the solution from that field, which is a differential equation for the time-warping curve $v = \mathbf{f}(u)$ in the (u, v) plane, and then proceeds to discuss choice of $\widetilde{W}$. He drops the dependence of $\widetilde{W}$ on u and $\mathbf{f}(u)$ "since all points in the $[(u, v)]$ plane should be weighted equally," and proposes as one logical choice for W the form

$$\widetilde{W}(\dot{\mathbf{f}}(u)) = \sqrt{1 + \dot{\mathbf{f}}(u)^2},$$

since $\widetilde{W}(\dot{\mathbf{f}}(u)) \, du$ then becomes the differential of arc length along the time-warping curve. Thus he obtains

$$\int_0^U w[\mathbf{a}(u), \mathbf{b}(\mathbf{f}(u))] \sqrt{1 + \dot{\mathbf{f}}(u)^2} \, du.$$

He points out that this can be thought of as the line integral of w with respect to arc length over the time-warping curve (and thus obtains an integral very similar to that of Velichko and Zagoruyko, though he does not make the connection or cite their paper). He attempted to find a numerical solution to the differential equation for his choice of $\widetilde{W}$, but indicates that this turned out to be difficult. He does not make any detailed connection between the continuous and discrete versions of the time-warping problem.

Myer's integral turns out to be symmetric in the two utterances $\mathbf{a}$ and $\mathbf{b}$, as we can see from the arc-length formulation, although his definition is not phrased in a symmetric manner and he does not consider the matter of symmetry. His integral above can easily be transformed, using the elementary change of variables $u = \mathbf{u}_0(t)$, into the symmetric form

$$\int_0^T w[\mathbf{a}(\mathbf{u}_0(t)), \mathbf{b}(\mathbf{v}_0(t))] \left[\frac{\mathbf{u}_0'(t)^2 + \mathbf{v}_0'(t)^2}{2} \right]^{1/2} dt,$$

which was one of the symmetric forms we described above.

5. DISTANCE AND LENGTH IN THE DISCRETE CASE

As in the continuous case, the distance between two sequences is defined as the minimum length of any time-warping between the two sequences. Thus the only question is how to form the length of a discrete time-warping $(\mathbf{i}_0, \mathbf{j}_0)$ by analogy with the length of a continuous time-warping. We shall explore several ways of making this analogy. In one approach, the infinitesimal intervals such as $d\mathbf{u}_0(t) = \mathbf{u}_0'(t) \, dt$ correspond to intervals from one sampling point to another, such as $[u_{i-1}, u_i]$. In another approach, each infinitesimal interval corresponds to an interval that surrounds one sampling point, so that each u_i is near the center of its interval. We shall explore several versions of the first approach, and one version of the second approach. It is not clear whether or not the differences among these versions have any substantive importance, but in some cases they do have computational importance. Throughout this section, we assume that sequences $\mathbf{a}$ and $\mathbf{b}$ have m and n points, respectively, and are drawn from trajectories $\mathbf{a}(u)$ and $\mathbf{b}(v)$ extending over the time intervals $[0, U]$ and $[0, V]$, respectively. We shall sometimes assume that the sampling times $u_1, \ldots, u_m$ and $v_1, \ldots, v_n$ are regularly and identically spaced, i.e., that $u_i - u_{i-1} = \tau$ and $v_j - v_{j-1} = \tau$ for all i and j.

We start with the first approach. For the time being, we assume that $u_m = U$ and we introduce nonsampling points $u_0 = 0$ and $v_0 = 0$, so that the first and last intervals for $\mathbf{a}(u)$ are $[u_0 = 0, u_1]$ and $[u_{m-1}, u_m = U]$, and similarly for $\mathbf{b}(v)$. To form the analogy, we use the following correspondence for $\mathbf{a}(u)$, and extend it in the obvious way to $\mathbf{b}(v)$:

$$dt \leftrightarrow 1 = h - (h - 1)$$

$$[u - du, u] \leftrightarrow [u_{i_0(h-1)}, u_{i_0(h)}]$$

$$d\mathbf{u}_0'(t) = \mathbf{u}_0'(t) \, dt \leftrightarrow \Delta u_{i_0(h)} = u_{i_0(h)} - u_{i_0(h-1)}$$

$$\int_0^T [\ldots] \mathbf{u}_0'(t) \, dt \leftrightarrow \sum_{h=1}^{H} [\ldots] \Delta u_{i_0(h)}$$

(As a check on the validity of the correspondence, we can apply both the integral and the summation to the function that is identically equal to 1. We obtain $\mathbf{u}_0(T) - \mathbf{u}_0(0) = U$ for the integral and $u_m - u_0 = U$ for the sum.) To complete the analogy, we decide that in the summation we will evaluate the summand $[\ldots]$ using the sampling point at the end rather than the beginning of its interval, i.e., we will use $u = u_{i_0(h)}$ rather than $u = u_{i_0(h-1)}$ in connection with the hth term of the summation. (The opposite decision would not be tenable, because it would involve use of $u = u_0$ when $h = 1$, which is not a sampling point.) Then letting

$$w(h) \equiv_{\text{def}} w[a_{i_0(h)}, b_{j_0(h)}],$$

the definition above for length of a continuous time-warping corresponds to the following, which is our first definition for length of a discrete time-warping:

$$\bar{d}(\mathbf{i}_0, \mathbf{j}_0) = \sum_{h=1}^{H} w(h) \frac{[\Delta u_{i_0(h)} + \Delta v_{j_0(h)}]}{2}.$$

If two sequences are regularly spaced, as described above, then $\Delta u_{i(h)} = \tau \Delta i_0(h)$ and the preceding formula reduces to

$$\bar{d}(\mathbf{i}_0, \mathbf{j}_0) = \tau \sum_{h=1}^{H} w(h) \frac{[\Delta \mathbf{i}_0(h) + \Delta \mathbf{j}_0(h)]}{2}.$$

For the first continuity constraint, the expression in brackets is 1 or 2 or 1 depending on which case occurs, so

$$\bar{d}(\mathbf{i}_0, \mathbf{j}_0) = \tau \sum_{h=1}^{H} z_h w(h),$$

where $z_h = 1$ for a diagonal step and $\frac{1}{2}$ for a vertical or horizontal step. We shall refer to the z_h as the *time weights*, though, properly speaking, it is the products $z_h \tau$ that are the true time weights. The values of z_h are illustrated for this formula in Case 1 of Fig. 8. This length formula is essentially identical to one that is well known in the time-warping literature, and the minimum-length time-warping can be calculated by standard methods. In particular, if $D_{ij} = $ distance $= $ minimum length between the incomplete sequences $a_1 \ldots a_i$ and $b_1 \ldots b_j$, then using recursion to find the values of D_{ij} is the main part of the calculation. Using

$$w(i, j) \equiv_{\text{def}} w[a_i, b_j],$$

the necessary recurrence equation is

$$D_{ij} = \min \begin{cases} D_{i-1, j} & + \frac{1}{2} \tau w(i, j), \\ D_{i-1, j-1} & + \tau w(i, j), \\ D_{i, j-1} & + \frac{1}{2} \tau w(i, j), \end{cases}$$

which can be evaluated recursively using the computational array of Fig. 5.

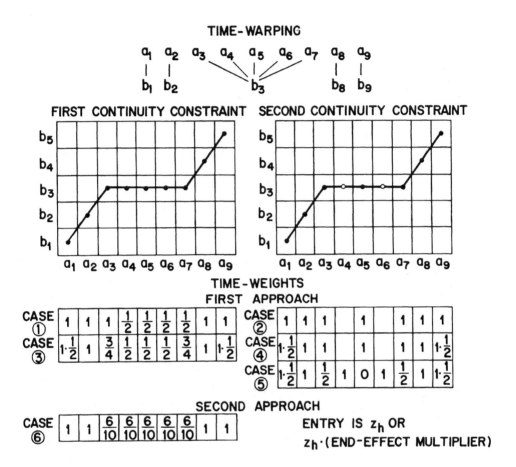

Figure 8. Illustration of time-weights for a given time-warping.

For the second continuity constraint, the expression in brackets is always 2, so the time weights $z_h = 1$ always (see Case 2 of Fig. 8). Using this yields

$$\bar{d}(\mathbf{i}_0, \mathbf{j}_0) = \tau \sum_{h=1}^{H} w(h).$$

This length formula is the same as that used in Chapter 5 by Hunt, Lennig, and Mermelstein. The minimization can easily be carried out by methods virtually identical to the standard ones, as illustrated in that chapter (of course, only half of the cells of the computational array are used, namely cells (i, j) with $i + j$ even). In particular, the recurrence equation is

$$D_{ij} = \min \begin{cases} D_{i-2,\,j} & + \tau w(i,\,j), \\ D_{i-1,\,j-1} & + \tau w(i,\,j), \\ D_{i,\,j-2} & + \tau w(i,\,j). \end{cases}$$

Suppose we proceed as above, but with one change. Instead of using $w(h)$, which means evaluating the summand at the end of the interval, suppose we use $[w(h-1) + w(h)]/2$, which means evaluating at both ends and taking the average. This requires a slight change of convention to avoid evaluating the summand at u_0 and v_0, which are not sampling points. Thus we set $u_1 = v_1 = 0$ (so the first interval for $a(u)$ is $[u_1, u_2]$). This leads to

$$\bar{d}(\mathbf{i}_0, \mathbf{j}_0) = \tau \sum_{h=2}^{H} \frac{w(h-1) + w(h)}{2} \cdot \frac{\Delta \mathbf{i}_0(h) + \Delta \mathbf{j}_0(h)}{2}.$$

For the first continuity constraint, we find that

$$\bar{d}(\mathbf{i}_0, \mathbf{j}_0) = \tau \left\{ \tfrac{1}{2} z_1 w(1) + \sum_{h=2}^{H-1} z_h w(h) + \tfrac{1}{2} z_H w(H) \right\},$$

where the time weights

$$z_h = \tfrac{1}{2} \quad \text{or} \quad \tfrac{3}{4} \quad \text{or} \quad 1$$

(see Case 3 of Fig. 8) depending on the steps which end with h and start with h (with a special rule for $h = 1$ and $h = H$). Again, the minimization can be carried out as usual. An appropriate equation is

$$D_{ij} = \min \begin{cases} D_{i-1,\,j} & + \tfrac{1}{4}\tau[w(i-1,\,j) + w(i,\,j)], \\\\ D_{i-1,\,j-1} & + \tfrac{1}{2}\tau[w(i-1,\,j-1) + w(i,\,j)], \\\\ D_{i,\,j-1} & + \tfrac{1}{4}\tau[w(i,\,j-1) + w(i,\,j)]. \end{cases}$$

For the second continuity constraint, we find that

$$\bar{d}(\mathbf{i}_0, \mathbf{j}_0) = \tau \left\{ \tfrac{1}{2} w(1) + \sum_{h=2}^{H-1} w(h) + \tfrac{1}{2} w(H) \right\},$$

which is reminiscent of the trapezoid rule for integration. Here $z_h = 1$ always (see Case 4 of Fig. 8). A recurrence equation for the minimization is

$$D_{ij} = \min \begin{cases} D_{i-2,\,j} & + \ \tfrac{1}{2}\tau[w(i-2,\,j) + w(i,\,j)], \\[2ex] D_{i-1,\,j-1} & + \ \tfrac{1}{2}\tau[w(i-1,\,j-1) + w(i,\,j)], \\[2ex] D_{i,\,j-2} & + \ \tfrac{1}{2}\tau[w(i,\,j-2) + w(i,\,j)]. \end{cases}$$

As an interesting side note, when using the second continuity constraint it is possible to use $w(h-\tfrac{1}{2})$ in place of $[w(h-1) + w(h)]/2$ for a vertical or horizontal step, because such a step moves two places along one of the sequences. Using such evaluation where possible leads to still another length formula, which we omit (but see Case 5 of Fig. 8), and the following recurrence equation:

$$D_{ij} = \min \begin{cases} D_{i-2,\,j} & + \ \tau w(i-1,\,j), \\[2ex] D_{i-1,\,j-1} & + \ \tfrac{1}{2}\tau[w(i-1,\,j-1) + w(i,\,j)] \\[2ex] D_{i,\,j-2} & + \ \tau w(i,\,j-1). \end{cases}$$

Consider the second approach to forming the intervals. We assume that sequences **a** and **b** were formed by dividing the trajectories into m and n pieces lasting time τ each and placing a sample point centrally in each interval. Then the definition of length for a continuous time-warping corresponds to

$$\bar{d}(\mathbf{i}_0, \mathbf{j}_0) = \sum_{h=1}^{H} w(h) z_h \tau$$

if we choose $z_h \tau$ analogous to $(\mathbf{u}_0'(t)dt + \mathbf{v}_0'(t)dt)/2$. Now $\mathbf{u}_0'(t)dt$ indicates time spent in trajectory $\mathbf{a}(u)$, and similarly for $\mathbf{v}_0'(t)dt$. Thus $z_h \tau$ should be the average of the time spent corresponding to h in the sequence **a** and the time spent corresponding to h in sequence **b**. One way to give this specific meaning relies on the constraint (see above) forbidding the presence of an "N"-shaped configuration in the discrete time-warping diagram. With this constraint, the diagram divides naturally into connected component groups, which are of three types (and see Case 6 of Figure 8):

i) A single a_i joined to a single b_j. This group contains one value of h, and it corresponds to time τ in each sequence, so the average is τ and we set $z_h = 1$.

ii) A single a_i joined to k terms from $\mathbf{b}$ (with $k \geq 2$). This group contains k values of h. It corresponds to time τ in the $\mathbf{a}$ sequence, and to time $k\tau$ in the $\mathbf{b}$ sequence. Taking the average, and dividing the amount of time evenly into k parts, we get $\tau(k + 1)/2k$, so we set $z_h = (k + 1)/2k$ for each of the k time-weights in the group.

iii) A single b_j joined to k terms from $\mathbf{a}$ (with $k \geq 2$). In a similar manner, we find that $z_h = (k + 1)/2k$ for each of the k time-weights in the group.

(Note that if we drop the requirement $k \geq 2$, then the latter two cases are consistent with the first one.) The recurrence equation for this definition of length is computationally slower than the recurrence equations above:

$$
D_{ij} = \min \begin{cases}
\min_{2 \leq i_1 < i} \left[D_{i-i_1, j-1} + \tau \dfrac{i_1 + 1}{2i_1} \sum_{i_2=1}^{i_1} w(i + 1 - i_2, j) \right], \\[2ex]
D_{i-1, j-1} + d(i, j), \\[2ex]
\min_{2 \leq j_1 < j} \left[D_{i-1, j-j_1} + \tau \dfrac{j_1 + 1}{2j_1} \sum_{j_2=1}^{j_1} w(i, j + 1 - j_2) \right].
\end{cases}
$$

6. HOW COMPRESSION–EXPANSION DIFFERS FROM DELETION–INSERTION

We have already noted one difference between compression–expansion and deletion–insertion in Sec. 3, when we contrasted the concepts of linking and trace. There is, however, a more important difference.

Consider the following bit from a time-warping diagram, and a very similar bit from a trace diagram:

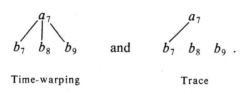

Time-warping   Trace

The former expands a_7 to match $b_7 b_8 b_9$; the latter inserts b_8 and b_9. By redescribing compression–expansion systematically in this manner, it is converted into deletion–insertion, and the time-warping problem can be thought of as the deletion–insertion problem. It is through this relationship that the two sequence-comparison problems have often been considered the same.

Despite this conversion, the problems are not the same. The difference between expansion and insertion lies in the costs that we wish to assign to these operations. In the trace, the cost assigned to the substitution of a_7 by b_7 is treated very differently from the cost of the insertions of b_8 and b_9. The cost of the substitution will be small if b_7 equals or resembles a_7, and will be larger otherwise. By contrast, there is no reason for the weight of the insertions to be small if b_8 or b_9 equals or resembles a_7; nor is there even any reason to select a_7 for the comparison over, say, a_8. The ultimate reason for this treatment of the costs is the basic physical processes we have in mind, namely, substitution or modification of a_7 to yield b_7, but insertion of a new element rather than modification of an existing one to yield b_8 and b_9.

On the other hand, in time-warping the costs assigned to each of the three comparisons (a_7 with each of b_7, b_8, b_9) are all treated in similar or identical fashion, and for each of them the cost should be smaller if b_j equals or resembles a_7. The ultimate reason for this treatment of the weights is again the basic physical process we have in mind, namely, that the trajectory moves more slowly through the region around a_7 on the second replication, so this region is represented by three points instead of one, and the difference between a_7 and b_j is due to additive random error.

7. COMBINING DELETION–INSERTION AND COMPRESSION–EXPANSION IN A SINGLE METHOD

One problem in applying time-warping to speech processing is that speech utterances may differ not only by time-distortion and additive random error but also by interpolated or deleted sounds. This can happen for a variety of reasons: extraneous sounds from the ambient environment (door slamming, footsteps, etc.), speaker-generated nonspeech sounds (lip smacks, coughs, breath noises, etc.), and more or less full pronunciation of a word (the dictionary pronunciation of "probably" may be reduced to "prob'ly" or even "pro'lly," the dictionary pronunciation "offen" may be expanded to the spelling pronunciation "often," etc.). One step towards dealing with such additional difficulties is to perform the comparison in a way that allows for deletion–insertion as well as compression–expansion. (In the case of an extraneous sound that does not delay the normal speech but merely conceals a bit of it, deletion–insertion permits the concealed bit to be deleted and the extraneous sound to be inserted, which is a more realistic and perhaps more desirable explanation than that permitted by additive random error.) Although this appears not to have been done before, it can be done in a simple and computationally tractable manner. Of course, there are even more alternative versions in this case than there are when only compression–expansion is permitted, but we restrict our discussion here. While these simple versions may not in themselves be adequate to handle the kind of difficulties referred to, they do indicate one approach to dealing with them. A more sophisticated approach is briefly mentioned.

The time warping is defined by the functions $i_0(h)$ and $j_0(h)$ as before, and we use the first continuity constraint,

$$(\Delta i_0, \Delta j_0) = \begin{cases} (1, 0) & \text{or} \\ (1, 1) & \text{or} \\ (0, 1), \end{cases}$$

(where $\Delta i_0(h) = i_0(h) - i_0(h-1)$), but the case $(1, 0)$ can refer either to compression as before or, instead, to deletion of $a_{i_0(h)}$; likewise, the case $(0, 1)$ can refer either to expansion or to insertion of $b_{j_0(h)}$. We introduce deletion and insertion weights $w_{del}[a_i]$ and $w_{ins}[b_j]$ in addition to the substitution weights $w[a_i, b_j]$. We think of these as weights *per unit time*, so that the cost for a deletion of a_i over an interval of τ time units is $\tau w_{del}[a_i]$.

The simplest recurrence that can be used is

$$D_{ij} = \min \begin{cases} D_{i-1, j} & + & \tfrac{1}{2}\tau w[a_i, b_j], \\ D_{i-1, j} & + & \tau w_{del}[a_i], \\ D_{i-1, j-1} & + & \tau w[a_i, b_j], \\ D_{i, j-1} & + & \tau w_{ins}[b_j], \\ D_{i, j-1} & + & \tfrac{1}{2}\tau w[a_i, b_j]. \end{cases}$$

However, the justification for time weights of $\tfrac{1}{2}$ for compression and expansion steps does not seem valid if such a step immediately follows a deletion or an insertion step. For that matter, it is probably more realistic to forbid a compression or expansion step immediately following a deletion or insertion step, and doing so improves the time-reversal symmetry of the linking concept. If we choose to do this, two coupled recurrences may be used. Let D_{ij}^{di} ("di" for deletion–insertion) be the length of the shortest possible time-warping between $a_1 \ldots a_i$ and $b_1 \ldots b_j$ that ends with a deletion or an insertion, and let d_{ij}^{o} ("o" for other) be the length of the best possible time-warping that ends in some other step. Then

$$D_{ij}^{di} = \min \begin{cases} \min (D_{i-1, j}^{di}, D_{i-1, j}^{o}) + & \tau w_{del}[a_i], \\ \min (D_{i, j-1}^{di}, D_{i, j-1}^{o}) + & \tau w_{ins}[b_j], \end{cases}$$

$$D_{ij}^{\circ} = \min \begin{cases} D_{i-1,\,j}^{\circ} & + \tfrac{1}{2}\tau w[a_i,\ b_j], \\ D_{i,\,j-1}^{\circ} & + \tfrac{1}{2}\tau w[a_i,\ b_j], \\ \min(D_{i-1,\,j-1}^{\mathrm{d\,i}}, D_{i-1,\,j-1}^{\circ}) & + \tau w[a_i,\ b_j], \end{cases}$$

are the desired recurrences, and $\min(D_{mn}^{\mathrm{di}}, D_{mn}^{\circ})$ gives the distance between **a** and **b**.

It is possible to extend these versions in a much more general approach that offers the possibility of considerable computational saving, though we mention this only briefly. By making the template utterance a network and generalizing the comparison problem as described in the section on Directed Networks in Chapter 10, we can treat normal speech-sound deletion–insertion (e.g., "probably" to "prob'ly") as an alternative path in a network, rather than as a long series of deletions or insertions of individual sequence elements. One computational advantage of this comes from the fact that deletion-insertion need not be considered as a possibility at every element in the sequence (which is, in any case, unrealistic for speech sounds), but only at a few specified places. It is still feasible and perhaps helpful to permit arbitrary deletion–insertion to handle extraneous sounds when using this network approach, because the weights used for the two different types of deletion–insertion may be quite different. We omit a more concrete description, since it would take us too far afield.

8. INTERPOLATION BETWEEN THE SAMPLING POINTS

The sampling procedure that converts trajectories $\mathbf{a}(u)$ and $\mathbf{b}(v)$ into sequences $\mathbf{a} = a_1 \ldots a_m$ and $\mathbf{b} = b_2 \ldots b_n$ can sometimes cause a problem when the curves swept out by the trajectories are very close together. (This problem has been encountered in x-ray pellet-tracking measurements of tongue and jaw movements during speech.) Suppose, as in Fig. 9, that the distance from one curve to the other in some region is small compared to the typical distances between successive sampling points (such as a_i to a_{i+1}, and b_j to b_{j+1}) in the same region. If the sampling in the two trajectories happens by chance to be nearly in phase, as in Fig. 9(a), then the optimum time-warping between the sequences may give a satisfactory description of the relationship between the two underlying trajectories. If, however, the sampling happens to be substantially out of phase, as in Fig. 9(b) or Fig. 9(c), then the values like $w[a_i,\ b_j]$ that enter into the length of the time-warping are much larger than the distance between the curves. In this case, the discrete time-warping gives an unduly pessimistic result. In addition, the correspondence between the trajectories is substantially less accurate than is possible by more sophisticated analysis of the same data.

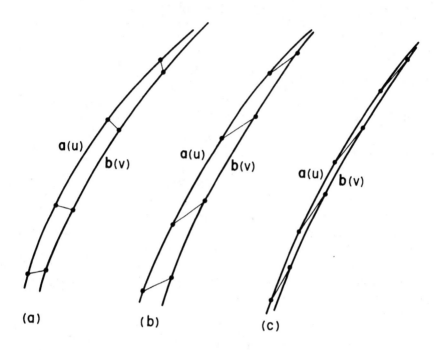

Figure 9. Close trajectories with sampling in or out of phase.

To overcome these problems, we have devised a new type of discrete time-warping, which we call *interpolation time-warping*. (A very different method for a somewhat similar purpose may be found in Burr (1979).) It makes use of two polygonal paths in feature space: the a-path that connects the a_i in order, and the **b**-path that connects the b_j in order. Each a_i is matched not to some b_j but instead to some point on the **b**-path, and each b_j is matched to some point on the a-path.

To describe an interpolation time-warping, we can use a diagram like Fig. 10(a), whose meaning is indicated by Fig. 10(b). Each $r(h)$ is a number between 0 and 1, and the notation (a_i, r) indicates an *interpolation point* on the segment from a_i to a_{i+1}. Specifically, (a_i, r) is the point that is r of the way from a_i to a_{i+1}, that is,

$$(a_i, r) \equiv_{\text{def}} (1 - r)a_i + ra_{i+1}.$$

Diagram 10(a) shows that a_1 is matched to the interpolation point $(b_1, r(1))$ between b_1 and b_2, and that b_2 is matched to an interpolation point between a_1 and a_2, etc. If the sampling rate is the same in the two trajectories, sampling points and interpolation points will tend to alternate along each row of the

$$\begin{bmatrix} a_1 & (a_1, r(2)) & a_2 & (a_2, r(4)) & (a_2, r(5)) & a_3 & \cdots \\ (b_1, r(1)) & b_2 & (b_2, r(3)) & b_3 & b_4 & (b_4, r(6)) & \cdots \end{bmatrix}$$

(a)

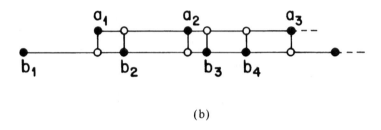

(b)

Figure 10. Interpolation time-warping.

diagram, but alternation need not occur, as illustrated by adjacent sampling points b_3 and b_4, which are both matched to interpolation points in the same segment.

More formally, an *interpolation time-warping between* $\mathbf{a} = a_1 \ldots a_m$ and $\mathbf{b} = b_1 \ldots b_n$ consists of functions $\mathbf{I}_0(h)$, $\mathbf{J}_0(h)$, $\mathbf{i}_0(h)$, $\mathbf{j}_0(h)$, and $\mathbf{r}(h)$. Each of them is defined for $h = 1$ to H, where $H = m + n - 2$. For each h, either

i) $\mathbf{I}_0(h)$ is the sampling point $a_{\mathbf{i}_0(h)}$ and $\mathbf{J}_0(h)$ is the interpolation point $(b_{\mathbf{j}_0(h)}, \mathbf{r}(h))$, or

ii) $\mathbf{I}_0(h)$ is the interpolation point $(a_{\mathbf{i}_0(h)}, \mathbf{r}(h))$ and $\mathbf{J}_0(h)$ is the sampling point $b_{\mathbf{j}_0(h)}$.

The functions $\mathbf{I}_0$ and $\mathbf{i}_0$ are required to satisfy the following continuity constraint, and $\mathbf{J}_0$ and $\mathbf{j}_0$ are required to satisfy a similar one.

a) If $\mathbf{I}_0(h)$ and $\mathbf{I}_0(h + 1)$ are both sampling points, then they are adjacent in $\mathbf{a}$, that is, $\mathbf{i}_0(h) + 1 = \mathbf{i}_0(h + 1)$.

b) If $\mathbf{I}_0(h)$ is a sampling point and $\mathbf{I}_0(h + 1)$ is an interpolation point, then the sampling point is the beginning point of the segment containing the interpolation point, that is, $\mathbf{i}_0(h) = \mathbf{i}_0(h + 1)$.

c) If $\mathbf{I}_0(h)$ is an interpolation point and $\mathbf{I}_0(h + 1)$ is a sampling point, then the line segment containing the interpolation point ends at the sampling point, that is, $i_0(h) + 1 = i_0(h + 1)$.

d) If $I_0(h)$ and $I_0(h + 1)$ are both interpolation points, they belong to the same line segment, that is, $i_0(h) = i_0(h + 1)$.

To insure that the time-warping covers the entire sequences, we can use the following constraint:

$$i_0(1) = j_0(1) = 1$$

and

$$\begin{cases} i_0(H) = m & \text{and } j_0(H) = n - 1, & \text{if } I_0(H) \text{ is a sampling point,} \\ i_0(H) = m - 1, & \text{and } j_0(H) = n, & \text{if } J_0(H) \text{ is a sampling point.} \end{cases}$$

If we assume that the sampling interval is the constant 2τ for both trajectories, then it is very simple to define length appropriately for an interpolation time-warping:

$$d(\mathbf{I}_0, \mathbf{J}_0, i_0, j_0, \mathbf{r}) \equiv \tau \sum_{h=1}^{H} w[\mathbf{I}_0(h), \mathbf{J}_0(h)].$$

The reason this is appropriate is that every h can be considered to correspond to time 2τ in the trajectory where a sampling point is used, and to time 0 in the trajectory where an interpolation point is used, so h corresponds to the average, τ. Although generalizing the standard recursion to handle this case involves some novel features, the resulting recurrence equation is quite simple and easy to calculate. Let A_{ij} be the set of incomplete interpolation time-warpings that start from the beginning of $\mathbf{a}$ and $\mathbf{b}$, that is, $i_0(1) = j_0(1) = 1$, and that end at (i, j), that is, $i_0(h_{\text{last}}) = i$ and $j_0(h_{\text{last}}) = j$. (Here $h_{\text{last}} = i + j - 1$.) In other words, the time-warpings in A_{ij} are incomplete because they end at (i, j), which means that a_i is matched to an interpolation point between b_j and b_{j+1}, or b_j is matched to an interpolation point between a_i and a_{i+1}. Let D_{ij} be the minimum length of any time-warping in A_{ij}. Then the recurrence for D_{ij} is

$$D_{ij} = \min \begin{cases} D_{i-1, j} + \min_{0 \le r \le 1} \tau w[a_i, (b_j, r)], \\ \\ D_{i, j-1} + \min_{0 \le r \le 1} \tau w[(a_i, r), b_j], \end{cases}$$

and the minimum length of any complete interpolation time-warping is given by

$$d(\mathbf{a}, \mathbf{b}) = D_{m-1, n-1} + \min \begin{cases} \min_{0 \le r \le 1} \tau w[a_m, (b_{n-1}, r)], \\ \\ \min_{0 \le r \le 1} \tau w[(a_{m-1}, r), b_n]. \end{cases}$$

(Note the differences in subscript pattern between this formula and the preceding one.) Each minimization over r corresponds to finding the point on a given line segment that is nearest to a given point, and this is easy to calculate. For example, the first minimization over r in the formula for D_{ij} corresponds to finding the point on the segment from b_j to b_{j+1} that is nearest to a_i. To find it, project a_i perpendicularly onto the complete line between b_j and b_{j+1}. If the projection is between b_j and b_{j+1}, it is the desired point. If the projection is beyond b_j, then b_j is the desired point, and if the projection is beyond b_{j+1}, then b_{j+1} is the desired point. In algebraic terms, this can be written as follows,

$$\bar{r} = \frac{(a_i - b_j) \cdot (b_{j+1} - b_j)}{(b_{j+1} - b_j) \cdot (b_{j+1} - b_j)} ,$$

$$r = \begin{cases} 1, & \text{if } \bar{r} > 1, \\ 0, & \text{if } \bar{r} < 0, \\ \bar{r}, & \text{otherwise,} \end{cases}$$

where the dot indicates the scalar product of two vectors.

We can generalize interpolation time-warpings to permit insertion and deletion in a way that is appropriate for use in speech processing (e.g., to get a good comparison between a precise pronunciation of "twenty" and the slurred pronunciation "twenny," by permitting deletion of "t"). The recurrence equation is straightforward, but the algorithm requires a three-dimensional array and time proportional to n^3 instead of n^2.

9. AVERAGE OF TWO TRAJECTORIES

It is sometimes useful to take the "average" of several trajectories, as illustrated in Rabiner and Wilpon (1979, 1980). The prime application occurs in speech processing, where several utterances of a single word are combined into a single average utterance to provide a "template" for use in word recognition. The Rabiner–Wilpon method has the advantage of being relatively simple, and of permitting a simple extension to the average of many trajectories (with one of them playing a special master role). It treats the trajectories in an asymmetric manner, however, and in this paper we are interested in developing a fully symmetric method, for reasons described earlier. We will give a natural symmetric definition for the weighted average of two trajectories with respect to a given time-warping between them. Of course, the optimum time-warping would normally be used. In principle, our definition could be extended to averaging N trajectories, but such an extension would rest on a simultaneous time-warping of N trajectories. We do not follow this approach, in part because of the great computation time such methods require. Instead, to combine N trajectories into a single average, $N-1$ repetitions of the two-trajectory average may be used, each with respect to the optimum time-warping between the two trajectories involved. For example, various pairs may be combined, then some of the resulting trajectories may be combined, either with each other or with original trajectories that have not yet been combined, and so on. When combining two trajectories that represent k_1 and k_2 original trajectories, respectively, presumably we would use weights $k_1/(k_1 + k_2)$ and $k_2/(k_1 + k_2)$. While the final average would not, unfortunately, be independent of the order of combination, it would probably not be very sensitive to the order, in realistic applications. Incidentally, the combining process described bears a strong relationship to widely used methods of clustering known as "pair-group" methods, and the rules used in clustering to determine the order of combination are probably quite suitable for use here also.

Now assume we are given the trajectories $a(u)$ and $b(v)$, and weights p and q with $p + q = 1, p \geq 0, q \geq 0$. Also assume that we are given a time-warping (u_0, v_0) between the trajectories. Presumably, this would usually be the optimum time-warping, though the following discussion does not rely on that assumption. Suppose u and v are linked, so that $a(u)$ and $b(v)$ are corresponding points in the two trajectories. The weighted average of these points is

$$p\mathbf{a}(u) + q\mathbf{b}(v).$$

Obviously the weighted-average trajectory should run along the curve formed by all such points.

What has not been so clearly set forth in the literature is the time pattern that should be used with this curve. We propose that the time assigned to the point shown above should be the weighted average of the two times involved, u and v, that is,

$$w = pu + qv$$

is the time that should be assigned to the point shown above. (It may appear that the time pattern chosen for the average is not important, because the distances we use are deliberately chosen to be insensitive to time pattern. However, it is in fact important, for two reasons. First, a time pattern is needed to use the procedures we discuss. Second, other ways of using the average trajectory, not discussed in this chapter, are sensitive to time pattern.)

This can all be wrapped up into one succinct definition, as follows. The *weighted average* $\mathbf{c}(w)$ of trajectories $\mathbf{a}(u)$ and $\mathbf{b}(v)$, with respect to the time-warping (u_0, v_0), using nonnegative weights p and q that sum to 1, is defined by

$$\mathbf{c}(\mathbf{w}_0(t)) = p\mathbf{a}(\mathbf{u}_0(t)) + q\mathbf{b}(\mathbf{v}_0(t))$$

where

$$\mathbf{w}_0(t) = p\mathbf{u}_0(t) + q\mathbf{v}_0(t).$$

If equivalent time-warpings $(\mathbf{u}_0, \mathbf{v}_0)$ and $(\mathbf{u}_1, \mathbf{v}_1)$ between $\mathbf{a}(u)$ and $\mathbf{b}(v)$ are each used to form the average, then it is easy to show that the two resulting average trajectories are the same.

10. AVERAGE OF TWO SEQUENCES

In the previous section, a reason for averaging trajectories was explained, but in practice, of course, it is *sequences* that are averaged. In this section, we define the average of two sequences with respect to a time-warping. As above, there are many ways to make the analogy with the continuous definition, and we give two alternative definitions. One difference between our definitions and the earlier definitions due to Rabiner and Wilpon (1979, 1980) is that we treat the two sequences in a fully symmetric manner.

Suppose that we are given two sequences $\mathbf{a} = a_1 \ldots a_m$ and $\mathbf{b} = b_1 \ldots b_n$ with sampling times u_i and v_j, and weights p and q with $p + q = 1, p \geq 0, q \geq 0$. Also assume that we are given a time-warping $(\mathbf{i}_0, \mathbf{j}_0)$ between the sequences, where $\mathbf{i}_0(h)$ and $\mathbf{j}_0(h)$ are defined for $h = 1$ to H. Presumably, the optimum time-warping would normally be used, but the following discussion is valid for any time-warping. Suppose i and j are linked by h (that is, $i = \mathbf{i}_0(h), j = \mathbf{j}_0(h)$), so that a_i and b_j are corresponding points in the two sequences. The weighted average of these points is

$$c_h = pa_i + qb_j,$$

which corresponds to h. Let the corresponding sampling time be defined by

$$w_h = pu_i + qv_j.$$

Our first definition of the average sequence is simply $\mathbf{c} = c_1 \ldots c_H$.

If $\mathbf{a}$ and $\mathbf{b}$ were both formed by sampling at constant time intervals τ, then we might want the average to have the same property. Our second definition achieves this, though there are many alternative versions of it, and we shall not spell out the details for any one of them. First we put a polygonal path (or more generally, a spline) through the points c_h. We label the points with w_h, and interpolate along the path to find points at the appropriate sampling times. The points yielded by this process constitute our second definition for the average.

REFERENCES

Bridle, J. S., and Brown, M. D., Connected word recognition using whole-word templates, *Proceedings of the Institute for Acoustics*, pages 25–28 (1979).

Burr, D. J., A technique for comparing curves, pages 271–277, in *Proceedings of the IEEE Conference on Pattern Recognition and Image Processing, 1979, Chicago*, IEEE: New York (1979).

Burr, D. J., Designing a handwriting reader, pages 715–722, in *International Conference on Pattern Recognition, 5th, 1980, Miami Beach, Florida*, IEEE: New York (1980).

Burr, D. J., Elastic matching of line drawings, *IEEE Transactions on Pattern Analysis and Machine Intelligence* PAMI-3:708–713 (1981).

Dixon, N. R., and Martin, T. B. (eds.), *Automatic speech and speaker recognition*, IEEE Press: New York City (1979).

Fujimura, O., Temporal organization of articulatory movements as a multidimensional phrasal structure, *Phonetica* **38**:66–83 (1981).

Fujimoto, Y., Kadota, S., Hayashi, S., Yamamoto, M., Yajima, S., and Yasuda, M., Recognition of handprinted characters by nonlinear elastic matching, pages 113–119, *International Joint Conference on Pattern Recognition, 3rd 1976, Coronado, California*, IEEE: New York (1976).

Itakura, F., Minimum prediction residual principle applied to speech recognition, *IEEE Transactions on Acoustics, Speech, and Signal Processing*, ASSP-23:67–72 (1975).

Levinson, S. E., Structural pattern recognition applied to automatic speech recognition, SCAMP Working Paper No. 13/81, in Proceedings of a Symposium on Acoustics Phonetics and Speech Modeling, June 1981, Volume 2, published by Communications Division, Institute for Defense Analyses, Princeton, N.J. (A.S. House, editor) (1981).

Myers, C., A comparative study of several dynamic time-warping algorithms for speech recognition, Master of Science thesis, Electrical Engineering and Computer Science Department, Massachusetts Institute of Technology, February 1980.

Myers, C. S., and Rabiner, L. R., A dynamic time-warping algorithm for connected word recognition, *IEEE Transactions for Acoustics, Speech, and Signal Processing* ASSP-29:284–297 (1981a).

Myers, C. S., and Rabiner, L. R., Connected-digit recognition using a level-building DTW algorithm, *IEEE Transactions on Acoustics, Speech, and Signal Processing* ASSP-29:351–363 (1981b).

Myers, C. S., Rabiner, L. R., and Rosenberg, A. E., Performance tradeoffs in dynamic time-warping algorithms for isolated-word recognition, *IEEE Transactions on Acoustics, Speech, and Signal Processing* ASSP-28:622–635 (1980).

Rabiner, L. R., Rosenberg, A. E., and Levinson, S. E., Considerations in dynamic time-warping for discrete-word recognition, *IEEE Transactions on Acoustics, Speech, and Signal Processing* ASSP-26:575–582 (1978).

Rabiner, L. R., and Schmidt, C. E., Application of dynamic time-warping to connected-digit recognition, *IEEE Transactions on Acoustics, Speech, and Signal Processing*, ASSP-28:337–388 (1980).

Rabiner, L. R., and Wilpon, J. G., Considerations in applying clustering techniques to speaker-independent word recognition, *Journal of the Acoustical Society of America*, **66**(3):663–673 (1979).

Rabiner, L. R., and Wilpon, J. G., A simplified, robust training procedure for speaker-trained, isolated-word recognition systems. *Journal of the Acoustical Society of America,* **68**(5): 1271–1276 (1980).

Reiner, E., and Bayer, F. L., Botulism: A pyrolysis-gas-liquid chromatographic study. *Journal of Chromatographic Science* **16**(12):623–629 (1978).

Reiner, E., and Kubica, G. P., Predictive value of pyrolysis-gas-liquid chromatography in the differentiation of mycobacteria. *American Review of Respiratory Disease* **99**:42–49 (1969).

Reiner, E., Abbey, L. E., Moran, T. F., Papamichalis, P., and Schafer, R. W., Characterization of normal human cells by pyrolysis-gas-chromatography mass spectrometry. *Biomedical Mass Spectrometry* **6**(11):491–498 (1979).

Sakoe, H., and Chiba, S., Dynamic-programming algorithm optimization for spoken word recognition, *IEEE Transactions for Acoustics, Speech, and Signal Processing,* ASSP-26:43–49 (1978).

Sakoe, H., and Chiba, S., A dynamic-programming approach to continuous speech recognition, *1971 Proceedings of the International Congress of Acoustics, Budapest, Hungary,* Paper 20 C13 (1971).

Velichko, V. M., and Zagoruyko, N. G., Automatic recognition of 200 words. *International Journal of Man–Machine Studies,* **2**:223–234 (1970).

White, G. M., and Neely, R. B., Speech-recognition experiments with linear prediction, bandpass filtering, and dynamic programming, *IEEE Transactions on Acoustics, Speech, and Signal Processing,* ASSP-24:183–188 (1976).

White, G. M., Dynamic programming, the Viterbi algorithm, and low-cost speech recognition, pages 413–417 in *Proceedings of the 1978 IEEE International Conference on Acoustics, Speech, and Signal Processing* (1978).

Yasuhara, M., and Oka, M., Signature-verification experiment based on nonlinear time alignment: feasibility study. *IEEE Transactions on Systems, Man, and Cybernetics,* SMC-7:212–216 (1977).

CHAPTER FIVE

USE OF DYNAMIC PROGRAMMING IN A SYLLABLE-BASED CONTINUOUS SPEECH RECOGNITION SYSTEM

Melvyn J. Hunt, Matthew Lennig, and Paul Mermelstein

1. INTRODUCTION

Speech is often said to be the fastest and most natural way for a person to communicate with a machine. Devices that can recognize isolated words have been available for some time now and are finding wide application (Bridle and Brown, 1974). However, the input required is often not a single word but a phrase or sentence. In that case, much of the speed and naturalness of voice input is lost if the words have to be spoken in isolation with periods of silence between them.

This chapter is concerned with the recognition of natural, continuously spoken sentences. The task is much more difficult than the isolated-word problem, and the few systems currently available are comparatively simple and not yet in widespread use. The problem becomes tractable only when the syntax and vocabulary of the phrases that the machine is expected to recognize are restricted. This may not be a serious handicap in practice: Most applications being considered are concerned with specific tasks and do not require free-ranging discussion. It is desirable, however, that the phrases that can be recognized should comprise a clean and natural subset of the user's language in order that he or she can easily understand which phrases are acceptable to the machine.

We are taking a syllable-based approach, where dynamic-programming methods are applied both to the recognition of syllables and to the evaluation of competing sentence hypotheses viewed as sequences of syllables. In this chapter we first point out some of the ways in which speech differs from many other sequence-matching problems. We look briefly at how some workers have addressed the problem and then try to explain why we feel that a syllable-based approach may be worth pursuing. In discussing our own system, we first describe the organization of the syntax and sentence-evaluation routines and then provide details of the time-warping algorithm used for syllable recognition. Results of experiments comparing variants of the syllable-recognition algorithm are presented.

2. SPEECH RECOGNITION AS A STRING-MATCHING PROBLEM

From an acoustical point of view, speech, like other sounds, consists of a sequence of small, rapidly varying perturbations in air pressure. These perturbations can be recorded as a waveform on an analogue medium such as a gramophone record.

The most direct way of making a digital recording is to sample the waveform at regular intervals and assign the pressure variation at each instant to the closest of a number of discrete levels. In order for good intelligibility and telephone-like quality to be preserved, the sampling rate needs to be at least 8000 times a second, and at least 128 levels are needed to represent the pressure changes. This is an enormously high information rate: The amount of speech that can be recorded in this way is severely limited and any digital processing applied to the waveform is computationally expensive. Moreover, because of the insensitivity of the ear to moderate changes in phase, two waveforms can look quite different and yet sound identical. The digitized waveform is therefore not a suitable representation of speech for recognition purposes.

Various methods have been developed that exploit the redundancy in the speech waveform, to reduce by at least an order of magnitude the amount of storage needed to reproduce speech of a given quality. Instead of a single-dimensional function sampled very frequently, the speech is represented by a vector of up to twenty components updated around a hundred times a second. For most of the time, the values of the vector components change slowly between updates, and they can be pictured as tracing out piecewise continuous curves in time. It is primarily between sequences of these vectors that string matching takes place.

String-matching for continuous speech recognition differs in several ways from many of the other string-matching problems discussed in this volume. First, speech is best described as a multilevel sequence: A sentence can be considered to be a sequence of words, and the words a sequence of speech sounds (loosely, "phonemes"). One might prefer to use other elements, perhaps syllables rather than words, or basic features of speech sounds such as the presence of voicing or frication rather than phonemes; but whatever elements are chosen, at least two levels are necessary if a reasonably compact description of allowed speech sequences is to be obtained. The recognition process itself may, as in our case, carry out string matching at both levels.

The underlying elements of a speech sequence do not have completely consistent acoustic realizations: No two utterances of the same word in the same context by the same speaker will ever be absolutely identical. Consequently, it is not possible to say that an element is present or that another element has been substituted for it; one tries instead to have a measure of the *probability* of a particular element being present, usually in terms of its closeness to some stored model form of the element. The model forms themselves have to be constructed in some way from previous utterances. There are no ideal forms in the sense that, for instance, the ideal form of some manmade signal might be known.

Finally, the elements making up the speech sequence have very variable durations, and in general they are difficult to isolate. There are no clear acoustic cues to word boundaries in speech corresponding to spaces between words in written language. At the phoneme level the situation is worse; although speech is widely transcribed by a linear sequence of phonemes, one could argue that trying to divide a spoken word into discrete phonemes by analogy with letters in a written word is nonsensical. The acoustic features that characterize a phoneme overlap in time and interact with those of adjacent phonemes. There is, for example, no way of cutting a piece of magnetic tape containing a recording of the word "do" so that a recognizable "d" sound can be heard in isolation from the following vowel; and the major acoustic difference between "once" and "ones" lies in the length of the "n" sound rather than in the form of the fricative sound that follows it. Even speech sounds that can be cleanly separated from their neighbours are usually influenced to some extent by the sounds around them.

3. APPROACHES TO THE PROBLEM

In tackling the problem of continuous speech recognition, a wide variety of approaches have been adopted, many of them involving dynamic programming. This section does not purport to be a survey of the field of continuous speech recognition (see Reddy (1976) for example), but is simply intended to set our own work in some perspective.

One of the major divisions in approaches to speech recognition is between systems that try to segment the speech into phoneme-like units and those that do not. The nonsegmenting systems usually describe the speech by a series of frames that are "isochronous," i.e., equally spaced in time (typically 10 milliseconds apart). There are generally several of these frames per phoneme and the changes between frames are in general fairly gradual. Thus, the nonsegmenting systems are less economical in their descriptions of words, and they are less well suited to pronunciation variations that involve the substitution, deletion, or insertion of whole phonemes. Systems that do attempt segmentation into speech sounds group varying numbers of isochronous frames together to form units of unequal duration intended to represent phonemes or phoneme subunits (such as stop gaps). The segmenting systems are faced with the difficulties in isolating phonemes that we mentioned in the previous section.

In the early 1970s the U.S. Department of Defense Advanced Research Projects Agency (ARPA) initiated a five-year program for the development of speech understanding systems. All of the major ARPA systems, namely HWIM, Hearsay II, Harpy, and the SDC system (see the review by Klatt, 1977) attempt segmentation into phonemes, though HWIM carries out a whole-word template-matching process on the isochronous frames in order to verify words hypothesized from the phoneme string. Complex logical rules have been proposed to deal with context problems in phoneme recognition (De Mori, 1979), and inevitable segment-identification ambiguities are often handled by holding an ordered list of likely phoneme candidates for subsequent higher-level analysis. Harpy and HWIM are examples of two different approaches to segmentation, in that Harpy first segments and then tries to recognize the phonemes (with the result that one phoneme often corresponds to several segments), while HWIM makes its segment-boundary decisions dependent on its phoneme hypotheses, holding several hypotheses with different boundaries in a "phoneme lattice." With its freedom to choose the optimum boundaries for each phoneme-sequence hypothesis, the HWIM approach presumably leads to better phoneme identification, but the Harpy method is more suited to the string-matching algorithms described in this volume.

Harpy demonstrates a further advantage of systems that classify sounds into phonemes: speaker adaptation can be achieved by collecting enough material from the new speaker to modify the reference phonemes. A system that does not classify into phonemes must either collect much more material and adapt at the higher (word or syllable) level, or else it must attempt some general acoustic transformation.

Some workers have carried out segmentation into purely acoustic units, such as regions of constant spectrum, without classification of the segments into phonemes. The primary advantage of this over nonsegmenting methods is in computational efficiency. Bridle and Sedgwick (1978) have described a dynamic-programming algorithm for the optimum division of a speech sequence into a given number of acoustic segments.

Equally, some systems that do not segment speech into phonemes nevertheless attempt to assign each individual frame to a phoneme-like class. This formulation of the speech-recognition task is the one that casts it in a form most similar to other sequence-matching problems such as macromolecule comparison. Substitution errors are inevitably very common. Kashyap (1979) recently described a system of this kind in which substitution costs were related to the estimated confusability of the phoneme classes.

It is, however, much more common for nonsegmenting systems to describe each frame by a set of continuous parameters. Work at IBM (Bahl *et al.,* 1979) is particularly interesting here because they have evaluated examples of the complete range of possibilities. Their first system contained a module (MAP) which segmented the speech and assigned a single phoneme label to each segment. Their next two systems (CSAP–1 and CSAP–2) divided the speech into centisecond (that is, 10 millisecond) frames and assigned a phoneme-like label to each frame, CSAP–2 differing from CSAP–1 in having more labels. Their most recently described system, WBAP, is also nonsegmenting, but it describes each frame by continuous parameters. There has been a uniform improvement in performance throughout this sequence of systems. Their results suggest that it is preferable to preserve the continuous spectral information for higher-level string matching rather than make early hard decisions on phoneme identities and locations, with a consequent loss of information.

Most nonsegmenting methods store acoustic models, "templates," representing whole words. A template consists of a sequence of frames containing information on the short-term power spectrum of the word as a function of time. It is aligned in time to a portion of the unknown speech and the spectral information is compared. Since different utterances of the same word can vary widely in their total duration and in the relative durations of parts of the word, the alignment process generally allows parts of the template and unknown speech to be stretched or compressed with respect to each other. Dynamic-programming template-matching algorithms are described in Sect. 5.

The great advantage of the word-matching approach is that since most phonetic context effects occur within words, whole-word templates automatically contain most of the required context information. The great problem is the difficulty of segmenting into words: Since there are no general acoustic cues to word boundaries, it is difficult to know which portion of speech should be matched against a template. Usually, the templates are matched consecutively across the speech, so that when a template has fitted part of the speech well, it is assumed that the next word starts at the point in the speech where the match to the previous word ended (Bridle and Brown, 1979). Thus the decision as to where to put the word boundary depends only on the match to the end of the previous word and not on the match to the beginning of the next word, although in some formulations (Levinson and Rosenberg, 1979), "dithering" of the word boundary is allowed in matching the next word.

4. THE SYLLABLE-BASED SYSTEM

In the following sections we describe our own syllable-based approach to the problem of continuous speech recognition.

4.1 Motivation for Using Syllables

The energy profile of a speech signal shows a modulation due to the syllable structure. This may make syllables the only elements of speech that can be consistently isolated independently of recognition, and our system in fact determines the syllable boundaries before any recognition is attempted.

The syllables to be recognized are matched against stored syllable templates. Syllable templates share the advantages of words in being substantially free of phonetic context effects, but in addition to being more compact for storage they are free of many of the disadvantages of word templates. It is possible to know, before recognition begins, how many templates need to be matched to the speech and between which points the matching must take place. Most importantly, alternative sentence hypotheses have the same syllable boundaries; thus they can be easily compared using dynamic-programming string-matching methods.

We do not pretend that our kind of syllable-based approach is the ultimate answer to the problem of continuous speech recognition. Future large-scale systems will almost certainly incorporate phonemic knowledge. However, the advantages outlined above make a syllable-based system attractive for the many practical applications where only a limited field of discourse is required.

The syllable method depends critically, of course, on there being an acceptable division into syllables, and this may set the ultimate limit to the performance of our method. Happily, sentences with a syllabification error (at present, roughly 10% of all sentences) almost always result in the recognition phase rejecting the sentence rather than misunderstanding it. For most applications a rejection and request for the sentence to be repeated will be less serious than a misrecognition.

The details of how a sentence is split into syllables (Mermelstein, 1975) are not pertinent to this volume, and we will describe only the most general features of the method here. It works primarily on the energy in the speech signal. Sufficiently deep local minima in the energy are candidates for syllable boundaries. The potential syllables produced by these boundaries then have to satisfy criteria concerning their length and loudness and degree of voicing. Note that the "syllables" produced in this way to not have to correspond to the conventional idea of what the syllables should be: there is no objection to "twenty" being taken to be a single syllable provided that this happens sufficiently often for the system to be aware of the possibility.

Moreover, absolute consistency is not required. For example, the system can store both single-syllable and two-syllable reference versions of "twenty" and so recognize either form. Difficulties can, however, arise when a "syllable" spans a syntactic boundary, such as when "twenty-eight" is split into "twen" and "ty-eight," and when syllable boundaries are placed in completely unexpected places.

4.2 System Overview

The sequence of operations involved in recognizing a sentence is shown in Fig. 1. The sentence is first low-pass filtered at 3.7 kHz, sampled at 8 kHz, and digitized.

The preliminary processing seeks to represent the speech in a compact manner by eliminating redundancy and features that provide information about the speaker or his mood rather than about the words he is using. We believe that the representation chosen should also reflect some of the properties of the human ear. As we mentioned in Sec. 2, the ear is relatively insensitive to phase, so it is appropriate to compute the short-term power spectrum and discard the phase information. The perceived loudness of a sound is roughly proportional to the logarithm of the intensity, so we work with the log power spectrum. Because the capacity of the ear to resolve two close frequencies decreases at frequencies above 1 kHz, we make use of a standard scale of frequencies known as the "mel" scale, in which equal differences in mel values correspond to equal perceived pitch differences. The varying resolving power of the ear is reflected in our system by dividing the spectrum into twenty channels of equal width on the mel scale.

Because speech spectra are fairly smooth, the energies in adjacent channels are highly correlated. Such correlation can be substantially removed by expanding the channel energies in a Fourier series. Since the spectrum is considered to be symmetric about 0 Hz, the only nonzero terms in the expansion are the cosine terms. The first few of these coefficients, which are known as "mel-scale cepstrum coefficients," form a very compact description of the spectrum. We compute the coefficients every 6.4 milliseconds. At the same time we estimate the perceptual loudness, which we use in the syllabifier. The syllabifier separates speech from silence and divides the speech regions into syllables. The rest of the system is concerned with trying to recognize the sequence of syllables.

In speech-recognition jargon, our system is "top-down," "left-to-right." That is to say, the syllables are considered in the order in which they occur in the sentence. Each syllable to be recognized is compared by the syllable comparator only with those templates that would be consistent with the syntax and the system's knowledge of the sentence so far. The output from the comparator is a measure of similarity between the template and the unknown and is referred to as a (squared) distance, although it would not always satisfy the requirements of a true distance measure.

To compare a syllable with a template, string-matching is used. During this process, as we shall see in Sec. 5, each frame making up the unknown syllable may be compared to a frame in a reference template, or it may be skipped over, or it may be involved in a comparison with two or more consecutive reference frames. The comparison rests entirely on a similarity measure, and does not involve any classification.

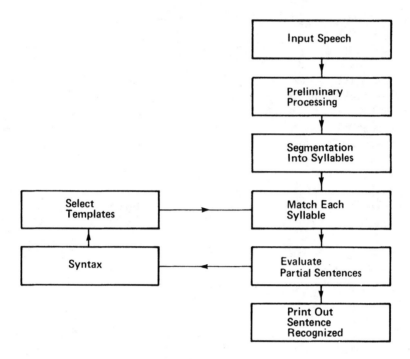

Figure 1. Processes involved in recognizing a sentence.

Frequently, the best fit to a syllable does not correspond to the correct interpretation, and it is only one or more syllables later that this becomes apparent. Some method of exploiting hindsight is therefore needed in interpreting a sentence. We do this by what is usually known as a "beam search" (Lowerre, 1976). At each stage, the most promising sentence hypothesis is held, together with a number of other possible sentence hypotheses. A fixed number of alternatives could be held, but it seems better to let the number held depend on the situation. The beam-search method keeps all hypotheses that are no more than a given threshold value worse than the current best one. The threshold value, which is a parameter in the system, therefore controls the number of hypotheses being held, i.e., the "width" of the beam. One seeks to set the beam width to be as narrow as possible while being fairly certain that a hypothesis that will turn out to fit the whole sentence best does not stray outside of the beam.

The squared distances returned by the syllable comparator are taken to be a measure of the negative log probability that the template and unknown syllable have the same interpretation. By summing the squared distances between the unknown syllables and the string of templates involved in a particular partial-sentence hypothesis, we obtain a measure of how promising that hypothesis is. If this sum exceeds that for the current best hypothesis by more than the threshold value, then the hypothesis under consideration is dropped.

When the end of the sentence is reached, the sentence with the lowest total distance is selected. If that distance is below a threshold, the sentence is printed out; otherwise the sentence is rejected and has to be spoken again.

To construct a sentence out of a sequence of syllables, string-matching is also used, though the process is quite different from that used at the lower level in matching unknown syllables against the syllable templates. By constrast with that process, the construction of a particular sentence hypothesis requires each unknown syllable to be assigned, with a degree of confidence derived from the syllable-comparison process, to a certain syllable class. The assignment is on a strictly one-to-one basis with no possibility of deletion or insertion. The distances output by the syllable comparator can be considered to be the substitution cost incurred when an unknown syllable is replaced by a particular reference form.

4.3 Syntax

The sentences we are currently trying to recognize specify dates and times. Two examples are: "September twenty-seventh at eight thirty-two P.M." and "First May at noon." Figure 2 shows the general structure of acceptable sentences. Slightly more than a million such sentences are possible. The rules that are used to specify these sentences must satisfy two main requirements:

i) The description has to be compact (to permit use of a small computer);

ii) The syntax should be written in terms of possible transitions between states such that the possible transitions out of any state do not depend on how that state was reached (to permit use of dynamic programming in evaluating sentence hypotheses).

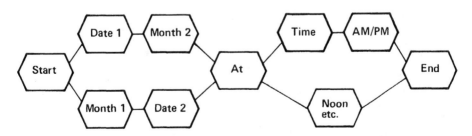

Figure 2. General structure of the date and time sentences used in this study.

Two further requirements are these:

iii) The syntax description should be as easy to read as possible;
iv) It should be easy to make changes to the general rules of the syntax.

These four requirements are not immediately compatible. To take the months as an example, it is clear that the list of months and the sequence of syllables making up each month will be the same whether the month occurs before or after the day number. It would therefore be very wasteful to specify all the month information both for the case of the month coming in first position and for its appearance in second position. On the other hand, the sequence of syllables allowed to *follow* the month depends on what has preceded it.

Conflicts of this kind have been resolved by having a four-level hierarchical syntax network. The sequence of syllables in a word forms the lowest level, and the units shown in Fig. 2 form the highest level. The levels in the hierarchy have been labeled syllable, word, word group, and phrase. In a sentence hypothesis each syllable is assigned a "state." A state is specified by four indices, which correspond to the four levels of the hierarchy and thus define the particular syllable, the word in which the syllable is embedded, and so on.

Given a syntactic state, the syntax network is used to determine all states that can grammatically follow that state. This involves horizontal and vertical searches. For example, if the state contained the first syllable of the word "seven," the network would point to the second syllable. If, however, the state contained the second syllable of the same word, the network would show that the word was complete and it would be necessary to move up one level in the hierarchy and test whether the network pointed to further words in the same phrase. The network might indicate that the phrase was definitely complete, in which case the network search would have to be continued at the phrase level, or that the phrase might or might not be complete, in which case the search would have to be continued at both the word and phrase levels.

The level designated as "word group" needs some explanation. It is used principally for the numbers in dates and times. Certain groups of numbers behave in similar ways; for example, when we specify minutes, the numbers one to nine have to be preceded by "oh," while higher numbers do not have this property. It is therefore convenient to treat certain groups of words within a phrase as a single unit.

The time taken to recognize a sentence depends on the number of possibilities considered. We therefore want to prune away impossible sentence hypotheses as quickly as possible. Information on the number of syllables in the sentence allows us to do this. For example, if a sentence hypothesis has reached the word "at" and only one syllable remains, that syllable can only be "noon," while if there are several syllables left it is pointless to consider "noon" as a possibility for the next syllable. In general, whenever the syntax proposes a state, there is a maximum and a minimum number of syllables that will be required to complete the sentence if that state is accepted. If the number of syllables remaining to be recognized lies outside the acceptable range, then the state must be rejected. The task of determining the range for any syllable in any position does not impose large storage or computational requirements. Each "unit" (e.g., phrase, syllable, etc.) has associated with it the maximum and minimum number of syllables that can follow it before it is necessary to rise to a higher level or, in the case of phrases, before the sentence is completed. Then, for any state, the maximum and minimum number of syllables to the sentence end is given by summing these four integers over the four units making up the state.

We have so far assumed that the acoustic form of a syllable is independent of its context. While we believe this to be substantially true, there is one sense within our system in which it is decidedly not true: The syllabifier frequently misplaces a boundary so that an initial or final consonant is assigned to the adjacent syllable. For example, the words "ninth October" may have a syllable boundary before, after, or during the "th" fricative, resulting in "nine thoct," "ninth oct," and "ninth thoct," respectively. It also occasionally happens that a weak fricative is not recognized as speech, giving the fourth possibility "nine oct." The syntax copes with this by allowing the "deep-structure" form "ninth" to be realized as "ninth" or "nine," and the deep-structure form "oct" to be realized as "oct" or "thoct" (or several other forms with different transferred consonants). All forms are proposed by the syntax, but before a match against "thoct" is tried, the system checks whether the deep-structure form of the previous syllable ended in "th."

Readers familiar with Harpy may have noticed several similarities between that system and the one being described here. In particular, both systems use a hierarchical syntax description and a left-to-right beam search. One might say that our system does at a syllable level what Harpy does at a phonemic level. There is, however, one major practical difference: In our system the number of fundamental recognition units—syllables—in a sentence is quite small, but the comparison process for identifying each syllable is computationally expensive and dominates the recognition time. By comparison, time taken to explore the four-level syntax network to find all the syllables that can grammatically follow the latest syllable of each hypothesis is quite negligible, so the network can be left in its compact, hierarchical form. In Harpy, on the other hand, the phonemic matching process is a quick table lookup, but there are many phonemic units in a sentence and each one requires a reference to the syntax network. The speed of the syntax-reference process is therefore very critical, and a multilevel search would be impossibly slow. Consequently, Harpy's syntax table has to be compiled out to a single level with sections that appeared once in the hierarchical network having to be specified many times over in the single-level network. This results in a very large structure that will not easily fit into the memory of a small computer.

5. THE SYLLABLE COMPARATOR

The syllable comparator determines the distance between an unknown syllable and a stored template. As described in Sec. 4.2, the comparator is invoked by the syntax as it advances through the unknown utterance, one syllable at a time, exploring different recognition hypotheses. To the comparator, the unknown syllable is represented as a sequence of acoustic-parameter vectors, each vector corresponding to one time frame. Syllable templates are represented in the same form as syllables.

This section describes how the distance between an unknown syllable and a syllable template can be defined in such a way as to be useful for speech recognition. Distance, in this sense, is a measure of acoustic dissimilarity between two syllables or between a syllable and a template.

5.1 The Need for Time-Warping

The fundamental source of difficulty in automatic speech recognition and, in fact, what makes speech recognition a more difficult task than speech synthesis, is variability. Not only does speech vary from speaker to speaker, but even within a single speaker's output appreciable variation occurs in pronouncing the "same" utterance. One aspect of speech variation is that of timing: Syllables may be pronounced faster or slower and the relative durations of their different portions may change. The syllable comparator is designed to take into account such timing variation in speech.

Given a situation in which we are required to find the distance between an unknown syllable and a template of different duration, one possible approach would be to linearly interpolate between time frames of the shorter syllable so that the result would have the same number of frames as the longer syllable. This is equivalent to a linear dilation of the time axis of the shorter syllable. Syllable-to-template distance could then be calculated by summing the vector distances between frames of the two syllables over their duration.

The problem with this linear-dilation approach is that it is not an accurate model of what goes on when speaking rate changes. In fact, certain phonetic segments are more elastic than others and hence change more with a change in speaking rate. For example, there is evidence that vowels are more elastic than consonants (Kozhevnikov and Chistovich, 1965). Although little quantitative data are available on the subject, it is reasonable to assume that many other factors besides segment type may affect the detailed timing variations that occur in speech.

In the absence of detailed quantitative data on how the time scales of syllables are dilated and compressed, we allow nonlinear distortion of the time scales of the unknown syllable and template, choosing the frame-to-frame correspondence that gives rise to the smallest distance between the two syllables. This technique, known as *time-warping*, has been employed in

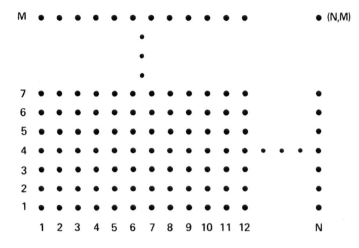

Figure 3. Lattice defined by the time frames of two syllables.

various forms (Velichko and Zagoruyko, 1970; Bridle and Brown, 1974; White and Neely, 1976; Sakoe and Chiba, 1978). In all these applications, a computationally efficient dynamic-programming algorithm simultaneously finds the optimum frame-to-frame correspondence between the template and the unknown speech segment and calculates the distance between them over that optimum frame correspondence.

In our speech-recognition system, the same time-warping algorithm used to calculate syllable-to-template distance is also used during system training to combine training syllables of the same phonetic class into composite templates. Templates are formed by averaging corresponding frames together, the correspondence being determined again by minimizing the summed vector distance over corresponding frames. The use of composite templates, generated in this way, results in better recognition performance than the use of a single typical syllable to represent the entire class. This empirical result is discussed in Sec. 5.5.

5.2 Details of the Time-Warping Algorithm

The time-warping problem may be represented graphically as a two-dimensional lattice in which the dimensions are the time axes of the two syllables being warped together, as shown in Fig. 3. Each point in the lattice has associated it with a *local distance* equal to the vector distance, appropriately defined, between the pair of parameter vectors defined by the point. The goal is to find a path through the lattice, beginning at $(1, 1)$ and ending at (N, M), which is continuous and monotonic and which minimizes the sum of the local distances of the points it traverses. The monotonicity constraint insures that if

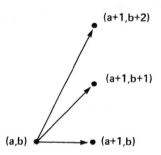

Figure 4. Rabiner, Rosenberg and Levinson's formulation of the continuity constraint, showing possible successors of point (a, b).

point (a, b) precedes point (c, d) in a path, then $a \leq c$ and $b \leq d$. This constraint follows from the physical requirement that warping of time scales must preserve the relative ordering of time points.

The continuity constraint has been formulated differently by different experimenters. A formulation by Rabiner, Rosenberg, and Levinson (1978), which has been widely reported on in the recent literature, allows the following three possible successor points to a point (a, b):

$$(a + 1, b), \quad (a + 1, b + 1) \quad \text{and} \quad (a + 1, b + 2).$$

This continuity constraint may be schematized as in Fig. 4. This formulation results in an asymmetric, syllable-to-syllable distance function since the distance so defined from syllable A to syllable B is not, in general, equal to the distance from syllable B to syllable A.

Velichko and Zagoruyko (1970), Bridle and Brown (1974), White and Neely (1976), and Sakoe and Chiba (1978) all propose a symmetric formulation of the continuity constraint, allowing point (a, b) to have as successors points $(a + 1, b), (a + 1, b + 1)$, and $(a, b + 1)$, as illustrated in Fig. 5. Sakoe and Chiba found experimentally that the symmetric formulation performed better in recognition than did the asymmetric one, although it should be noted that their data were isolated words and not syllables of continuous speech.

It is seen that, while in the Rabiner *et al.* formulation the number of local distance terms being summed to calculate total syllable-to-syllable distance is always equal to the number N of time frames on the abscissa, the same is not true in the symmetric formulation. In the symmetric method, the number of terms being summed depends upon the path chosen through the space. It can vary between $\max(N, M)$ and $N + M - 1$. Since all the local distances are positive, the algorithm is biased in favor of paths involving fewer local distance terms, i.e., those paths involving fewer nondiagonal arcs.

Sakoe and Chiba remove this bias by counting a local distance *twice* if it is reached via a diagonal arc, but only once if it is reached via a horzontal or vertical arc. In this way, every path through the lattice from $(1, 1)$ to (N, M) has the same effective number of local distance terms, $N + M$.

Sakoe and Chiba's method of eliminating the bias toward diagonal arcs causes problems when the time-warping algorithm is used for template generation. In template generation, we calculate one average parameter vector for every local distance term in the optimal warp path. If we were to count diagonally reached nodes twice, as in Sakoe and Chiba's algorithm, two syllables of length N and M frames would produce a template of length $N + M$ frames when warped together. This is undesirable; we would prefer the template length to approximate the *average* of the lengths of its constituent syllables rather than the sum of their lengths.

We have therefore taken a different approach to insure that all paths contain effectively the same number of local distances. We use the continuity constraint, first formulated by Mermelstein (1978), that a point (a, b) may be followed by points $(a + 1, b + 1)$, $(a, b + 2)$, or $(a + 2, b)$, as illustrated in Fig. 6. It is seen that any path from $(1, 1)$ to $N, M)$ will contain the same number of local distances, $(N + M)/2$. In the case where $N + M$ is odd, we seek the minimum path from $(1, 1)$ to either $(N, M - 1)$ or $(N - 1, M)$, resulting in $(N + M - 1)/2$ local distance terms. This formulation is desirable since the resulting length (number of frames) of warping two syllables together will always be the integer average of the two syllable lengths.

It may appear from Fig. 6 that we are losing information by ignoring points $(a, b + 1)$ and $(a + 1, b)$. In fact, since spectra are generally derived from overlapping windows (in our work the overlap is 75%), successive frames are sufficiently redundant that omitting single frames rarely leads to the loss of useful information.

5.3 Time-Warping as Sequence Comparison

It will be helpful at this point to draw some comparisons between the problem of sequence comparison and the problem of time-warping. Whereas sequence

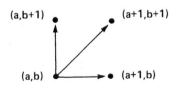

Figure 5. Symmetric formulation of the continuity constraint, showing possible successors of point (a, b).

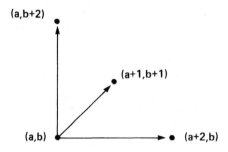

Figure 6. Continuity constraint used in this study, showing possible successor points of point (a, b).

comparison is a basically discrete problem, time-warping of speech is a discrete approximation to a continuous problem. This leads to different definitions of the cost functions.

In sequence comparison, the objects to be compared are strings of symbols from a fixed alphabet. Therefore, it makes sense to speak of deletion of a symbol, insertion of a symbol, and substitution of one symbol for another, assigning fixed costs for each. The symbols themselves are discrete units.

In speech, as we noted in Sec. 2, there are no discrete units on the acoustic level. Rather, we observe a continuous signal in the time domain from which we derive short-term spectra at arbitrary intervals (frames). Once parametrization has been accomplished, it is possible to think of each parameter vector, which denotes a point in continuous parameter space, as belonging to an infinite alphabet of parameter vectors. Substitution cost is then the appropriately defined vector-difference function we have called local distance.

In sequence comparison, where strings of discrete symbols are being compared, the operations of deletion and insertion are well defined. In speech, where the "symbols" are really samples from a continuous process, the notions of time compression and dilation are more appropriate. Deletion of speech frames is a discrete approximation to the continuous operation of time compression, just as insertion of frames is a discrete approximation to time dilation.

In order to approximate time dilation by insertion of extra frames, the inserted frames should be interpolations: their values should depend upon the values of surrounding frames. In the algorithm described in Sec. 4.2, this is accomplished by repeatedly matching the same frame in one syllable against several frames in the other syllable. Figure 6 illustrates how this is accomplished. The arrival of the warp path at point (a, b) implies that frame a of the abscissa is matched against (substituted for) frame b of the ordinate. If the path then proceeds vertically to point $(a, b + 2)$, frame a of the abscissa is also matched against frames $b + 1$ and $b + 2$ of the ordinate. Only one of the two resulting substitution costs (local distances) is summed into the total distance, however: the one corresponding to frame $b + 2$ of the ordinate. The local distance corresponding to frame $b + 1$ of the ordinate is discarded so that the number of local distances added to determine the total distance will be independent of the warp path chosen.

This formulation captures the desired notion of time compression and expansion better than a fixed cost for deletion would. A fixed cost for deletion of a frame might be well suited to handling bursts of noise in the speech signal, but does not seem appropriate to modelling the nonlinear time translation we are trying to characterize.

What we really want to calculate in speech recognition is the likelihood that an unknown syllable belongs to a given syllable class. In speech there is no absolute standard to which an unknown syllable can be compared. We can estimate a standard for a syllable class by constructing a reference template for that class. An effective way of constructing a template is to average together through time-warping several examples of syllables belonging to the same class. Time-warping is then used to compare unknown syllables to the template in order to measure the likelihood that the unknown belongs to the class represented by the template. The fact that we calculate the distance between a template and an unknown by summing local distances implies independence between frames, which may not be perceptually warranted.

5.4 Constraints on Time Elasticity: A Preliminary Experiment

The time-warping algorithm as described so far, while not permitting changes in the ordering of parameter frames, allows unconstrained compression and dilation of the time scales of the two syllables. While speech is elastic to a certain extent, timing can also carry meaningful information used to distinguish phonetically distinct syllables. For example, the burst of noise following a [t] release, if sufficiently prolonged, might be a good match for an [s] at the beginning of a template. The capacity of timing to carry phonetically relevant information leads to the consideration of methods of constraining the warp path in such a way as to limit the amount of time compression and dilation that occurs.

Constraints on time elasticity may be defined in various ways. We experimented with two types of constraints, which in the lattice representation are interpreted as constraints on the slope of the warp path. These two types were *hard slope constraints* and *slope-cost penalties* (or *soft slope constraints*). Hard constraints place absolute limits on the slope of the path; this also has the effect of narrowing the region of the lattice through which the path can pass. This reduces computation, an important consideration when many phonetic comparisons must be made in a short time. Slope-cost penalties (soft constraints) assign a penalty or added cost to every vertical or horizontal (rather than diagonal) arc in the path (cf. Fig. 6).

We did a preliminary experiment to test the combined effect of hard and soft slope constraints on the performance of our speech-recognition system. The hard slope constraint we implemented prevents the warp path from containing two consecutive vertical or two consecutive horizontal arcs. The soft slope constraint we implemented is based on minimizing a sum over all *arcs* in the path (instead of a sum over all nodes in the path), where the value associated with each arc is the local distance of its destination node times the following factor:

$$\begin{cases} 1 & \text{if the arc is diagonal,} \\ 1.5 & \text{if the arc is not diagonal.} \end{cases}$$

Since these slope constraints cause the path to tend toward the 45° line, they are correct only when the grid is square (i.e., when the syllables contain an equal number of time frames). In the experiment described in this section, the lengths of the syllables were equalized by extending the ends of the shorter syllables with *dummy frames*. Two different types of dummy frame were tried and their effects on recognition rate were compared:

1. *Mean end frames.* The initial frames of all syllables in the training set were averaged together to create an average initial frame to use for preceding dummy frames. The average final frame was used for following dummy frames. During template generation these dummy frames were averaged into the template at the ends whenever the syllables were of unequal length.
2. *Constant differences.* These can be thought of as abstract frames whose distance to any other frame is a constant. The constant chosen was approximately halfway between the average interframe distances for a correct and for an incorrect identification. During template generation these frames were not averaged into the template. Instead, the frames of the longer syllable were simply copied into the template.

To determine how many dummy frames should be placed before, versus how many after, the shorter syllable, the total distance between the syllables was minimized as a function of this split, where distance was based on the straight-line path which is the diagonal of the square (i.e., no warping).

The two different types of dummy frame were tested with and without slope constraints; the results are shown in Table 1. For each experimental condition, a set of composite templates was created from a set of 59 training sentences, using the sequential method of template generation (see Sec. 5.5). A total of 135 templates were created from 495 training syllables. Twelve of the 59 training sentences were chosen as the test set. These particular sentences were chosen because most of them were segmented correctly by the syllabifier. They contain 95 test syllables.

Table 1. Error rates for syllable recognition without syntax under two constraint conditions and two dummy-frame conditions ($N = 95$).

	Slope constraints			
Dummy frames	Both constraints		No constraints	
	No.	%	No.	%
Mean end frames	13	(14%)	11	(12%)
Constant difference	31	(33%)	28	(29%)

In order to focus on the deficiencies in the syllable comparator, the recognition system was tested without using any syntactic information. In particular, each test syllable was treated as an unknown and its distance to each of the 135 templates was calculated using the time-warping algorithm. The identification was treated as correct only if the single correct template had the smallest distance. Of course, this error estimate is pessimistic both because no syntax is used, and because some of the errors reflect confusion between templates for different pronunciations of the same syllable, which would not cause any error in practice.

It can be seen from Table 1 that the type of dummy frame used has a much larger effect on performance than the absence or presence of slope constraints. At the same time we see that the effect of slope constraints, although small and probably not statistically significant, is to degrade performance in both dummy-frame conditions.

We can explain the large difference in performance between the two dummy-frame conditions by calling attention to a phenomenon we have not yet mentioned but which has consistently obtained in the data we have examined: When we compare different tokens of the same syllable, we find greater variability among tokens near the edges of the syllables than near the middle. By generating templates from syllables extended with mean end frames, we are averaging the most variable vectors of longer syllables with constant vectors. When shorter syllables are then compared with the template, the ends are extended, with the same constant vectors giving rise to small local distances at the edges of the syllable. The use of a constant vector as a dummy frame therefore attenuates the effect of the high variability of end frames. Since abstract constant-difference frames are never averaged into the template and no actual distance is calculated, they cannot have this effect.

We conclude from the results in Table 1 that the slope constraints proposed above do not improve recognition performance, at least in conjunction with the dummy-frame extension procedure used here. Future work should explore more elaborate slope constraints applicable to rectangular grids.

Initial tests on these data of a rectangular grid algorithm without slope constraints produced 13 misrecognitions (14%), a performance one percentage point worse than the best performance condition of Table 1. This unconstrained, rectangular time-warping algorithm, because it combines good performance with simplicity and intuitive appeal, was chosen as the standard algorithm for use in the recognition system. The experiment described in the following section uses this algorithm.

5.5 Template Generation

Composite templates are generated from a set of training sentences. The training sentences are preprocessed and automatically segmented by the syllabifier. Each sentence is then transcribed by a human operator using the ARPAbet, an ASCII phonetic alphabet: Sentences are played back one syllable at a time and the phonetic transcription of each syllable is entered at the terminal. After verification, the system combines all the syllables corresponding to a distinct ARPAbet transcription into a single composite template associated with that transcription. This section describes and compares three alternative processes for creating one composite template from a set of phonetically equivalent syllables. The first template-generation procedure that was implemented combines syllables sequentially. In this procedure, the time-warping algorithm is used to determine an optimal frame-to-frame correspondence between the first two syllable tokens of the phonetic class. Corresponding parameter vectors are averaged to form a template of weight = 2 (the *weight* of a template is the number of syllable tokens from which it was derived). The warping algorithm is invoked again to determine the optimal frame correspondence between the template of weight = 2 and the third token of the phonetic class. Parameter vectors of corresponding frames are averaged as before, except that vectors belonging to the template are weighted twice as heavily as those belonging to the token, since they contain information from two tokens. The resulting template has weight = 3. The process is continued until all the syllable tokens in the phonetic class have been combined.

When we combine a new syllable token with a template, we weight the vectors of the template more heavily so that each training token in the phonetic class contributes equally to the final template. Because of the symmetric nature of the time-warping algorithm, however, the token and template contribute equally to the warp path. This means that the last token combined with the template affects the temporal structure of the final template as much as all the other tokens combined. Since the temporal structure of the syllable is to some extent neutralized by the warping process anyway, this problem is less serious than it may appear at first, although it should not be ignored.

Conceptually, we would like to make the template proportionally less flexible as its weight increases, although it is not clear how this could be done. It is theoretically possible to warp together all the syllable tokens in a phonetic

Table 2. Performance of the recognizer without syntax, using three different template-generation procedures.

Template generation procedure	Number of misrecognitions ($N = 495$)	Percent misrecognition
Sequential-combination	92	19%
Binary-combination	84	17%
Single-token templates	159	32%

class simultaneously by formulating the problem in a lattice space of dimensionality equal to the number of tokens involved. This solution was considered too computationally expensive for our purpose. Instead, we approximate this generalized averaging process by sequential combinations. The approximation seems fairly good: Davis (1979) found that for the slope-constrained, symmetric warping algorithm, differences in the order of sequential combination had no significant effect on recognition performance.

In order to eliminate temporal distortion due to sequential combination, a second template-generation procedure was implemented, which first combines tokens in pairs. Then the resulting templates of weight two are combined in pairs, resulting in a set of templates of weight four. This process of *binary combination* is continued until a single template results. If the original number of tokens is a power of two, no warp between templates of different weights occurs. Otherwise, at least one warp between templates differing in weight by unity must occur. This is still a much smaller weight difference than in the sequential-combination procedure.

A third approach to template generation, which avoids problems of combination, is to choose a single typical token to represent a phonetic class. An experiment was performed to see whether composite templates performed better or worse than single token templates. One syllable token from each phonetic class was selected as a template. In each class, the syllable selected was the one whose summed distance to the other syllables in the class was minimum.

Sequential-combination, binary-combination, and single-token template-generation procedures were tested on the set of 59 training sentences as in the previous section, except that, instead of choosing 12 test sentences, all 59 sentences were included in the test set, giving 495 test syllables. The results from this experiment are shown in Table 2.

The table shows that composite templates perform dramatically better than single-token templates, even when the single-token templates are chosen optimally. This large improvement is consistent with results from a similar experiment in which automatically generated segmentation boundaries were verified and hand-adjusted where necessary (Mermelstein, 1978). We also see

Table 3. Recognition performance and number of matches carried out as a function of beam-width threshold.

Beam width threshold (Arbitrary squared-distance units)	No. of incorrect sentences (out of 53)	No. of matches	No. of matches per syllable
1	16	5631	10.2
50	9	5841	10.6
100	5	6069	11.0
300	3	6783	12.3

that binary combination, which gives approximately equal weight to the time information of all its constituent tokens, performs somewhat better than does sequential combination. This small difference, which is probably not significant, is consistent with Davis's findings on order dependence.

6. SYSTEM PERFORMANCE

In order to test the performance of the system, 59 date-and-time sentences comprising approximately 550 syllables were recorded to form the training set, and two months later 59 similar sentences were recorded by the same speaker to form the test set. In each set the syllabifier generated syllable-boundary positions in six sentences that were incompatible with the syntax. Of the remaining 53 sentences, 50 were fully correctly recognized when a wide beam was used in the beam search.

On a per-syllable basis, there was a 99% acceptable syllabification, namely six errors out of 550 but, because a single serious syllabification error makes recognition of a sentence impossible, a 1% error rate led to roughly 10% of the sentences being lost to the system because of the syllabifier. Since an acceptable sequence of templates can seldom be found to fit an incorrectly syllabified sentence, most such sentences will be rejected rather than misrecognized. In this experiment it turned out that five out of the six incorrectly syllabified sentences in the test set could be rejected. Moreover, the probability of unacceptable syllabification is primarily a function of sentence length rather than of task difficulty and would not be expected to increase with increasing vocabulary size.

We carried out some experiments to see how performance depended on beam width. The simplest sentence-hypothesis strategy consists of selecting and retaining at each stage the single best match among the syllables proposed by the syntax. Strategies that are able to use hindsight in selecting their sentence hypotheses should correctly recognize a larger proportion of sentences but at the cost of greater processing times. An upper limit to the performance of the hypothesis-evaluation strategy is reached when the only errors occurring are single-syllable substitutions (e.g., "sixteen" recognized as "fifteen"). Table 3 shows the performance of our system on the 53 correctly syllabified sentences. The three errors remaining at large values of the threshold (i.e., using a wide beam) are all single-syllable substitutions. Consequently, no other evaluation strategy could produce better recognition performance on our data using our syllable-recognition algorithm.

The processing time is approximately proportional to the number of syllable-to-template matches that have to be carried out. It can be seen in the table that the number of matches is not very sensitive to beam width. The beam width at which performance improvement saturates requires only 20% more syllable matches than the fastest strategy using the narrowest possible beam width, yet the number of errors is reduced by a factor of more than five.

Recognition was correct for 546 out of 549 syllables, i.e., 99.5%, leading to correct recognition of 50 out of 53 acceptably syllabified sentences, i.e., to a recognition performance of 94%.

7. CONCLUSIONS

In providing an example of how dynamic programming is being applied to continuous speech recognition, we feel that the ideas should stand on their own; detailed consideration of the overall system performance is not particularly helpful. Overall performance depends critically on many factors not discussed here: the syllabifier, the spectrum range and spectrum parametrization used, and the form of local distance measure, among others. Moreover, our work is at an early stage, and our system performance has neither been optimized nor adequately evaluated. Nevertheless, our results so far at least give us grounds to believe that our approach may be practical.

The vocabulary we used is small by comparison with those used by some other groups and our sentences are relatively short. However, this is balanced b / a relatively free syntax and by the fact that many word pairs (e.g., "sixteen" and "fifteen") differ in only one syllable and share the same vowel. Experience so far suggests that close to real-time performance would be possible on fairly modest hardware.

We have aimed in this article to show how one group among many others is applying string-matching techniques to speech recognition. Such techniques have contributed much to speech recognition already and will no doubt continue to do so. We hope, however, that it has become clear that speech has special properties that have to be taken into account. Rather than mutilating the speech to fit some particularly attractive mathematical model, it is important to select suitable techniques carefully and adapt them to the speech signal.

REFERENCES

Bahl, L.R. Bakis, R., Cohen, P.S., Cole, A.G., Jelinek, F., Lewis, B.L., and Mercer, R.L. Recognition Results with Several Experimental Acoustic Processors, *Proc. IEEE International Conference on Acoustics, Speech, and Signal Processing,* Washington, pp. 249–251, 1979.

Bridle, J.S., and Brown, M.D., An Experimental Automatic Word-Recognition System, *Interim Report, JSRU Report No. 1003,* Joint Speech Research Unit, Ruislip, England, 1974.

Bridle, J.S., and Brown, M.D., Connected-Word Recognition using Whole-Word Templates, *British Institute of Acoustics Autumn Conference,* November 1979.

Bridle, J.S., and Sedgwick, N.C., A Method for Segmenting Acoustic Patterns, with Applications to Automatic Speech Recognition, *Proc. IEEE International Conference on Acoustics, Speech, and Signal Processing,* Hartford, pp. 656–659, 1978.

Davis, S.B., Order Dependence in Templates for Monosyllabic Word Identification, *Proc. IEEE International Conference on Acoustics, Speech, and Signal Processing,* Washington, pp. 570–573, 1979.

De Mori, R., Automatic Phoneme Recognition in Continuous Speech: A Syntactic Approach, *Proc. of NATO Advanced Study Institute on Spoken-Language Generation and Understanding,* Bonas (Gers), France, 1979.

Kashyap, R.L., Syntactic Decision Rules for Recognition of Spoken Words and Phrases Using a Stochastic Automaton, *IEEE Transactions on Pattern Analysis and Machine Intelligence,* PAMI-1(2), pp. 154–163, April 1979.

Klatt, D.H., Review of the ARPA Speech-Understanding Project, *J. Acoust. Soc. Am,* 62(6), pp. 1345–1366, December 1977.

Kozhevnikov, V.A., and Chistovich, L.A., *Speech: Articulation and Perception.* Washington: Joint Publications Research Service, 1965.

Levinson, S.E., and Rosenberg, A.E., A New System for Continuous Speech Recognition—Preliminary Results, *Proc. IEEE International Conference on Acoustics, Speech, and Signal Processing,* Tulsa, pp. 239–246, 1979.

Lowerre, B.T., The Harpy Recognition System, Ph.D. Thesis, Dept. of Computer Science, Carnegie-Mellon Univ., Pittsburgh, PA, 1976.

Martin, T.B., Practical Applications of Voice Input to Machines, *Proc. IEEE,* 64(4), pp. 487–500, April 1976.

Mermelstein, P., Automatic Segmentation of Speech into Syllabic Units, *J. Acoust. Soc. Am.,* 58, pp. 880–883, 1975.

Mermelstein, P., Recognition of Monosyllabic Words in Continuous Sentences using Composite Word Templates, *Proc. IEEE International Conference on Acoustics, Speech and Signal Processing,* pp. 708–711, Hartford, 1978.

Rabiner, L.R., Rosenberg, A.E., and Levinson, S.E., Considerations in Dynamic Time-Warping Algorithms for Discrete-Word Recognition, *IEEE Transactions on Acoustics, Speech, and Signal Processing,* ASSP-26 (6), pp. 575–586, 1978.

Reddy, D.R., Speech Recognition by Machine: A Review, *Proc. IEEE,* 64(4), pp. 501–531, April 1976.

Sakoe, H., and Chiba, S., Dynamic-Programming Algorithm Optimization for Spoken-Word Recognition, *IEEE Transactions on Acoustics, Speech and Signal Processing,* ASSP-26 (1), pp. 43–49, Feb. 1978.

Velichko, V.M., and Zagoruyko, N.G., Automatic Recognition of 200 Words, *International Journal of Man–Machine Studies,* 2, pp. 223–234, 1970.

White, G.M., and Neely, R. , Speech Recognition Experiments with Linear Prediction, Bandpass Filtering, and Dynamic Programming, *IEEE Transactions on Aco , Speech, Signal Processing,* ASSP-**24** (2), pp. 183–188, April 197 .

CHAPTER SIX

APPLICATION OF SEQUENCE COMPARISON TO THE STUDY OF BIRD SONGS

David W. Bradley and Richard A. Bradley

1. INTRODUCTION

Scientific study of bird song began with the advent of lightweight portable tape-recording equipment in the early 1950s. Since that time the volume of published work has increased dramatically. Early work on bird vocalizations has been reviewed by several authors (Thorpe (1961), Hinde (1969), Armstrong (1973), Thielcke (1976)). Each year several hundred new scholarly papers appear that deal with bird song. Studies have concentrated on the social functions of vocalizations, their ontogeny, geographic variation, and evolution.

The study of bird vocalizations is of interest to biologists for several reasons. Bird song is a complex, learned behavioral pattern. The development of bird song shows parallels with language acquisition in humans (Marler, 1970). As a learned behavior, the patterns are passed from one generation to the next by cultural transmission. Changes in subsequent generations represent an example of cultural evolution, rare among the nonprimates. The divergence of bird songs through evolutionary time gives the biologist a record of past geographic isolation of populations. In many species of birds, variation takes the form of song dialects, with members of each isolated population sharing certain variants on the characteristic species-specific theme.

The analysis of animal sounds makes use of much elaborate equipment. The basic tool of quantitative analysis is the audiospectrograph, or "voice print" machine. This unit produces a frequency-versus-time display (audiospectrogram, or sonagram). Such tracings allow for detailed study of the structure of sounds. For relatively simple utterances, presence (or absence) of specific acoustic features, maximum and minimum frequency, length of sound, etc., are used. More complex sounds are often studied by meticulous qualitative comparison of sonagram tracings. The song structure is then delineated by encoding the song as a sequence of basic notes, which are identifiable as separate traces on the sonagrams. Variability in the note sequences is described and evaluated. Many studies have used this approach (e.g., Thompson (1970, 1976), Lemon (1971), Williams and MacRoberts (1977)).

The process has recently been extended to incorporate quantitative techniques of data analysis by Payne (1978), and by Payne and Budde (1979). They employed principal components analysis and cluster analysis to explore microgeographic variation in songs of splendid sunbirds and Acadian fly-catchers. A prerequisite for their method of analysis was development of a distance measure applicable to comparisons between pairs of songs. Payne and Budde used correlations between songs based upon several measurements of pitch and timing of notes within the songs. These species have a very simple song structure well suited to this approach. Unfortunately, differences between complex songs are not readily adapted to this form of measurement.

The primary objective of this paper is to develop a strategy for measuring inter-song distances that can be used with complex songs. A secondary objective is to illustrate some applications of inter-song distances. Our method uses both similarity of the notes and similarity of their arrangement to provide a sensitive measure of inter-song similarity. This measure can be used for intra- and inter-individual repertoire analysis, population phenology, geographic variation studies, and interspecific comparisons. Songs are first encoded as sequences of notes. Then differences between pairs of notes are measured. Finally, these two types of data are utilized by a sequence-comparison algorithm that produces a matrix of inter-song distance scores. The results make it feasible to perform statistical analysis of bird song data.

We have employed several different options at each stage of the analysis, which are described in detail in the next three sections of this chapter. These alternatives are used to analyze data from an earlier study (Bradley, 1977) on the songs of a subspecies of the savannah sparrow (*Passerculus sandwichensis beldingi*). These birds live in salt-water marshes along the Southern California coast. Consequently, they are confined to isolated populations that provide a good resource for the study of song dialects. Performance comparisons between various methods for quantifying these data are presented, along with the results of an application of this method to the Beldings sparrow data.

2. METHODS FOR CODING THE SONGS

Initial coding of the sequences of notes for the Beldings data was published previously (Bradley, 1977). A sample sonagram along with its corresponding coding scheme is illustrated in Fig. 1. Note that all separate traces were coded

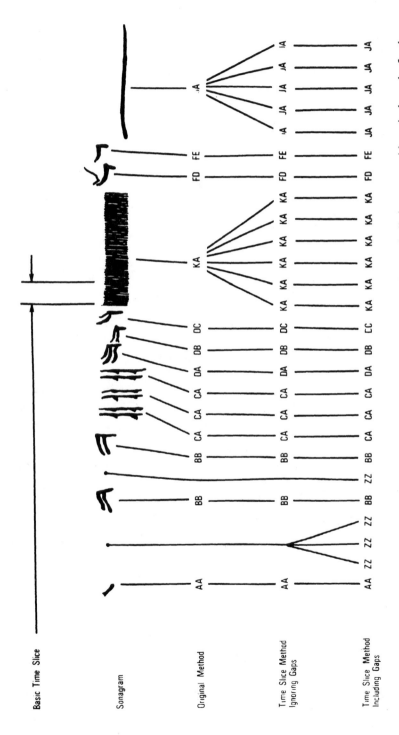

Figure 1. Examples of coding schemes for the songs. The original method matches each distinct note with a single code. In the second method, long notes are represented with repeated codes; the number of repetitions is the number of basic time slices covered by the note. The final coding method adds another code to denote gaps between notes.

as single notes. This is common practice but has the unfortunate effect of assigning equal weight to notes that differ greatly in duration. We devised a scheme for refinement of the original coding by choosing a basic time interval and then counting a note more than once if it spans more than one basic time slice. The second code sequence in Fig. 1 shows the resulting list of notes for the sample song.

The revised method also has the advantage that the number of notes in the sequence provides an approximate estimate of the length of the song. However, it does not take into account the time intervals between successive notes. These gaps may play an important role in the character of a song. Consequently, we considered a third approach which incorporates a special code to designate inter-note gaps. These are counted using the same scale for time slices. We ignore gaps that are shorter than one basic time slice (Fig. 1). These sequences should provide a good representation of the temporal structure and duration of the songs.

The precision of this representation depends upon the size of the basic time slice. Investigators are faced with a dilemma: Reduction in the duration of time slices improves accuracy but results in great increases in the amount of storage and computer time required for the analysis. Since analysis of a typical sample containing two hundred songs would require 19,900 inter-song comparisons, the computer time usage may be an important consideration. We chose a time interval that makes almost every short note fit within one time slice.

One possible alternative to our approach would be to use extremely short time slices so that each note would consist of many slices. Digital signal-analysis techniques could then be employed for direct comparison of time slices. This strategy has been used extensively in the context of automated human-speech recognition (Rabiner et al., 1977). They consider methods for matching the sounds by optimal warping of the time scale. We see two difficulties with their approach: First, a tremendous amount of computer time would be required for processing all of the thousands of comparisons that arise in bird song samples. Secondly, while bird song researchers frequently work with finite-sample audiospectrographs, they rarely have access to the much more expensive real-time digital signal-processing equipment that is needed for the alternative strategy.

3. DETERMINING INTER-NOTE DISTANCES

Constructing a quantitative gauge of the degree of similarity between notes is in itself a complex task. Digital signal-processing of individual notes combined with time-warping for inter-note comparison might be useful for this purpose except for the cost limitations indicated above. Bertram (1970) proposed that sonagraphs, when viewed as grids of binary data, could be used to directly calculate an overlap measure between notes. This requires tremendous coding effort when large numbers of notes are considered. It has the additional disadvantage of high sensitivity to the quality of the original recording and sonagram. Neither of the above approaches seemed practical for our study of the Beldings sparrow data.

Use of human judges provides an alternative strategy for determining inter-note distances. There are many ways in which this could be accomplished. For example, each judge could rate every possible pair of notes on a scale with zero representing perfect matches and large values corresponding to very different notes. Rating sets for several judges could then be averaged together, producing a matrix of inter-note distances. The primary disadvantage of this method is the tremendous number of pairwise comparisons that must be made when there are many notes. For example, with the Beldings sparrow data we categorized the notes into 85 distinct types, so each judge would need to make more than three thousand comparisons.

We adopted a less laborious scheme. Each judge draws a dendrogram indicating his or her impression of how the notes cluster together, based both on a general familiarity with how the notes sound and on examining the sonagrams for the notes. This dendrogram tree construction is a relatively easy task to perform since each step in building the tree requires only that the judge select the most similar pair of clusters and connect them together. Once a dendrogram has been developed, the distance between any pair of notes can be calculated by tracing the path along the dendrogram between those notes and observing the highest level it attains (see Fig. 2). This is essentially the same method as that employed in cluster analysis for computing cophenetic values (Sneath and Sokal, 1973). The data from all of the judges are pooled to produce the final inter-note distance matrix.

Direct measurement (with a ruler on the sonagram traces) of note characteristics such as medial pitch, range of pitch, and duration provides another way of comparing notes. Although this information is too crude to be of much value alone, it could be incorporated along with other data to obtain refined values to inter-note distance. We constructed alternative inter-note distances for the Beldings data in this way by averaging human dendrogram judgment values together with Euclidean distances between notes based upon these three sonagram characteristics. We also experimented with one other method for note-distance refinement using a smaller data set of songs of white crowned sparrows (*Zonotrichia leucophrys*). This was accomplished by fitting the judges' raw dendrogram data to a two-dimensional Euclidean plane with nonmetric multidimensional scaling (Lingoes (1965), Guttman (1968)). Our objective in these efforts was to discover whether additional refinements of the inter-note distances resulted in any appreciable change in the final inter-song distance values.

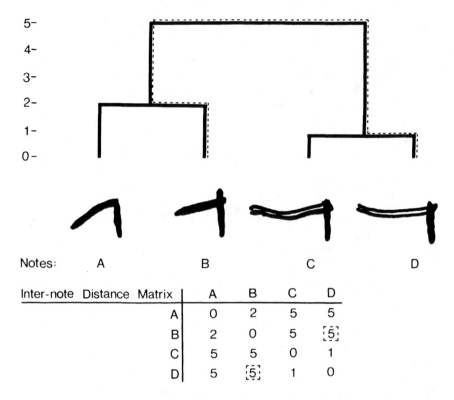

Figure 2. Illustration of the dendrogram method for generating inter-note distances. In this example, notes C and D are seen as extremely similar to each other, while A and B are slightly less similar. These two pairs of notes are then clustered together at a higher level. The distance matrix is generated directly from the dendrogram. The dotted lines illustrate this for the comparison of B with D. The connecting line between them reaches level five, resulting in the inter-note distance of five as shown in the matrix.

4. ALGORITHMS FOR SEQUENCE COMPARISON

The primary goal in the design of our sequence-comparison algorithms is to produce some measure that will accurately reflect the birds' impressions of song similarities. Having no direct way of assessing the birds' opinions, we use the human impression of similarity as a guide and seek a technique that is sensitive to both the selection and the ordering of notes within the songs, and hope that this is sufficient to effectively represent audible sound differences. Emlen (1971) used synthesized songs with altered sequences of notes in an experiment that provides field evidence that note sequence is an important factor in the bird's response to songs. In any case, a successful application of the distance function that reveals useful data patterns automatically provides *ex post facto* evidence of its validity. Ideally, songs that incorporate similar notes in approximately the same order would have a very low inter-song distance, while increasing disparity between the songs would result in higher distance values. Divergence of one song from another could arise either from re-ordering of notes within the song, insertion or deletion of notes, or substitution of one note for another. We have investigated two major algorithms for making comparisons between note sequences. The first is based upon weighted averages of inter-note distance, while the second determines an optimal method for matching the songs by time-warping. Variations on each of these methods are presented.

The first algorithm that we developed computes a weighted average of inter-note distances for all pairs of notes from two songs. The weights place the greatest emphasis on pairs of notes that occur at the same relative position in the songs. The following formula provides a precise description of how the algorithm operates:

$$S^{ab} = \max \left[\left(\sum_{i=1}^{m} \sum_{j=1}^{n} W_{ij}^{mn} D_{ij}^{ab} \right) - \tfrac{1}{2} \left\{ \left(\sum_{K=1}^{m} \sum_{L=1}^{m} W_{KL}^{mm} D_{KL}^{aa} \right) \right. \right.$$

$$\left. \left. + \left(\sum_{K=1}^{n} \sum_{L=1}^{n} W_{KL}^{nn} D_{KL}^{bb} \right) \right\}, 0 \right]$$

where:

S^{ab} = the distance from song a to song b,
D_{ij}^{ab} = the distance from note i in song a
　　　to note j in song b,
m = the length of song a,
n = the length of song b,
W_{ij}^{mn} = the weight for item pair (i, j)
　　　from songs of length m and n,

$$W_{ij}^{mn} = k \left(1 - \left| \frac{i-1}{m-1} - \frac{j-1}{n-1} \right| \right),$$

with k chosen to normalize the weights so that

$$\sum_{i=1}^{m} \sum_{j=1}^{n} W_{ij}^{mn} = 1.$$

Note that the raw weighted averages are adjusted by intra-song terms that compensate for differences in within-song variability in note selection and guarantee that identical songs will have a distance value of zero. As an additional precaution, negative distances are explicitly truncated to zero, although we have never seen negative values arise in practice. The weights in this algorithm incorporate a linear time-warp to match notes at similar relative positions in the songs, so the algorithm has no sensitivity to differences in song length. The latter limitation was addressed in an alternative version of the algorithm where the distance based upon the weighted average was multiplied by the ratio of the length of the longer song to that of the shorter song.

The second major algorithm uses the method of dynamic programming to find an optimal matching scheme for items in the sequences. This approach was previously applied in studies of nucleotide-sequence homology (Needleman and Wunsch (1970), Sankoff et al. (1976)) and error correction in communications (Okuda et al., (1976). The strategy involves accumulating the total cost of conversion from one sequence of notes to another where each step in the conversion (insertion, deletion, or substitution of notes) has its own associated cost. Substitution of similar notes costs little, while substitution of very different notes adds a large amount to the overall cost. The distance between songs is defined by the minimum possible total cost by which one of the songs can be converted into the other. This minimum is then normalized by dividing it by the length of the longer sequence in the pair, in order to produce the actual inter-song distance value. This algorithm can be interpreted as adjusting the time scales for the two songs to put them into the best possible alignment. Unlike the weighted-average algorithm it allows nonlinear time warps. The details of the algorithm are contained in this recursive definition:

$$S^{ab} = \frac{S^{ab}_{mn}}{\max(m, n)}$$

where

$$S^{ab}_{ij} = \min\left[S^{ab}_{i-1,j-1} + D^{ab}_{ij}, \quad S^{ab}_{i-1,j} + C^a_i, \quad S^{ab}_{i,j-1} + C^b_j \right]$$

with initial conditions

$$S^{ab}_{00} = 0, \qquad S^{ab}_{i0} = \sum_{k=1}^{i} C^a_k, \qquad S^{ab}_{0j} = \sum_{k=1}^{j} C^b_k,$$

and

$$S^{ab} = \text{the distance from song } a \text{ to song } b,$$
$$D^{ab}_{ij} = \text{the distance from note } i \text{ in song } a$$
$$\text{to note } j \text{ in song } b \text{ (substitution cost)},$$
$$C^a_k = \text{the insertion cost and deletion cost}$$
$$\text{for note } k \text{ of song } a,$$
$$m = \text{the length of song } a,$$
$$n = \text{the length of song } b.$$

The insertion costs were set proportional to the duration of the note when the original coding scheme for notes was employed. A constant value applicable to all notes was used with the time-slice coding schemes. Appendix 1 contains APL functions that precisely describe these algorithms. Actual implementation of the algorithms was carried out with FORTRAN programs that used the dynamic-programming approach instead of explicit recursion, in order to increase computing efficiency.

5. COMPARATIVE EVALUATION OF THE METHODS

One concern that arose immediately with use of the optimal matching algorithm was the selection of the insertion-cost values. If the cost was small in comparison with any of the inter-note distances, then those distances would become ineffective in the matching process. On the other hand, very large insertion costs would place undue emphasis on differences in the lengths of the sequences.

We sought a more precise picture of the relationship between insertion-cost values and differences in sequence lengths using Monte Carlo simulation. Performance of the algorithms was measured with two conflicting criteria. One criterion measured degree of sensitivity to differences in sequence length, while the other responded to differences in sequence structure and was used to guard against overemphasis on length differences. The criteria were actually success rates from a series of 100 random trials, where each trial involved three pairs of sequences. The pairs were generated by a design scheme which specified, based on our own judgment, which of the three pairs should show the largest distance.

For the first criterion, all six sequences were built from the same basic random generating process, differing systematically only in length: one pair contained sequences of length 5; the second pair contained sequences of length 10; and the target pair contained one sequence of length 5 and one of length 10. The trial was counted as a success if the target pair had a greater distance value than the other two pairs.

For the second criterion, the target pair was one that had relatively large differences in item selection, although the sequences were of the same length. In each comparison pair, the two sequences had unequal length but were generated with smaller likelihood of differences in item selection.

The optimal matching algorithm was tested with five choices for the insertion cost which ranged from 0.5 to 2.0 relative to a maximum inter-note distance value of one. Two versions of the weighted-average algorithm were also included. The first employed the basic algorithm as it was presented earlier. The second used the method which incorporates some sensitivity to length differences by adjusting the weighted average by the ratio of sequence lengths.

Table 1. Monte Carlo simulation study of sequence length effects

Algorithm	Criterion	
	1. Sufficient sensitivity to sequence lengths	2. Not too much sensitivity to sequence lengths
Weighted average: original	1*	85*
Weighted average: adjusted for length	0	100
Optimal matching: insertion cost = 0.50	0	100
Optimal matching: insertion cost = 1.00	24	90
Optimal matching: insertion cost = 1.12	76	80
Optimal matching: insertion cost = 1.50	100	9
Optimal matching: insertion cost = 2.00	100	0

*Entries show per cent of trials successful (out of 100 trials).

The results appear in Table 1. The weighted-average algorithm was unable to distinguish sequences of different lengths even after adjustment. The optimal matching algorithm performed more effectively. Increased insertion cost improved the sensitivity to sequence length, eventually overemphasizing that factor. The only version that performed well on both criteria had insertion costs set at 1.1176. This value had been chosen so that it exceeded the maximum inter-note distance by an amount equal to the gap between this maximum and the second highest inter-note distance value.

Another value that needed specification was the relative distance assigned in the inter-note-distance matrix to comparisons between inter-note gaps and actual notes. Monte Carlo simulation again helped clarify the situation. Here one criterion measured the effectiveness of the algorithms at distinguishing sequences that differ only in the probabilities that control the positioning of gaps. As in the earlier case, a second criterion was used to verify that differences in gap positioning did not overwhelm more important differences in the underlying note structure.

Table 2 shows the results for three different sequence-comparison options. The weighted-average algorithm performed very well. Optimal matching was slightly less successful. Each of these techniques was evaluated with several choices for the distance between gaps and actual notes. All cases where this value was set equal to or slightly higher than the maximum inter-note distance produced satisfactory results. When inter-note distances are derived from judges' dendrograms, this can be easily accomplished by clustering gaps in as a final outlier, slightly above the remainder of the tree.

The previous discussion provided some consideration for the effects of changes in inter-note distances, but only with reference to gaps in the songs. The sensitivity of the algorithms to changes among the remaining inter-note distances was also of concern. We first looked at this with the white crowned sparrow data. Refinement of the original inter-note distances with multidimensional scaling using the Guttman–Lingoes program SSAl (Lingoes (1965), Guttman (1968)) reduced the notes to three clusters located at the vertices of an equilateral triangle. We generated inter-song distances with the weighted-average technique using both the original and the scaled note distances. The values we obtained showed a nearly perfect correlation ($r = 0.95$).

Comparison of raw with refined inter-note distances was also performed using the Beldings sparrows data. We tested both the weighted-average and optimal-matching algorithms with both the original and time-slice coding schemes for sequences. Inter-song distances computed by these techniques based upon refined inter-note distances correlated highly with their counterparts which used the raw values ($r = 0.98, 0.95, 0.96, 0.98$). Thus the sequence-comparison strategies seem to be quite robust in the face of perturbations of the inter-note distances.

As a preliminary evaluation of the different methods for calculating inter-song distances, we used our own judgment on randomly selected triads of songs from the Beldings data. Each author independently reviewed the same set of triads and judged which pair of songs would sound most similar. Those triads for which our subjective judgments were in agreement were then used to evaluate

Table 2. Monte Carlo simulation study of inter-note gap effects

Criterion 1: Sufficient sensitivity to gaps

	Ratio between gap-to-note distance and maximum inter-note distance			
	1.00	1.12	1.50	2.00
Weighted average: original	79*	79*	80*	80*
Optimal matching: insertion cost = 1.00	76	82	85	90
Optimal matching: insertion cost = 1.12	76	79	81	84

Criterion 2: Not too much sensitivity to gaps

	Ratio between gap-to-note distance and maximum inter-note distance			
	1.00	1.12	1.50	2.00
Weighted average: original	97*	96*	94*	87*
Optimal matching: insertion cost = 1.00	88	81	52	39
Optimal matching: insertion cost = 1.12	83	73	48	35

*Entries show percent of trials successful (out of 100 trials).

sequence-comparison strategies. Trials were considered successful whenever the algorithm agreed with us as to which pair of songs was most similar. Each of the three coding schemes for the songs described earlier (see Fig. 1) was tried with both the weighted-average and the optimal-matching algorithms. This resulted in six combinations of sequence-comparison methods which were evaluated with the triads.

Inspection of Table 3 reveals that the time-slice coding schemes performed better than did the original coding scheme. Inclusion of gaps helped very slightly for the optimal-matching algorithm but reduced the quality of fit for the weighted-average approach. Analysis-of-variance of these data using an arcsin transformation showed that the optimal-matching algorithm performed significantly better than the weighted-average technique ($p = 0.046$). There were also differences between the coding schemes ($p = 0.057$) but less evidence ($p = 0.142$) for an interaction (see Table 4). We feel, considering how difficult the subjective evaluation process was, that the performance of the algorithms was quite satisfactory.

The correlation matrix among the six methods of producing inter-song distances is shown in Table 5. All of the values are positive, indicating a general consistency among the different methods for measuring song distance. There is no further discernible pattern.

6. APPLICATION OF THE METHOD

In actual use of the sequence-comparison algorithms to study 279 Beldings sparrow songs from 14 locations, we selected the method that appeared to be the most successful in the various evaluations presented above. Thus the songs were coded with the time-slice scheme including gaps, while the inter-song distances were computed by optimal matching with the insertion cost set about 12 per cent above the highest inter-note distance value. We wished to identify the structure of the interrelationships among songs and to see whether this showed any consistency with the distribution of sparrows among populations. In particular we sought quantitative evidence for the existence of the local dialects in Beldings sparrow songs, which were described in the previous qualitative study (Bradley, 1977).

The first technique that we employed to analyze the inter-song distances was group-average cluster analysis (Anderberg, 1973). This gave us a dendrogram that indicated the way in which the various songs related to each other. There were nine primary clusters that include 92 percent of the songs. The remaining songs were outliers. Table 6 shows how the songs from each population are partitioned into the clusters. Within-population homogeneity of song patterns appears to be strong, since for every population the majority of its songs fall into one or two clusters. However, the primary clusters often combine songs from different populations, and it is interesting that these combinations frequently fail to follow geographic associations. Locations that were widely

Table 3. Triad analysis of Belding's data

Algorithm†	Original coding	Time-slice coding	Inclusion of gap codes
Weighted-average	67*	82*	67*
Optimal-matching	74	82	84

*Entries show percent trials successful (out of 103 trials).
† The weighted-average algorithm is not adjusted for lengths. The optimal-matching algorithm used relative insertion cost 1.1176.

Table 4. Analysis-of-variance of triads data

Effect	Chi-squared	Degrees of freedom	Probability
Algorithm	4.0	1	0.046
Coding scheme	5.7	2	0.057
Interaction	3.9	2	0.142

Table 5. Product-moment correlations between inter-song distance measures.

		Algorithm					
		Weighted average† with:			Optimal matching* with:		
Algorithm		Original coding	Time-slice coding	Inclusion of gaps	Original coding	Time-slice coding	Inclusion of gaps
Weighted average† with:	Original coding	1.00	0.78	0.54	0.66	0.54	0.47
	Time-slice coding	0.78	1.00	0.68	0.66	0.71	0.63
	Inclusion of gaps	0.54	0.68	1.00	0.46	0.59	0.50
Optimal matching* with:	Original coding	0.66	0.66	0.46	1.00	0.46	0.56
	Time-slice coding	0.54	0.71	0.59	0.46	1.00	0.71
	Inclusion of gaps	0.47	0.63	0.50	0.56	0.71	1.00

*For optimal-matching versions, relative insertion cost = 1.1176
† Distances derived from the weighted-average approach were transformed to logarithms in order to remove skewness.

Table 6. Tabulation of songs by populations and clusters

Population*	1	2	3	4	5	6	7	8	9	Outliers	Total
Goleta Slough	—	—	—	—	—	—	—	7	—	—	7
El Estero	—	—	—	—	—	—	—	28	—	—	28
Point Mugu	15	—	—	—	—	—	—	—	7	6	28
Playa del Rey	—	—	—	—	—	—	—	—	4	—	4
Anaheim Bay	—	—	1	—	—	—	58	—	1	1	61
Huntington Harbour	—	—	—	—	—	—	5	—	—	—	5
Upper Newport Bay	—	—	—	—	—	—	—	19	—	10	29
Santa Margarita Lagoon	21	—	—	8	—	—	—	—	—	—	29
Agua Hedionda Lagoon	10	—	—	—	—	—	—	—	—	—	10
San Elijo Lagoon	2	1	—	—	—	—	—	—	—	1	4
Los Penasquitos Estuary	1	—	—	—	—	—	—	19	—	1	21
Imperial Beach	—	—	8	—	—	—	—	—	—	—	8
Bahia de San Quintin	—	—	—	—	9	21	—	3	—	1	34
Laguna El Rosario	—	11	—	—	—	—	—	—	—	—	11
Total	49	12	9	8	9	21	63	76	12	20	279

*The populations are listed in order from north to south along the coastline.

separated occasionally showed quite similar songs, while nearby marshes possessed sparrows that sang quite differently. Some examples of sonagrams for each cluster are included in Fig. 3 to permit direct examination of the range of variation in song patterns within and between clusters and locations.

Our next analysis involved comparison of the 4,340 distances *within* populations (i.e., locations) and the 34,441 distances *between* populations. (See Appendix 2 for details). Presumably, if the songs formed distinct dialects, the distances between populations would be generally higher than the distances within populations. The ratio between the "average" distances of each type can be employed in a manner analogous to analysis-of-variance to test for the existence of dialects. We considered two ways of performing the significance test.

The first is the randomization test, which works by randomly permuting the population identifications of the songs and recomputing the ratio after each permutation (Sokal and Rolhf, 1969). The proportion of random permutations for which the recomputed ratio is as large as the ratio observed with the true data produces a probability level for assessing significance of the inter-group distances. After 150 trials with the Beldings data, the p-value was exactly zero. In fact, the largest ratio obtained from the 150 random trials (1.005) was far smaller than the observed ratio (1.141) from the Beldings data, so the songs clearly fall into local dialects.

The second significance-testing procedure requires much less computer time and is an application of the Jackknife technique (Mosteller and Tukey,

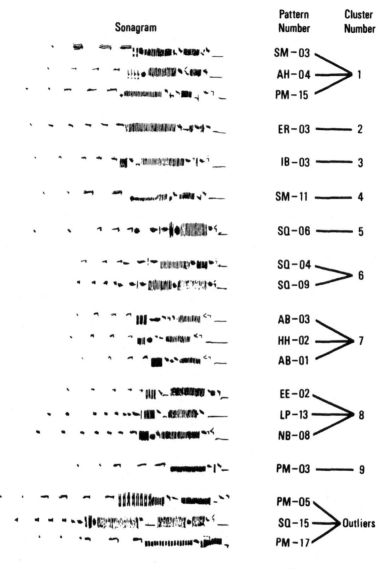

Figure 3. Samples of sonagraphs from each major cluster. These examples were selected from songs which were frequently sung and appeared to be typical of most songs in the cluster. Large clusters are represented by several songs, which indicate the range of variation within the cluster. The pattern numbers correspond to Bradley (1977). The cluster numbers refer to Table 6.

1977). We transform the distances by a modified logarithmic function to reduce skewness of the distribution (see Appendix 2 for details), and then calculate

$$f = \left(\begin{array}{c} \text{Average value} \\ \text{within populations} \end{array} \right) - \left(\begin{array}{c} \text{Average value} \\ \text{between populations} \end{array} \right).$$

By the Jackknife technique (see Appendix 2), we estimate the standard deviation s of f, and find that $f/s = 16.4$. If distances between populations are not systematically different from distances within populations, f/s should be distributed approximately like the standard t variable with 278 degrees of freedom, hence, like the standard normal variable. Obtaining a value as large as 16.4 from such a distribution is enormously unlikely (the p-value is extremely small), so this approach confirms that the songs fall into local dialects.

The final analysis that we performed was to compute the correlation of the $91 = 14 \times 13/2$ average song distances between the 14 populations with the geographic distances between the locations. This correlation was very low (0.081), confirming the impression gleaned from Table 6 that there was little tendency for geographic and song distances to correspond (when different locations are compared). This would lead us to suspect that the populations show a great deal of isolation and that dispersal of these sparrows from one location to another is a rare occurrence. The ability to examine these interpopulation patterns is the primary new information that we gained from the quantitative analysis. Such patterns are difficult to discover in qualitative exploration of raw sonagraphs.

7. DISCUSSION

Selection of the most appropriate method for encoding the sequences of notes requires careful consideration of the song structure under study. In the case of the Beldings sparrows, songs included a mixture of very brief sounds with long buzzing notes. This led us to expect that the time-slice coding scheme would provide a better measure of song characteristics than the original approach. The analysis of triads confirms this.

On the other hand, we did not anticipate any improvement to appear when inter-note gaps were included among the codes. The songs of Beldings sparrows show a consistent pattern of widely spaced introductory notes followed by a closely packed series of sounds that make up the main body of the song. Two songs from any single individual are almost always identical except for variation in the number of introductory notes. Inclusion of codes for inter-note gaps puts added emphasis on the least stable portion of the songs. This accounts for the lack of significant improvement in triad successes when gaps were incorporated as codes.

The insensitivity of the sequence-comparison algorithms with respect to minor changes in the inter-note distances is comforting. Apparently any technique that provides a reasonable measure of differences between notes would perform adequately. The use of judges' clustering patterns yielded satisfactory results with the Beldings data.

The optimal-matching algorithm seems to be superior to the weighted-average method for several reasons. It is sensitive to sequence lengths while the weighted-average procedure is not. Optimal matching also performed better in the triads analysis. One defect in the weighted-average approach is the fact that it is restricted to a linear time warp. Greater versatility could be achieved with a nonlinear fit. The optimum-matching algorithm accomplishes this by providing a monotonic fitting strategy.

A series of FORTRAN programs that implements all of these options is available from the authors upon request. For the optimal-matching algorithm, the default choice for insertion cost is a value that exceeds the maximal inter-note distance by an amount equal to the gap between this and the second highest distinct inter-note distance value. This is consistent with the results of the Monte Carlo simulation runs. In addition to measuring inter-song distance, the programs also provide several methods for analyzing the data. These include cluster analysis, multidimensional scaling, and the randomization and Jackknife tests for differences in song patterns between groups of birds.

There are many different ways in which these methods could be applied in the study of bird songs. The analysis of Beldings sparrows exemplifies but one possibility: the study of song dialects. For those species where one individual sings several different songs, comparative analysis of individual repertoires could be performed. Studies of geographic variation on a much smaller scale (mapping out individual birds' territories) is also possible. Temporal effects may also be of interest, such as investigation of year-to-year changes in patterns of intra- and inter-population variation in songs. Interspecific comparisons in search of convergence or divergence of song patterns for several species living in the same habitat would be facilitated by these quantitative techniques. Questions that have arisen along these lines include the presence of song convergence to encourage inter-specific territoriality and divergence of songs to help partition the audible space. Various researchers are currently engaged in study of these kinds of patterns in vocalization.

8. CONCLUSION

Sequence-comparison analysis of bird songs produces a useful measure of distance between pairs of songs. Calculation of these inter-song distance values permits cluster analysis, multidimensional scaling, or other multivariate data-analysis techniques to be employed in the study of complex song patterns. In addition, significance tests for presence of local dialects can be performed.

Several different measurement strategies led to similar inter-note distance values. The best results were obtained by application of the optimal-matching algorithm in conjunction with time-slice coding of the songs (incorporating codes for gaps).

The methods presented here provide a new tool that may assist in the study of bird vocalizations, as well as other biological phenomena where behavior can be viewed as sequences of actions over a finite time interval.

9. ACKNOWLEDGMENTS

We would like to thank Joseph Kruskal, James Coggins, and David Sankoff for introducing us to the optimal-matching algorithm and suggesting how it might be applied to bird songs. Cary Daniels read an earlier draft of this chapter and provided many helpful comments.

REFERENCES

Anderberg, M. R., *Cluster Analysis for Applications*. New York: Academic Press, 1973.

Armstrong, E. A., *A Study of Bird Song* (Second ed.), New York: Dover Publications Inc., 1973.

Bertram, B., The vocal behavior of the Indian Hill Mynah, *Gracula religiosa*. *Animal Behavior Monographs* 3:80–192, 1970.

Bradley, R. A., Geographic variation in the song of Belding's Savannah Sparrow (*Passerculus sandwichensis beldingi*). *Bulletin of the Florida State Museum, Biological Sciences* 22:57–100, 1977.

Emlen, S. T., An experimental analysis of the parameters of bird song eliciting species recognition. *Behaviour* 41: 130–171, 1972.

Guttman, L., A general nonmetric technique for finding the smallest coordinate space for a configuration of points. *Psychometrika* 33: 469–506, 1968.

Hinde, R. A., *Bird Vocalizations*. London: Cambridge University Press, 1969.

Lemon R. E., Analysis of song of red-eyed vireos. *Canadian Journal of Zoology* 49: 847–854, 1971.

Lingoes, J. C., An IBM-7090 program for Guttman–Lingoes smallest-space analysis. *Institute for Behavioral Sciences* 10: 183–184, 1965.

Marler, P., Birdsong and speech development: Could there be parallels? *American Scientist* 58: 669–673, 1970.

Mosteller, F., and Tukey, J. W., *Data Analysis and Regression*. Reading, Massachusetts: Addison-Wesley Publishing Company, 1977.

Needleman S. B., and Wunsch, C. D., A general method applicable to the search for similarities in the amino-acid sequence of two proteins. *Journal of Molecular Biology* 48: 443–453, 1970.

Okuda, T., Tanaka, E., and Kasai, T., A method for the correction of garbled words based on the Levenshtein metric. *IEEE Transactions on Computers* C25: 172–177, 1976.

Payne, R. B., Microgeographic variation in songs of splendid sunbirds (*Nectarinia coccinigaster*): Population phenetics, habitats, and song dialects. *Behaviour* 65: 282–308, 1978.

Payne, R. B., and Budde, P., Song differences and map distances in a population of Acadian flycatchers. *Wilson Bulletin* 91: 29–41, 1979.

Rabiner, L. R., Rosenberg, A. E., and Levinson, S. E., Considerations in dynamic time-warping algorithms for discrete-word recognition. Unpublished manuscript, 1977.

Sankoff, D., Cedergren, R. J., and Lapalme, G., Frequency of insertion–deletion, transversion, and transition in the evolution of 5S ribosomal RNA. *Journal of Molecular Evolution* **7**: 133–149, 1976.

Sneath, P. H., and Sokal, R. R., *Numerical Taxonomy*. San Francisco: W. H. Freeman and Company, 1973.

Sokal, R. R., and Rohlf, F. J., *Biometry*. San Francisco: W. H. Freeman and Company, 1969.

Thielcke, G., *Bird Sounds*. Ann Arbor: University of Michigan Press, 1976.

Thompson, W. L., Song variation in a population of indigo buntings. *Auk* **87**: 58–71, 1970.

Thompson, W. L., Vocalizations of the lazuli bunting. *Condor* **78**: 195–207, 1976.

Thorpe, W. H., *Bird Song: The Biology of Vocal Communication and Expression in Birds*. London: Cambridge University Press, 1961.

Williams, L., and MacRoberts, M. H., Individual variation in songs of dark-eyed juncos. *Condor* **79**: 106–112, 1977.

APPENDIX 1. APL PROGRAMS

Weighted Average Algorithm:

```
      DISTANCE←SEQ1 WTD_AVG SEQ2;D11;D22;D12;N1;N2
[ 1]  N1←ρSEQ1
[ 2]  N2←ρSEQ2
[ 3]  D11←INTER_ELEMENT_DISTANCE_MATRIX[SEQ1;SEQ1]
[ 4]  D22←INTER_ELEMENT_DISTANCE_MATRIX[SEQ2;SEQ2]
[ 5]  D12←INTER_ELEMENT_DISTANCE_MATRIX[SEQ1;SEQ2]
[ 6]  DISTANCE←0⌈(WA D12)-((WA D11)+(WA D22))÷2
      ∇

      DIST←WA D;WEIGHTS;M1;M2;□IO
[ 1]  □IO←0
[ 2]  M1←N1-1
[ 3]  M2←N2-1
[ 4]  WEIGHTS←1-|((M2×ιN1)∘.-(M1×ιN2))÷M1×M2
[ 5]  DIST←(+/,D×WEIGHTS)÷+/,WEIGHTS
      ∇
```

Optimal Matching Algorithm:

```
      DISTANCE←SEQ1 MATCH SEQ2;N1;N2;D;W1;W2
[ 1]  N1←ρSEQ1
[ 2]  N2←ρSEQ2
[ 3]  D←INTER_ELEMENT_DISTANCE_MATRIX[SEQ1;SEQ2]
[ 4]  W1←INSERTION_COSTS[SEQ1]
[ 5]  W2←INSERTION_COSTS[SEQ2]
[ 6]  DISTANCE←(N1 DIST N2)÷N1⌈N2
      ∇

      DST←I DIST J;OPT1;OPT2
[ 1]  →2+2×(×I)+2×(×J)
[ 2]  DST←0
[ 3]  →0
[ 4]  DST←+/W1[ιI]
[ 5]  →0
[ 6]  DST←+/W2[ιJ]
[ 7]  →0
[ 8]  OPT1←W1[I]+(I-1) DIST (J)
[ 9]  OPT2←W2[J]+(I) DIST (J-1)
[10]  DST←⌊/OPT1,OPT2,(I-1) DIST (J-1)+,D[I;J]
      ∇
```

APPENDIX 2. STATISTICAL METHODS

For both the randomization and the Jackknife tests, the distances d_{ij} between songs were first transformed into values c_{ij} by

$$c_{ij} = \ln \left(0.618 + \frac{d_{ij}}{\max d_{ij}} \right).$$

The additive constant (0.618) was selected for symmetry, in the sense that the transformed value corresponding to a distance of zero is the negative of the largest transformed value in the sample. The intent of the transformation was to reduce skewness in the distribution of distances, and to avoid boundary problems in the Jackknife procedure.

Let $N = 279$ be the size of the whole sample, and let n_k, $k = 1$ to 14, be the size of the kth population (see Table 6 for values of n_k). Then the number of c_{ij}

$$within \text{ populations is } \sum \frac{n_k(n_k - 1)}{2} = 4340,$$

and

$$between \text{ populations is } \frac{N(N - 1)}{2} - 4340 = 34{,}441.$$

Let the ordinary arithmetic averages of the c_{ij} *within* and *between* populations be w and b. Let $f = w - b$, and $r = \exp(f)$. Note that r is equal to the ratio of the geometric means of $(0.618 + d_{ij}/\max d_{ij})$.

The values of r were used for the randomization test, and reported in the text. The Jackknife test was applied to f rather than r. For each song i, a value f_{-i} with this song excluded was obtained by recomputing f without using any of the $N - 1$ distances involving song i. The pseudo–value, pv_i, for song i was then obtained using the standard Jackknife technique:

$$pv_i = Nf - (N - 1)f_{-i}.$$

The t-value was then calculated as:

$$t = \frac{\text{the mean of the } N \text{ pseudo–values}}{\text{the standard error of the mean of the pseudo–values}}$$

Note that the ratio that indicates equality of averages, namely, $r = 1$, corresponds to $f = 0$.

PART THREE

VARIATIONS ON A THEME: ALGORITHMS FOR RELATED PROBLEMS

1. INTRODUCTION

Central to the literature on sequence comparison is one basic problem, which is essentially the same in speech processing, macromolecular biology and error-correcting compilers, and one dynamic-programming algorithm to solve it. Given two sequences, the basic comparison problem, roughly speaking, is to find a matching (or trace, or alignment, etc.) between the elements in the two sequences which requires the smallest number of changes, such as deletions, insertions, and substitutions. Many variations and generalizations of the basic problem have been introduced, however, to meet the needs of different applications.

2. FINDING A PATTERN IN A SEQUENCE

In some applications, including both molecular biology and speech processing, interesting similarities may involve not an entire sequence, but only some portion of it. Chapter 2 (in Part I), by Erickson and Sellers, presents algorithms to find the best matching between a shorter sequence and some unspecified portion of a longer sequence.

3. FINDING SIMILAR PORTIONS IN TWO SEQUENCES

In some applications, interesting similarities may involve only unspecified portions in *both* sequences. Chapter 10, by Kruskal and Sankoff, contains a method for this purpose based on Smith and Waterman (1981).

4. COMPARING SEVERAL SEQUENCES

Instead of comparing just two sequences, Chapter 9, by Sankoff and Cedergren, considers the simultaneous comparison of N sequences which are connected by an evolutionary tree. Other comparison problems involving more than two sequences are referenced in Chapter 12 (in Part IV), by Hirschberg. See also Itoga (1981).

5. COMPARING ONE SEQUENCE WITH ITSELF

Rather than the comparison of two distinct sequences, the analysis of RNA secondary structure in Chapter 3 (of Part I), by Sankoff, Kruskal, Mainville, and Cedergren, involves the comparison of different parts of a single sequence.

6. COMPARING TWO TREES

The linear ordering of terms in a sequence can be supplemented, in more complex structures, by other types of relation among the components of these structures. Chapter 8, by Noetzel and Selkow, describes the tree-to-tree editing problem, which deals with the comparison of ordered labelled trees, and presents algorithms that make such comparisons.

7. COMPARING TWO DIRECTED NETWORKS

Another type of structure more general than a linear sequence is studied in Chapter 10. The comparison of two directed networks is considered, with the goal of extracting the pair of source-to-sink paths, one through each network, that are most similar.

8. COMPARING TWO CONTINUOUS FUNCTIONS: PERMITTING COMPRESSION AND EXPANSION

The sequence-comparison problem originates, in some fields, through a discretization of the problem of comparing continuous trajectories. For such problems, it is important to permit the use of what is called "compression" and "expansion" in place of, or in addition to, deletion and insertion. Studies of this type are grouped together in Part II.

9. PERMITTING TRANSPOSITIONS

In order to deal with the kinds of errors that occur during manual keyboard operation and other similar situations, Chapter 7, by Wagner, permits transpositions (swaps) of adjacent letters when comparing sequences as well as the usual substitutions, insertions, and deletions. This single change in formulation enormously complicates the nature of the sequence-comparison problem.

Transpositions lead to a generalized concept of trace (see Chapter 1) in which trace lines can intersect each other. Nonintersection of trace lines has been a fundamental property of almost all sequence-comparison problems, and a property that has most characteristically distinguished them from other problems in operations research, algorithm design, and combinatorial theory. This property is basic to the remarkable convergence of techniques in different applied fields, specifically, the repeated independent invention of methods that can be described as dynamic programming. Permitting intersecting trace lines makes impossible any simple dynamic-programming algorithm, blurs the distinctive character of sequence-comparison problems, and, as Wagner elegantly shows, creates a class of problems closely related to well known topics in operations research and computational complexity theory.

10. PERMITTING GENERALIZED SUBSTITUTIONS

The limited class of edit operations allowed in the original formulations can be extended in various ways. An apparently simple extension is to allow swaps, but as Wagner has shown, this leads to profound complication of the problem. Another extension, mentioned above, is to "compression" and "expansion." Still another extension, discussed in Chapter 10, is to allow blocks of terms to be inserted or deleted in a sequence instead of just single terms. Ultimately, insertions, deletions, substitutions, and even transpositions and block insertion/deletion may be subsumed under the notion of a generalized substitution, whereby a block of m consecutive terms in the sequence is replaced by a block of n terms. Chapter 10 discusses the consequences of this approach for the design of algorithms and for the distinctions among traces, alignments, and listings.

11. COMPARING UNDER CONSTRAINTS

Constraints are often imposed in finding the alignment that optimizes some criterion, so that only certain alignments are permitted. These are motivated by pragmatic considerations, such as speeding up the calculation, and also by subject matter goals, e.g., not having too many compressions or deletions consecutively (when comparing two speech utterances) or, in the opposite spirit, having the necessary deletions or insertions occur consecutively as far as possible (when comparing two macromolecules). Algorithms for finding optimal matchings under many different constraints are presented in Chapter 10.

12. GENERALIZING LEVENSHTEIN DISTANCE

Chapter 11, by Coggins, is concerned with generalizing Levenshtein distance to meet certain objectives required for use in a grammatical application. These objectives include reducing the number of ties in the matrix of distances among some given sequences, but doing so in a way that avoids arbitrary choices.

REFERENCES

Itoga, S. Y., The string merging problem. *BIT* 21:20–30 (1981).
Smith, T. F., and Waterman, M. S., Identification of common molecular subsequences. *Journal of Molecular Biology* 147:195–197 (1981).

CHAPTER SEVEN

ON THE COMPLEXITY OF THE EXTENDED STRING-TO-STRING CORRECTION PROBLEM

Robert A. Wagner

1. INTRODUCTION

The insert, replace, and delete single-character operations permitted in the basic string-to-string correction problem (Wagner and Fischer, 1974) may be extended to include the operation of interchanging any two adjacent characters of the source string (Lowrance and Wagner, 1975; Wagner and Brown, 1974). Allowing this *swap* operation more accurately models the error-introduction process accompanying such actions as keyboarding text (typing), so that the search for the minimum-cost edit sequence may better identify errors.

If we diagrammatically align the input and output strings from a sequence of editing operations, and connect by a line each undeleted term in the input string to the term that results from it in the output string through the replace and swap operations, the set of lines obtained is called the *trace* of the edit sequence. Inversely, any set of lines connecting some terms in one string with terms in the other, such that no term is the endpoint of more than one line, is the trace of an edit sequence that transforms one string into the other.

In the original problem (no swaps allowed), the dynamic-programming method for finding minimal-cost traces or edit sequences depends strongly on the nonintersection of the lines in the trace. When swaps are permitted, traces can become rather tangled, as with the *knot* in Fig. 1(a) and the *plaid* in Fig. 1(b). This possibility of tangling makes the reconstruction of minimal-cost traces more difficult, and motivates extensions of the dynamic-programming

Figure 1(a): a 3-knot.

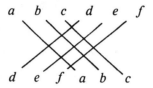

Figure 1(b): a 3-plaid.

scheme that have been incorporated (Wagner and Brown, 1974) into an algorithm called CELLAR. In the present chapter we analyze this algorithm and investigate its running time with respect to the length of the strings being compared and the relative costs of the various edit operations.

2. THE PROBLEM

For a and b, two distinct elements of an alphabet $\mathscr{A}$ not containing the null element λ, we denote by $a \to b$, $ab \to ba$, $a \to \lambda$, and $\lambda \to a$ the edit operations of replacing a by b, swapping a and b, deleting a, and inserting a, respectively. Each edit operation e has a nonnegative integer cost $w(e)$, and here we fix the cost of replacements, swaps, deletions, and insertions to be w_R, w_S, w_D, and w_I.

For $\mathbf{a}$ and $\mathbf{b}$, two strings over $\mathscr{A}$, an edit operation $e = (\mathbf{u} \to \mathbf{v})$ can transform $\mathbf{a}$ into $\mathbf{b}$ (written $\mathbf{a} \underset{e}{\Rightarrow} \mathbf{b}$) if there exist strings $\mathbf{x}$ and $\mathbf{y}$ over $\mathscr{A}$, such that $\mathbf{a}$ is of the form $\mathbf{x} \cdot \mathbf{u} \cdot \mathbf{y}$ and $\mathbf{b}$ is of the form $\mathbf{x} \cdot \mathbf{v} \cdot \mathbf{y}$. (The notation $\mathbf{s} \cdot \mathbf{t}$ indicates the *concatenation* of the two strings $\mathbf{s}$ and $\mathbf{t}$.) Further, for a sequence of edit operations $E = e_1, e_2, \ldots, e_{|E|}$, we say that $\mathbf{a}$ can be transformed into $\mathbf{c}$ by E (written $\mathbf{a} \underset{E}{\Rightarrow} {}^*\mathbf{c}$) if there exist strings $\mathbf{b}_0, \mathbf{b}_1, \ldots, \mathbf{b}_{|E|}$ such that $\mathbf{a} = \mathbf{b}_0$, $\mathbf{b}_{|E|} = \mathbf{c}$, and $\mathbf{b}_{i-1} \underset{e_i}{\Rightarrow} \mathbf{b}_i$, for $i = 1, \ldots, |E|$. (The notation $|\cdot|$ will serve without ambiguity for the cardinality of a set or the length of a sequence or a string. Similarly $\in$ will indicate membership in a sequence, string, or set.)

A formal statement of the extended string-to-string correction problem (ESSCP) is as follows: Given $\mathbf{a}$, $\mathbf{b}$, w_R, w_S, w_D, w_I, as well as $k \geq 0$, does there exist a sequence E of edit operations such that $\mathbf{a} \underset{E}{\Rightarrow} {}^*\mathbf{b}$ and such that

$$\sum_{i=1}^{|E|} w(e_i) \le k?$$

We will show that the running time of the CELLAR algorithm for answering this question is $O(|\mathbf{a}|\,|\mathbf{b}|\,|\mathbf{a}|\,|^s)$ where

$$s \le \min(4\,|\mathbf{a}|, \; w_I + w_D)/w_S + 1.$$

If $s = \infty$, either because $w_S = 0$ or because $w_R = w_I + w_D = \infty$, it is also shown that the algorithm runs in polynomial time, except for the cases

a) $w_R = w_I = \infty > w_D > 0 \quad w_S > 0,$

b) $w_R = w_D = \infty > w_I > 0, \quad w_S > 0.$

These cases will be shown to be NP-complete problems, in that a deterministic polynomial-time solution for them yields a deterministic polynomial-time solution of the minimum set cover problem.

3. CUTS OF TRACES

Let $I_n = \{1, 2, \ldots, n\}$. A *trace* from $\mathbf{a}$ to $\mathbf{b}$ may be defined as a triple $[U, \mathbf{a}, \mathbf{b}]$, where the set of *lines* $U \subset I_{|\mathbf{a}|} \times I_{|\mathbf{b}|}$ has distinct endpoints, that is, $(u_1, v_1), (u_2, v_2) \in U$ implies $u_1 = u_2$ iff $v_1 = v_2$. The line (u, v) is said to *touch* characters a_u of $\mathbf{a}$ and b_v of $\mathbf{b}$.

For any such set of ordered pairs of integers U, we define the projections $\Pi_1 U = \{u \,|\, (u,v) \in U\}$, $\Pi_2 U = \{v \,|\, (u, v) \in U\}$ and the number of pairs of lines of U that cross:

$$\chi(U) = |\{[(u_1, v_1), (u_2, v_2)] \in U \times U \,|\, u_1 < u_2 \text{ and } v_1 > v_2\}|.$$

The characters in strings $\mathbf{a}$ and $\mathbf{b}$ not touched by lines of U are found at the sets of positions

$$U_1 = I_{|\mathbf{a}|} - \Pi_1 U \quad \text{and} \quad U_2 = I_{|\mathbf{b}|} - \Pi_2 U, \quad \text{respectively.}$$

We may then define the cost $d(T)$ of a trace $T = [U, \mathbf{a}, \mathbf{b}]$ as follows:

$$d(T) = d([U, \mathbf{a}, \mathbf{b}]) = \sum_{(u,\,v) \in U} w(a_u \to b_v) + w_S \chi(U)$$

$$+ \sum_{u \in U_1} w(a_u \to \lambda) + \sum_{v \in U_2} w(\lambda \to b_v).$$

The four cost components in the above sum represent, respectively:

1. The cost of changing each character of **a** touched by a line (u, v) of U to that character of **b** touched by (u, v);
2. The number of pairs of lines of U that cross, times w_S;
3. The cost of deleting each character of **a** not touched by a line of U; and
4. The cost of inserting each character of **b** not touched by a line of U.

Lowrance and Wagner (1975) have shown that for each trace T from **a** to **b**, there exists a sequence E_T of edit operations such that

$$\sum_{e \in E_T} w(e) = d(T).$$

Furthermore, to each sequence E of edit operations such that $\mathbf{a} \underset{E}{\Rightarrow} {}^*\mathbf{b}$, there corresponds a trace T_E from **a** to **b** such that

$$d(T_E) \le \sum_{e \in E_T} w(e).$$

Thus the search for minimal-cost sequences of edit operations can be reduced to a search for a minimal-cost extended trace from **a** to **b**.

As mentioned in the Introduction (Sec. 1), the "tangling" of traces caused by the introduction of swaps means that the dynamic-programming method of reconstructing minimal-cost traces becomes more difficult. A generalization of the notion of trace, which we call a *cut*, is required.

Intuitively, a cut is a trace, extended to include lines that touch characters not in the strings joined by the trace. These characters form new strings **a'** and **b'**, which extend strings **a** and **b**, respectively, on the right. Every portion of **a'** (and of **b'**) is touched by a line of a cut whose other end also touches **b** (or **a**). One can then define a trace as a special case of a cut in which $\mathbf{a'} = \mathbf{b'} = \lambda$, the null string. One can also compute the weight of every cut from $\mathbf{a} \cdot \mathbf{a'}$ to $\mathbf{b} \cdot \mathbf{b'}$, and can determine that cut of least weight from $\mathbf{a} \cdot \mathbf{a'}$ to $\mathbf{b} \cdot \mathbf{b'}$, by an extension of the dynamic-programming scheme used in the string-to-string correction problem.

Let $T = [U, \mathbf{a}, \mathbf{b}]$ be a trace. For any integers i, j, define

$$U^{\le i} = \{(u, v) \in U \,|\, u \le i\},$$
$$U_{>j} = \{(u, v) \in U \,|\, v > j\},$$
$$U_{>j}^{\le i} = U^{\le i} \cap U_{>j},$$

and analogously for $U^{>i}$, $U_{\le j}^{\ge i}$, etc. Then

$$\Pi_1 \, U_{>j}^{\le i} = \{u \,|\, (u, v) \in U_{>j}^{\le i}\}$$

represents the set of character positions of **a** touched by a line of U that crosses (i, j) from "left to right" in going from **a** to **b**, and $u \in \Pi_1 U_{\leq j}^{\leq i}$ implies $1 \leq u \leq i$. Similarly $\Pi_2 U_{\leq j}^{\geq i}$ is the set of positions of **b** touched by lines that cross (i, j) from right to left and $v \in \Pi_2 U_{\leq j}^{\geq i}$ implies $1 \leq v \leq j$. Letting $x^k = x_1 \ldots x_k$, we define

$$T[i, j] = [U_{\leq j}^{\leq i}, \mathbf{a}^i, \mathbf{b}^j, \Pi_1 U_{\leq j}^{\leq i}, \Pi_2 U_{\leq j}^{\geq i}]$$

to be *the (i,j) cut of T*. The cost $d(T[i, j])$ will be defined to correspond with $d(T)$ when $i \geq |\mathbf{a}|$ and $j \geq |\mathbf{b}|$. Specifically, let

$$V_1 = U_{\leq j}^{\leq i}, \qquad V_2 = U - U_{> j}^{\geq i}, \qquad W_1 = I_i - \Pi_1 U^{\leq i}, \qquad W_2 = I_j - \Pi_2 U_{\leq j}$$

and

$$d(T[i, j]) = \sum_{(u,v) \in V_1} w(a_u \rightarrow b_v) + w_S \chi(V_2)$$

$$+ \sum_{u \in W_1} w(a_u \rightarrow \lambda) + \sum_{v \in W_2} w(\lambda \rightarrow b_v).$$

It can easily be shown that $d(T[0, 0]) = 0$ and $d(T[|\mathbf{a}|, |\mathbf{b}|]) = d(T)$.

4. CELLARS AND PATHS

The CELLAR algorithm is based on the construction of a directed graph G whose paths correspond to admissible sequences of cuts, i.e., each cut in the sequence introduces either one character of **a**, or one of **b**, but not both. The nodes of G are identified not only by the pair of integers (i, j) specifying the position of the cut, but also by the *cellar* of the (i, j) cut. We define

$$\text{cellar } (T[i, j]) = (\mathbf{x}, \mathbf{y}),$$

where

$$\mathbf{x} = \mathbf{a}\text{-cellar } (T[i, j]) = \text{the substring of } \mathbf{a}$$
$$\text{touched by the lines of } U_{> j}^{\leq i},$$

and

$$\mathbf{y} = \mathbf{b}\text{-cellar } (T[i, j]) = \text{the substring of } \mathbf{b}$$
$$\text{touched by the lines of } U_{\leq j}^{\geq i}.$$

From the following more formal characterization of G it will be seen that each of its paths determines an *admissible cut sequence* of any trace T from **a** to

b, that is, a sequence $C_0, \ldots, C_r$ of cuts of T such that if C_k is an (i_k, j_k) cut, then

$$(i_0, j_0) = (0, 0), \qquad (i_r, j_r) = (|\mathbf{a}|, |\mathbf{b}|),$$

and either

a. $(i_{k+1}, j_{k+1}) = (i_k + 1, j_k)$, or

b. $(i_{k+1}, j_{k+1}) = (i_k, j_k + 1)$

for each $k \in I_r$.

The nodes or vertices of G are the elements of the set

$$\{[i, j, \mathbf{x}, \mathbf{y}] \mid i \in I_{|\mathbf{a}|}, j \in I_{|\mathbf{b}|},$$

where $\mathbf{x}$ is a substring of $\mathbf{a}^i$, and $\mathbf{y}$ is a substring of $\mathbf{b}^j\}$.

Arcs lead from $[i, j, \mathbf{x}, \mathbf{y}]$ only to nodes having one of the following forms:

1. a) $[i + 1, j, \mathbf{x}, \mathbf{y}]$
   b) $[i + 1, j, \mathbf{x} \cdot a_{i+1}, \mathbf{y}]$
   c) $[i + 1, j, \mathbf{x}, \mathbf{s} \cdot \mathbf{s}']$
   where $\mathbf{y} = \mathbf{s} \cdot b \cdot \mathbf{s}'$
   for some $b \in \mathscr{A}$.

2. a) $[i, j + 1, \mathbf{x}, \mathbf{y}]$
   b) $[i, j + 1, \mathbf{x}, \mathbf{y} \cdot b_{j+1}]$
   c) $[i, j + 1, \mathbf{t} \cdot \mathbf{t}', \mathbf{y}]$
   where $\mathbf{x} = \mathbf{t} \cdot a \cdot \mathbf{t}'$
   for some $a \in \mathscr{A}$.

The *length* of an arc of type 1(a) or 2(a) is defined to be equal to $w(a_{i+1} \to \lambda)$ or $w(\lambda \to b_{j+1})$, respectively.

An arc of type 1(b) (2(b)) is termed a *placement* arc, which places a_{i+1} (b_{j+1}) in the cellar. Its length is $w_S |\mathbf{y}|$ ($w_S |\mathbf{x}|$).

An arc of the type 1(c) (2(c)) is termed a *removal* arc, which removes the character b (a) from the **b**-cellar (**a**-cellar). Its length is

$$w(a_{i+1} \to b) + w_S |\mathbf{s}| \qquad (w(a \to b_{j+1}) + w_S |\mathbf{t}|).$$

The arc is said to be at position $i + 1$ $(j + 1)$.

Let $C_0, \ldots, C_r$ be any admissible sequence of cuts of a trace T. The following lemma shows that the cost of T can be decomposed into a sum of arc lengths of a path whose nodes are defined by the successive cuts in the sequence.

Lemma 1. Let T by any trace, and let $C_0, \ldots, C_r$ be any admissible sequence of cuts of T, where

$$C_k = T[i_k, j_k] \quad \text{and} \quad \text{cellar }(C_k) = (\mathbf{x}_k, \mathbf{y}_k).$$

Then there exists a path P in G whose kth node is $p_k = [i_k, j_k, \mathbf{x}_k, \mathbf{y}_k]$. Furthermore, the length of the kth arc of P is just $d(C_k) - d(C_{k-1})$, so that

$$\text{length }(P) = d(C_r) - d(C_0) = d(T).$$

Proof: By induction on k:

Since $C_0, \ldots, C_r$ is an admissible sequence of cuts, C_k follows C_{k-1}. Thus, for each k, $k = 1, \ldots, r$, either

a. $(i_k, j_k) = (i_{k-1} + 1, j_{k-1})$, or

b. $(i_k, j_k) = (i_{k-1}, j_{k-1} + 1)$.

Consider case (a), and let cellar $(C_k) = (\mathbf{x}, \mathbf{y})$.

Case a(1): Suppose $(i_k, u) \notin T$, for any u. Then a_{i_k} is *deleted* by T, and cellar $(C_k) = $ cellar $(C_{k-1}) = (\mathbf{x}, \mathbf{y})$. In G, an arc of length $\ell = w(a_{i_k} \to \lambda)$ extends from p_{k-1} to $[i_k, j_k, \mathbf{x}, \mathbf{y}] = p_k$. In this case, $d(C_k) - d(C_{k-1}) = \ell$, as well.

Case a(2): Suppose $(i_k, u) \in T$, with $u > j_k$. Then $(i_k, u) \in U_{>j_k}^{\leq i_k}$, but $(i_k, u) \in U_{>j_{k-1}}^{\geq i_{k-1}}$. Hence cellar $(C_k) = (\mathbf{x}, \mathbf{y})$ is of form $(\mathbf{z} \cdot a_{i_k}, \mathbf{y})$ while cellar$(C_{k-1}) = (\mathbf{z}, \mathbf{y})$. In G, an arc of length $\ell = w_S|\mathbf{y}|$ extends from $p_{k-1} = [i_{k-1}, j_{k-1}, \mathbf{z}, \mathbf{y}]$ to $[i_k, j_k, \mathbf{x}, \mathbf{y}] = p_k$. Again, $d(C_k) - d(C_{k-1}) = \ell$, since the line (i_k, u) crosses each of the $|\mathbf{y}|$ lines of $U_{\leq j_k}^{\geq i_k}$.

Case a(3): Suppose $(i_k, u) \in T$, with $u \leq j_k$. Then $(i_k, u) \in U_{\leq j_k}^{\leq i_k}$, while $(i_k, u) \in U_{\leq j_{k-1}}^{\geq i_k}$. Hence, cellar (C_{k-1}) is of form $(\mathbf{x}, \mathbf{s} \cdot b_u \cdot \mathbf{s}')$ while cellar$(C_k) = (\mathbf{x}, \mathbf{s} \cdot \mathbf{s}')$. In G, an arc of length $\ell = w(a_{i_k} \to b_u) + w_S|\mathbf{s}|$ extends from p_{k-1} to p_k. Since the line $(i_k, u) \in U_{\leq j_k}^{\leq i_k}$, it crosses each line $(v, t) \in U_{\leq j_k}^{\geq i_k - 1}$ such that $u > t$, since $i_k \leq v$ for all $(v, t) \in U_{\leq j_k}^{\geq i_k - 1}$. By the definition of cellar, there are exactly $|\mathbf{s}|$ such lines. Thus $d(C_k) - d(C_{k-1}) = \ell$.

The preceding argument establishes case (a), and case (b) may be proven in similar fashion. $\square$

The following lemma is basically a converse of Lemma 1, in that it shows that the nodes on any path in G correspond to an admissible sequence of cuts of some trace, where the cost of each cut equals the path length up to the corresponding node.

Lemma 2: Let $P = [p_0, \ldots, p_r]$ be any path in G from $p_0 = [0, 0, \lambda, \lambda]$ to $p_r = [|\mathbf{a}|, |\mathbf{b}|, \lambda, \lambda]$, where $p_k = [i_k, j_k, \mathbf{x}_k, \mathbf{y}_k]$. Then there exists a trace T from $\mathbf{a}$ to $\mathbf{b}$, and a sequence $C_0, \ldots, C_k$ of cuts of T such that $C_k = T[i_k, j_k]$, C_k follows C_{k-1},

$$\text{cellar } (C_k) = (\mathbf{x}_k, \mathbf{y}_k), \text{ and}$$
$$d(C_k) = \text{length } ([p_0, \ldots, p_k]).$$

Proof: With each character a_i (b_i) placed by an arc of P into the a-cellar (b-cellar), associate the number i. Trace T then contains one line for each removal arc of P:

T contains the line (i, j) just when the removal arc is at position i, and removes b_j from the b-cellar, or when the removal arc is at position j, and removes a_i from the a-cellar. The sequence of cuts of T such that

$$C_k = T[i_k, j_k] \text{ then satisfies the lemma. } \square$$

Let R be some arbitrary predicate on (or property of) nodes of G. Trace T is said to have an *R-restricted cut sequence* just when an admissible cut sequence $C_0, \ldots, C_r$ of T exists such that $R([i_k, j_k, \mathbf{x}_k, \mathbf{y}_k])$ holds, for each $k \in \{0, \ldots, r\}$, where $C_k = T[i_k, j_i]$, and cellar $(C_k) = (\mathbf{x}_k, \mathbf{y}_k)$.

Theorem 1. Let T be a minimum-cost trace, with an R-restricted cut sequence, for some arbitrary predicate R. Let $G_R = (N_R, E_R)$ be that subgraph of $G = (N, E)$ such that

$$N_R = \{p \in N \,|\, R(p)\} \quad \text{and}$$
$$E_R = \{(p, q) \in E \,|\, p \in N_R \quad \text{and} \quad q \in N_R\}.$$

Then every minimum-length path P in G_R from $\sigma = [0, 0, \lambda, \lambda]$ to $\tau = [|\mathbf{a}|, |\mathbf{b}|, \lambda, \lambda]$ satisfies length $(P) = d(T)$. Furthermore, some minimum-cost trace T' may be recovered from each such P.

Proof: Lemma 1 shows that a path whose kth node is $[i_k, j_k, \mathbf{x}_k, \mathbf{y}_k]$ exists in G; since each node of P lies in N_R, P lies in G_R, and length $(P) = d(T)$. P must be a shortest path in G (and hence in G_R) from σ to τ, for if some shorter path P' existed from σ to τ, by Lemma 2, so would a trace T' satisfying

$$d(T') = \text{length}(P') < \text{length}(P) = d(T),$$

which contradicts the minimality of $d(T)$. Lemma 2 assures us that a minimal-cost trace T^* may be recovered from every path P^* from σ to τ in G_R such that $\text{length}(P^*) = \text{length}(P)$. $\square$

5. SUBGRAPHS OF G

In what follows, we make use of two predicates R on admissible cut sequences:

Let $R_1([i, j, \mathbf{x}, \mathbf{y}])$ be

$$\mathbf{y} = \lambda \qquad \text{and} \qquad |\mathbf{x}| < \frac{(w_1 + w_D)}{w_S} + 1 = s_1 + 1.$$

Let $R_2([i, j, \mathbf{x}, \mathbf{y}])$ be

$$0 \le |\mathbf{x}| - |\mathbf{y}| \le 1 \qquad \text{and} \qquad |\mathbf{y}| < \max\left(\frac{2w_R}{w_S}, 1\right) = s_2.$$

Nodes satisfying R_1 correspond to cuts with empty **b**-cellars and, depending on edit operation costs, small **a**-cellars. Nodes satisfying R_2 have **a**- and **b**-cellars about the same size, and these are also constrained not to be too large. From the construction of G, there are clearly $4 + |\mathbf{x}| + |\mathbf{y}|$ arcs, at most, leaving any given node $[i, j, \mathbf{x}, \mathbf{y}]$. Let $s = \min(2\lceil s_2 \rceil - 1, \lceil s_1 \rceil)$ and

$$R = \begin{cases} R_1, & \text{if } \lceil s_1 \rceil \le 2\lceil s_2 \rceil - 1, \\ R_2, & \text{otherwise.} \end{cases}$$

Then G_R has, for each distinct pair (i, j),

Case $R = R_1$:

$$\sum_{t=0}^{\lceil s_1 \rceil} (2 + t)|\mathscr{A}|^t = \mathcal{O}(s|\mathscr{A}|^s) \text{ arcs,}$$

Case $R = R_2$:

$$\sum_{t=0}^{\lceil s_2 \rceil - 1} \sum_{u=t}^{t+1} (4 + u + t)|\mathscr{A}|^u|\mathscr{A}|^t = \sum_{t=0}^{\lceil s_2 \rceil - 1} [(2t + 4)|\mathscr{A}|^{2t} + (2t + 5)|\mathscr{A}|^{2t+1}]$$
$$= \mathcal{O}(s|\mathscr{A}|^s) \text{ arcs.}$$

Since G_R is cycle-free, it is possible to order the nodes of G_R so that all nodes m that lie on any path from σ to any node n precede n in the ordering. This permits the determination of some shortest path from σ to every node of G_R in a number of operations proportional to the number of arcs of G_R. Algorithm CELLAR is one implementation of such a shortest-path calculation. CELLAR requires $\mathcal{O}(|\mathbf{a}||\mathbf{b}|s|\mathscr{A}|^s)$ time, and $\mathcal{O}(|\mathbf{b}||\mathscr{A}|^s)$ space, where $|\mathbf{b}| \leq |\mathbf{a}|$.

6. CELLAR ALGORITHM

1. Procedure MINSET(a, b); $a: = \min(a, b)$;
2. string procedure DELETE($\mathbf{a}, i$); return $(a_1 \ldots a_{i-1} a_{i+1} \ldots a_{|\mathbf{a}|})$
     comment: DELETE($\mathbf{a}, i$) deletes the ith character of string $\mathbf{a}$.
     The notation $\mathbf{s} \cdot \mathbf{t}$ indicates concatenation of strings
     $\mathbf{s}$ and $\mathbf{t}$;
3. XL: $= \max(\lfloor 2 * w_R/w_S \rfloor, 1)$;
4. YL: $= $ XL $- 1$;
5. ZL: $= \lfloor(w_I + w_D)/w_S\rfloor$;
6. if XL $+$ YL $>$ ZL then (YL, XL): $= (0, \text{ZL})$;
7. for $j: = 0$ to $|\mathbf{b}|$ do
8.   for each $\mathbf{x} \in \mathscr{A}*$ with $|\mathbf{x}| \leq$ XL do
9.   for each $\mathbf{y} \in \mathscr{A}*$ with $|\mathbf{y}| \leq$ YL do
10. $K[0, j, \mathbf{x}, \mathbf{y}]: = \infty$;
11.   end;
12. $K[0, 0, \lambda, \lambda]: = 0$;
13. for $i: = 0$ to $|\mathbf{a}|$ do
14.   for $j: = 0$ to $|\mathbf{b}|$ do begin
15.   for each string $\mathbf{x}$ with $|\mathbf{x}| \leq$ XL do
16.   for each string $\mathbf{y}$ with $|\mathbf{y}| \leq$ YL do begin
17. $K[i + 1, j, \mathbf{x}, \mathbf{y}]: = K[i, j, \mathbf{x}, \mathbf{y}] + w(a_{i+1} \rightarrow \lambda)$;
18.   MINSET($K[i, j+1, \mathbf{x}, \mathbf{y}], K[i, j, \mathbf{x}, \mathbf{y}] + w(\lambda \rightarrow b_{j+1})$);
19.   for $q: = 1$ to $|\mathbf{x}|$ do
20.   MINSET($K[i, j,$ DELETE($\mathbf{x}, q$), $\mathbf{y}]$,
                $K[i, j, \mathbf{x}, \mathbf{y}] + w_S * (q - 1) + w(x_q \rightarrow b_{j+1})$);
21.   for $q: = 1$ to $|\mathbf{y}|$ do
22.   MINSET($K[i + 1, j, \mathbf{x},$ DELETE($\mathbf{y}, q$)]$,
                $K[i, j, \mathbf{x}, \mathbf{y}] + w_S * (q - 1) + w(a_{i+1} \rightarrow y_q)$);
23.   if $|\mathbf{x}| <$ XL then
24.   MINSET($K[i + 1, j, \mathbf{x} \cdot a_{i+1}, \mathbf{y}], K[i, j, \mathbf{x}, \mathbf{y}] + w_S * |\mathbf{y}|$);
25.   if $|\mathbf{y}| <$ YL then
26.   MINSET($K[i, j + 1, \mathbf{x}, \mathbf{y} \cdot b_{j+1}], K[i, j, \mathbf{x}, \mathbf{y}] + w_S * |\mathbf{x}|$);
        end;
Result: $K[|\mathbf{a}|, |\mathbf{b}|, \lambda, \lambda] = d(T) = $ cost of minimal-cost trace from $\mathbf{a}$
to $\mathbf{b}$.

The CELLAR algorithm computes $d(T)$ in $K[\,|\mathbf{a}|,\,|\mathbf{b}|,\lambda,\lambda]$, by computing the shortest path P from $\sigma = [0,\,0,\,\lambda,\,\lambda]$ to each node $\nu = [i,\,j,\,\mathbf{x},\,\mathbf{y}]$ in G_R, and storing the length of P in $K[i,\,j,\,\mathbf{x},\,\mathbf{y}]$. CELLAR takes advantage of the acyclic nature of G_r, to ensure that $K[\mu]$ has been assigned correctly to each immediate ancestor μ of the node ν, before computing a final value of $K[\nu]$. This is performed by ensuring that $K[i-1,j,\mathbf{s},\mathbf{t}]$ and $K[i,j-1,\mathbf{s},\mathbf{t}]$ are known for all $\mathbf{s}$, $\mathbf{t}$, before $K[i,\,j,\,\mathbf{x},\,\mathbf{y}]$ is used. Because of the order in which this computation is carried out, only $K[i,\,\cdot,\,\cdot,\,\cdot]$ and $K[i+1,\,\cdot,\,\cdot,\,\cdot]$ need coexist in memory simultaneously; hence, the space needed by CELLAR as written can be reduced, by replacing each occurrence of $K[x,\,y,\,z,\,w]$ with $K[x \bmod 2, y, z, w]$, for any expressions x, y, z, w, appearing in the algorithm. This achieves a space utilization of

$$2\,|\mathbf{b}|\;\frac{(\,|\mathscr{A}|^{s+1} - 1)}{(\,|\mathscr{A}| - 1)} + \mathscr{O}\,(\,|\mathbf{a}| + |\mathbf{b}|\,)\ \text{cells.}$$

If $|\mathbf{a}| < |\mathbf{b}|$, space may be saved by interchanging strings $\mathbf{a}$ and $\mathbf{b}$, and computing

$$d'([U,\,\mathbf{b},\,\mathbf{a}) = d([U,\,\mathbf{a}',\,\mathbf{b}]),$$

where

$$(\mathbf{a}',\,\mathbf{b}') = (\mathbf{b},\mathbf{a}) \qquad \text{and} \qquad w'(a \to b) = w(b \to a) \quad \text{for all } a,\,b \in \mathscr{A} \cup \{\lambda\}.$$

7. PROOF OF THE ALGORITHM

To prove that CELLAR indeed computes the correct value of $d(T)$ in $K[\,|\mathbf{a}|$, $|\mathbf{b}|,\,\lambda,\,\lambda]$, we will show that some minimal-cost trace has an R_1-restricted cut sequence, and likewise, some minimal-cost trace has an R_2-restricted cut sequence. Theorem 1 then applies to show the correctness of CELLAR, which chooses the R_1- or R_2-restricted subsequence, depending on whether the number of arcs is less in G_1 or in G_2, for given values of w_I, w_D, w_R, and w_S.

Lemma 3 shows that for any trace there is an admissible cut sequence where each $\mathbf{b}$-cellar is empty and the size of each $\mathbf{a}$-cellar is essentially bounded by the size of the $\mathbf{a}$-cellar of a cut made at some line of T. The construction of this cut sequence is illustrated in Fig. 2.

Lemma 3. Every trace T has an admissible cut sequence $C_0, \ldots, C_r$ with $C_k = T\,[i_k, j_k]$ such that $|U_{\leq j_k}^{\geq i_k}| = 0$ and $|U_{>j_k}^{\leq i_k}| \leq |U_{>j}^{\leq i}| + 1$ for some $(i,\,j) \in T$.

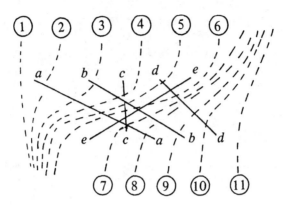

Figure 2. A cut sequence in which $|U_{\leq j}^{\geq i}| = 0$ and which satisfies Lemma 3. The crucial step is between cut-lines 6 and 7.

Proof: Let (i_{k+1}, j_{k+1}) be $(i_k, j_k + 1)$ if $\exists\, u$ such that $(u, j_k + 1) \in T$ and $u > i_k$, and $(i_k + 1, j_k)$ otherwise.

An easy induction on k shows that $|U_{\leq j_k}^{\geq i_k}| = 0$. Whenever $|U_{>j_k+1}^{\leq i_{k+1}}| > |U_{>j_k}^{\leq i_k}|$ it is clear that $i_{k+1} = i_k + 1$, and hence by construction that $\nexists\, u$ such that $(u, j_k + 1) \in T$ and $u > i_k$. Then

$$U_{>j_k}^{\leq i_k} \subseteq U_{>j_k+1}^{\leq u} \qquad \text{for } (x, y) \in U_{>j_k}^{\leq i_k} - U_{>j_k+1}^{\leq u}$$

implies

$$(x \leq i_k \text{ and } y > j_k) \qquad \text{and} \qquad (x > u \text{ or } y \leq j_k + 1).$$

Since $u > i_k$, $u > x$, so that $j_k + 1 \geq y > j_k$; that is, $y = j_k + 1$. But only $(u, j_k + 1)$ touches b_{j_k+1}, so that $x = u$, contrary to the requirement that $x \leq i_k$. Since

$$|U_{>j_k+1}^{\leq i_{k+1}}| \leq |U_{>j_k}^{\leq i_k}| + 1,$$

the lemma holds. $\square$

Lemma 4 shows that it is possible to construct a minimal-cost trace T such that if a cut were made at any line of T, the size of the **a**-cellar would be bounded by $s_1 + 1$, as in property R_1.

Lemma 4. If $w(a \to \lambda) + w(\lambda \to b) - w(a \to b) \leq nw_S$, for all a, $b \in \mathscr{A}$, then some minimal-cost trace T exists such that $|U_{>j}^{\leq i}| < n$, for every $(i, j) \in T$.

Proof: Let T be a minimal-cost trace with $|U_{>j}^{\le i}| = m \ge n$, for some $(i, j) \in T$. Consider $T' = T - \{(i, j)\}$. Now

$$d(T') = d(T) - mw_S - w(a_i \rightarrow b_j) + w(a_i \rightarrow \lambda) + w(\lambda \rightarrow b_j).$$

Hence, $d(T') \le d(T)$. The operation of removing a line from T may be repeated until no line $(i, j) \in T$ exists such that $|U_{>j}^{\le i}| \ge n$. $\square$

Theorem 2. Some minimal-cost trace T exists with an R_1-restricted cut sequence.

Proof: Apply Lemma 3 to a trace T satisfying Lemma 4. $\square$

We will now prove the existence of some trace with an R_2-restricted cut sequence, as illustrated in Fig. 3.

Lemma 5. Any trace T has an admissible cut sequence containing only cuts $T[i, j]$ satisfying $0 \le |U_{>j}^{\le i}| - |U_{\le j}^{\ge i}| \le 1$.

Proof: Let C_0 be $T[0, 0]$, C_k be $T[i, j]$. Then C_{k+1} is $T[i + 1, j]$ if $\Delta_k = |U_{>j}^{\le i}| - |U_{\le j}^{\ge i}| = 0$, and C_{k+1} is $T[i, j + 1]$ otherwise. $\Delta_0 = 0$. If C_{k+1} is the

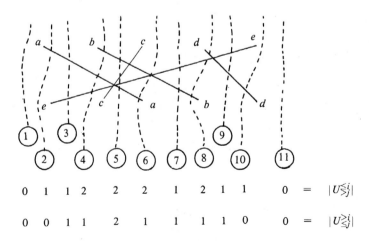

Figure 3. A sequence of cuts illustrating Lemma 5. The cut-lines are dotted, and numbered in sequence; the column of numbers under each cut-line gives the size of that cut's **a**- and **b**-cellars.

first cut in the sequence to violate the lemma, then $\Delta_k \in \{0, 1\}$. But $\Delta_k = 0$ implies $\Delta_k \leq \Delta_{k+1} \leq \Delta_k + 1$, for if $\exists u$ such that $(i + 1, u) \in T$, then $u > j$ implies that

$$U_{>j}^{\leq i+1} = U_{>j}^{\leq i} \cup \{(i + 1, u)\} \qquad \text{and} \qquad U_{\leq j}^{\geq i+1} = U_{\leq j}^{\geq i}.$$

Also, $u \leq j$ implies $(i + 1, u) \in U_{\leq j}^{\geq i}$. Now $(i + 1, u) \in U_{\leq j}^{\leq i+1}$ so that

$$U_{\leq j}^{\geq i} = U_{\leq j}^{\geq i+1} \cup \{(i + 1, u)\} \qquad \text{and} \qquad U_{>j}^{\leq i} = U_{>j}^{\leq i+1}.$$

In both cases, $\Delta_{k+1} = \Delta_k + 1$, while if $\nexists u$ such that $(i + 1, u) \in T$, $\Delta_{k+1} = \Delta_k$. Similarly, if $\Delta_k \neq 0$, $\Delta_k - 1 \leq \Delta_{k+1} \leq \Delta_k$. Since $\Delta_k \neq 0$ implies $\Delta_k = 1$, $\Delta_{k+1} \in \{0, 1\}$ in all cases, contradicting the supposed failure of the lemma at C_{k+1}. $\square$

Lemma 6: There exists some minimal-cost trace T, such that every cut $T[i, j]$ with $0 \leq |U_{>j}^{\leq i}| - |U_{\leq j}^{\geq i}| \leq 1$ satisfies $|U_{\leq j}^{\geq i}| < 2w_R/w_S$ or $|U_{\leq j}^{\geq i}| = 0$.

Proof: Suppose not. Let T be that minimal-cost trace such that $\chi(T)$ is least. Then T must have a cut $T[i, j]$ with $0 < |U_{\leq j}^{\geq i}| = s \geq 2w_R w_S$, and $|U_{>j}^{\leq i}| = t$, with $s \leq t \leq s + 1$. Choose $Q_1 \subseteq U_{>j}^{\leq i}$ such that $|Q_1| = s$, and take $Q = Q_1 \cup U_{\leq j}^{\geq i}$. $|Q| = 2s$, and $\chi(Q) \geq s^2$, for every line of Q_1 crosses each line of $U_{\leq j}^{\geq i}$. Consider trace $T' = (T - Q) \cup (\Pi_1(Q) \odot \Pi_2(Q))$, where

$$\mathbf{x} \odot \mathbf{y} = \begin{cases} \{(x_i, y_i) \mid i \in I_{|x|}\}, & \text{if } |\mathbf{x}| = |\mathbf{y}|, \\ \text{undefined}, & \text{otherwise} \end{cases}$$

(where x_i and y_i are the ith smallest elements of $\mathbf{x}$ and $\mathbf{y}$ respectively). No pair of lines of $\Pi_1(Q) \odot \Pi_2(Q)$ cross, and hence

$$\chi(T') \leq \chi(T) - s^2.$$

Furthermore, every position of $\mathbf{a}$ or of $\mathbf{b}$ touched by a line of T also is touched by a line of T'; however, the $2s$ lines of Q all may have satisfied

$$(u, v) \in Q \text{ implies } w(a_u \to b_v) = 0,$$

while each $(u', v') \in \Pi_1(Q) \odot \Pi_2(Q)$ satisfies

$$w(a_{u'} \to b_{v'}) \leq w_R.$$

Thus, $d(T') \leq d(T) - s^2 w_S + 2s w_R$. If $s \geq 2w_R/w_S$, $s^2 w_S \geq 2s w_R$, so that

$d(T') \le d(T)$. Furthermore, $\chi(T') < \chi(T)$. This contradicts the choice of T, since T' is a minimal-cost trace with $\chi(T') < \chi(T)$. $\square$

Theorem 3. There exists a minimal-cost trace with an R_2-restricted cost sequence.

Proof. Immediate, by Lemmas 5 and 6. $\square$

Finally, the implementation of CELLAR presented above correctly computes $K[\,|\mathbf{a}|, |\mathbf{b}|, \lambda, \lambda] = d(T)$. For, let the shortest distance from vertex $\sigma = [0, 0, \lambda, \lambda]$ in G_R to vertex v be denoted by $D(v)$. Let the variables i and j of CELLAR take on the values i_t and j_t, respectively, during the tth execution of the loop of lines 15 to 27, which we term "time t."

Theorem 4. At time t, $K[i_u, j_u, \mathbf{x}, \mathbf{y}] = D([i_u, j_u, \mathbf{x}, \mathbf{y}])$ for all $u \le t$, and all strings $\mathbf{x}$ and $\mathbf{y}$ such that $|\mathbf{x}| \le XL$ and $|\mathbf{y}| \le YL$.

Proof: By induction on t:

When $t = 1$, $(i_1, j_1) = (0, 0)$. The initialization of lines 7 through 12 has established that

$$K[0, 0, \mathbf{x}, \mathbf{y}] = \begin{cases} 0, & \text{if } \mathbf{x} = \mathbf{y} = \lambda, \\ \infty, & \text{otherwise.} \end{cases}$$

Since the vertices $[0, 0, \mathbf{x}, \mathbf{y}]$ with $\mathbf{x} \ne \lambda$ or $\mathbf{y} \ne \lambda$ are not reachable in G_R from $[0, 0, \lambda, \lambda]$, $K[0, 0, \mathbf{x}, \mathbf{y}] = D([0, 0, \mathbf{x}, \mathbf{y}])$. If $t > 1$, assume the theorem for all $t' < t$. Let $(i_t, j_t) = (i, j)$. By inspection of the loop structure, if

$$(i_u, j_u) = (i - 1, j) \qquad \text{and} \qquad (i_v, j_v) = (i, j - 1),$$

then $u < v < t$. By assumption, at time u,

$$K[i_u, j_u, \mathbf{x}, \mathbf{y}] = D([i_u, j_u, \mathbf{x}, \mathbf{y}]),$$

and the arc in G_R from $[i - 1, j, \mathbf{x}, \mathbf{y}]$ to $v = [i, j, \mathbf{x}, \mathbf{y}]$ of weight $w(a_i \to \lambda)$ is the first arc entering v in G_R to be explored by CELLAR. Thus the assignment of line 17 correctly initiates the computation of

$$D(v) = \min \left\{ \begin{array}{l} D(\mu) + \ell((\mu, v)_k) \mid (\mu, v)_k \text{ is the} \\ k\text{th arc in } G_R \text{ entering vertex } v \end{array} \right\}.$$

Subsequent calls on MINSET during times u and v complete the calculation in $K[i, j, \mathbf{x}, \mathbf{y}]$ of $D(v)$. $\square$

8. SPECIAL ALGORITHMS FOR SPECIAL SETS OF w_X VALUES

As has been shown, CELLAR requires time $\mathcal{O}(|\mathbf{a}|\,|\mathbf{b}|\,s|\,\mathscr{A}|^s)$, and space $\mathcal{O}(|\mathbf{b}|\,|\mathscr{A}|^s)$, where

$$s = \min\left(\left\lceil \frac{w_I + w_D}{w_S} \right\rceil,\ \max\left(1,\ 2\left\lceil \frac{2w_R}{w_S} \right\rceil - 1\right)\right),\ \text{and}\ |\mathbf{b}| \leq |\mathbf{a}|.$$

Two cases exist in which s is unbounded:

a) $w_S = 0$, and (b) $w_R = w_I + w_D = \infty$.

The case $w_S = 0$ seems trivial, for in this case we can consider $\mathbf{a}$ and $\mathbf{b}$ to be *multi-sets* of characters, rather than ordered strings. Then compute $\mathbf{x} = \mathbf{a} \cap \mathbf{b}$, $\mathbf{a}' = \mathbf{a} - \mathbf{x}$ and $\mathbf{b}' = \mathbf{b} - \mathbf{x}$. The ESSCP distance $d_{ESSCP}(\mathbf{a},\ \mathbf{b})$ equals $d_{WF}(\mathbf{a}',\ \mathbf{b}')$, where $d_{WF}(\mathbf{a}',\ \mathbf{b}')$ is the result obtained by the polynomial-time string-to-string correction algorithm of Wagner and Fischer (1974). In this case, $\mathbf{a}'$ and $\mathbf{b}'$ contain no characters in common, allowing $d_{WF}(\mathbf{a}',\ \mathbf{b}')$ to be computed particularly quickly:

$$d_{WF}(\mathbf{a}',\ \mathbf{b}') = |\mathbf{a}'|\,w_D + |\mathbf{b}'|\,w_I + \min(|\mathbf{a}'|,\ |\mathbf{b}'|)\ \min(0,\ w_R - w_I - w_D).$$

The case $w_R = w_I + w_D = \infty$ is slightly more interesting. Three subcases can be distinguished: $w_I = w_D = \infty$, and $w_I < \infty = w_D$ or, symmetrically, $w_D < \infty = w_I$.

We will see in the next section that $w_I < \infty = w_D = w_R$ or, symmetrically, $w_D < \infty = w_I = w_R$ leads to an NP-complete problem; but by means of a totally different algorithm, the ESSCP problem restricted to $w_S > 0$, $w_R = w_I = w_D = \infty$ can be solved in time $\mathcal{O}(|\mathbf{a}|^2 + |\mathbf{a}| + |\mathbf{b}|)$:

For in this case, if T is an optimal trace from $\mathbf{a}$ to $\mathbf{b}$ with $d(T) < \infty$, $\mathbf{a}$ must be a permutation of the characters of $\mathbf{b}$. The ESSCP then asks for the least costly sequence of swap operations that permutes $\mathbf{a}$ into $\mathbf{b}$. But we can show that the lines of T can be determined easily. Each line (i, j) of T must satisfy $a_i = b_j$ since otherwise $d(T) = \infty$. Hence the only remaining opportunity for "choice" occurs when more than one instance of a character of $\mathbf{a}$ appears in $\mathbf{b}$. However, a theorem of Wagner (1973) asserts that even this "choice" is denied us: For two lines (i_1, j_1) and (i_2, j_2) such that $a_{i_1} = a_{i_2}$ cannot cross in a minimal-weight trace ("reversing" them removes at least one line-crossing, and adds nothing to the weight of a trace). Algorithm PERMUTE, below, computes U in time $\mathcal{O}(|\mathbf{a}| + |\mathbf{b}|)$, where $[U, \mathbf{a}, \mathbf{b}]$ is an optimal trace.

PERMUTE: for $j = |\mathbf{b}|$ to 1 do
              push j onto $Z[b_j]$;

for $i = 1$ to $|\mathbf{a}|$ do begin
    $j: = \text{pop}(Z[a_i])$;
    output line (i, j) end;

Let the set of lines output by PERMUTE be U. Then $d(T) = \infty$ unless

$$|U| = |\mathbf{a}| = |\mathbf{b}|.$$

Otherwise, the cost $w([U, \mathbf{a}, \mathbf{b}])$ must be determined from the lines of U. This requires time at most $\mathcal{O}(|U|^2) = \mathcal{O}(|\mathbf{a}|^2)$, giving a total time for solution of ESSCP in this case of $\mathcal{O}(|\mathbf{a}| + |\mathbf{b}| + |\mathbf{a}|^2)$. It is interesting to note that U may be determined in less time than $d([U, \mathbf{a}, \mathbf{b}])$, in this case, contrary to the approaches elsewhere in this paper which assume that if U is to be constructed, information about the "choices" made by MINSET are needed, and an admissible cut sequence for $[U, \mathbf{a}, \mathbf{b}]$ is reconstructed by following some shortest path backward from $[|\mathbf{a}|, |\mathbf{b}|, \lambda, \lambda]$ to $[0, 0, \lambda, \lambda]$, using the remembered choice information.

9. AN NP-COMPLETE ESSCP

We will consider the system of weights

$$S = [w_S = 1, w_I = \infty, w_D < \infty, w_R = \infty],$$

and show that, for each Minimum Set-cover [MSC] problem g, there can be derived in polynomial time an ESSCP, h, in this system S such that g can be answered in polynomial time if and only if h can.

In system S, any minimum-cost trace $T = [U, \mathbf{a}, \mathbf{b}]$ has the following property, if $d(T) < \infty$:

For each $j \in I_{|b|}$ there exists $(i, j) \in U$ such that $a_i = b_j$. Also, if character c appears i times in $\mathbf{b}$ and j times in a, exactly $(j - i)$ occurrences of c must be deleted from a, if $d(T) < \infty$. Thus the deletion cost is independent of w_D, and also is independent of the set of positions of a actually touched by lines of T.

The MSC problem can be stated:
    Given finite sets $x^{(1)}, \ldots, x^{(n)}$, and integer ℓ. Does there exist a subset $J \subset \{1, \ldots, n\}$ such that $|J| \leq \ell$, and

$$\bigcup_{j \in J} x^{(j)} = \bigcup_{j=1}^{n} x^{(j)}?$$

Let $y = \bigcup_{j=1}^{n} x^{(j)}$, let $t = |y|$ and $r = t^2$, and choose symbols Q_i, $i = 1, \ldots, n$, R, and S not in y. Let $\mathbf{y}$ be the string made up of the elements of y in some order and $\mathbf{x}^{(i)}$ be the string made up of the elements of $x^{(i)}$ with their order corresponding to their order in $\mathbf{y}$.

Take string $\mathbf{a}$ to be $Q_1^r \, R \cdot \mathbf{x}^{(1)} \cdot Q_1^r \, S^{r+1} \ldots Q_n^r \, R \cdot \mathbf{x}^{(n)} \cdot Q_n^r \, S^{r+1}$, string $\mathbf{b}$ to be

$$RQ_1^r \ldots RQ_n^r \cdot \mathbf{y} \cdot \underbrace{S^{r+1} \ldots S^{r+1}}_{n \text{ times}},$$

and k to be

$$(\ell + 1)r - 1 + 2t(r + 1)(n - 1) + n(n - 1)\frac{(r + 1)^2}{2} + \delta w_\mathrm{D},$$

where

$$\delta = rn + \sum_{j=1}^{n} |x^{(j)}| - |y|.$$

The resulting ESSCP$_\mathrm{S}$ is:

Given strings $\mathbf{a}$ and $\mathbf{b}$ and integer k defined above, does there exist a trace $T = [U, \mathbf{a}, \mathbf{b}]$ such that $d(T) \le k$, for weights

$$w_\mathrm{S} = 1, \qquad w_\mathrm{I} = w_\mathrm{R} = \infty, \qquad w_\mathrm{D} < \infty \, ?$$

The derived ESSCP$_\mathrm{S}$ is satisfiable if and only if the source MSC problem is.

Our construction hinges on the object diagrammed below, which we call an *r-gate:*

Closed *r*-gate   Open *r*-gate

In system S, the cost, ignoring deletion and insertion costs, of an open *r*-gate trace is just r, while that of a closed *r*-gate trace is 0. If $q \ge 1$ noncrossing lines join $\mathbf{x}$ to $\mathbf{y}$, the cost of the resulting minimal trace is r, via the "open" *r*-gate trace; if no lines join $\mathbf{x}$ to $\mathbf{y}$, the "closed" *r*-gate trace becomes optimal, at cost zero. Furthermore, each line joining $\mathbf{y}$ to strings $\mathbf{z}$ and $\mathbf{z}'$ (considered "outside" the gate shown) crosses exactly $r + 1$ lines of this *r*-gate, independent of the "open" or "closed" trace form for this gate. Our constructed ESSCP$_\mathrm{S}$ essentially concatenates n *r*-gate "tops" containing the $\mathbf{x}^{(j)}$, while successively nesting $\mathbf{y}$ into the "bottoms" of those n gates.

Example:

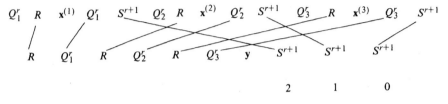

Number of gates crossed
by these lines

Note that in the "closed" positions, none of the structural Q- or R-lines cross, while regardless of gate position, the S-lines each cross $0, 1, 2, \ldots, n - 1$ entire gates (each containing $(r + 1)$ structural lines). Any line drawn from y must enter some one gate, and must therefore cross $(n - 1)$ other gates, each containing r Q-lines, one R-line, and $(r + 1)$ S-lines, a total of $2(r + 1)(n - 1)$ lines crossed. Since every character of y is touched by some line, there must be a total of

$$d(T) - \delta w_{\mathrm{D}} = 2t(r + 1)(n - 1) + n(n - 1)\frac{(r + 1)^2}{2} + f + gr$$

line crossings in the minimal-cost trace, where $g =$ the number of "open" gates and $f =$ the number of pairs of y-touching lines that cross. Since there are t y-touching lines,

$$0 \leq f < t\frac{(t - 1)}{2} < t^2 = r.$$

Thus

$$gr + c \leq d(T) \leq gr + c + t\frac{(t - 1)}{2} < (g + 1)r + c,$$

where

$$c = 2t(r + 1)(n - 1) + n(n - 1)\frac{(r + 1)^2}{2} + w_{\mathrm{D}}\delta.$$

Hence, if and only if $d(T) \leq (\ell + 1)r + c - 1 = k$, there exists a trace with ℓ or fewer open gates such that every position of y is touched by a line of T leading from **y** to the string inside an open gate. We obviously can identify the strings inside open gates with the sets indexed by J in the MSC problem.

10. A GENERALIZATION

The results of the previous sections can be generalized to the case where the cost of each edit operation $a \to b$, $a,\ b \in \mathscr{A} \cup \{\lambda\}$ depends on a and b, $w(a \to b) \geq 0$, while the cost of a swap operation remains constant at w_S, no matter what characters are being transposed. (The restriction to constant swap-cost seems necessary, if the "trace" model we rely on is to maintain its validity.) Specifically, we note that the Wagner–Fischer algorithm and the CELLAR algorithm are in fact stated for this more general problem. Thus, if

$$w_R = \max_{a,b \in \mathscr{A}} w(a \to b), \qquad w_I = \max_{a \in \mathscr{A}} (\lambda \to a),$$

and $$w_D = \max_{a \in \mathscr{A}} W(a \to \lambda), \text{ then}$$

Case: $w_S = \infty$:   Use Wagner–Fischer [1974] algorithm.

Case: $0 < w_S < \infty$:
  Case: $w_I = w_D = w_R = \infty$:   Polynomial complete (for our restrictions on w do not *force* exclusion of *all delete, replace,* and *insert* operations).

  Case: $w_I = w_R = \infty, w_D < \infty$, or   Polynomial complete.
      $w_D = w_R = \infty, w_I < \infty$:

    Case: $w_R < \infty$, or $w_I + w_D < \infty$:   Use the CELLAR algorithm.
Case: $w_S = 0$:

The case $w_S = 0$, in the generalized-weight ESSCP, turns out to be the classical "transportation problem" of linear programming [Edmonds and Karp, (1972)]. Specifically, let $q(x, \mathbf{s})$ be the number of occurrences of character x in string $\mathbf{s}$. Interpret c_{ij} as the number of characters "i" in **a** which should be changed to "j" in the optimal solution. Then solving the following transportation problem produces, as the optimal objective-function value, the cost of a minimal-weight trace T from **a** to **b**:

$$d(T) = \min \sum_{\substack{i,j \in \\ \mathscr{A} \cup \{\lambda\}}} w(i \to j)c_{ij}$$

such that

$$\sum_i c_{ij} = q(j, \mathbf{b}) \quad \text{for } j \in \mathscr{A},$$

$$\sum_i c_{i\lambda} = |\mathbf{a}|,$$

$$\sum c_{ij} = q(i, \mathbf{a}) \quad \text{for } i \in \mathscr{A},$$

$$\sum c_{\lambda j} = |\mathbf{b}|,$$

$$c_{ij} \geq 0, \quad w(\lambda \to \lambda) = 0.$$

The solution of this problem can be computed in time $\mathcal{O}(v^3 \log(|\mathbf{a}| + |\mathbf{b}|))$ by the method of Edmonds and Karp (1972), where $v = |\mathscr{A} \cup \{\lambda\}|$. The problem itself can be constructed in time $\mathcal{O}(|\mathbf{a}| + |\mathbf{b}|)$. Since $|\mathscr{A}| \leq |\mathbf{a}| + |\mathbf{b}|$, the solution time for this case is $\mathcal{O}(v^3 + |\mathbf{a}| + |\mathbf{b}|)$, including the time required for computation of the $q(i, \mathbf{a})$'s and $q(j, \mathbf{b})$'s.

REFERENCES

Edmonds, J., and Karp, R.M., Theoretical improvements in algorithmic efficiency for network flow problems, *Journal of the Association for Computing Machinery* **19**, 248–264 (1972).

Karp, R.M., Reducibility among combinatorial problems, in *Complexity of Computer Computations,* Miller, R.E. and Thatcher, J.W. (Eds.), Plenum Press, (1972); 85–104.

Lowrance, R., and Wagner, R.A., An extension of the string-to-string correction problem, *Journal of the Association for Computing Machinery* **22**, 177–183 (1975).

Wagner, R.A., Generalized correction of context-free languages. Technical Report, Systems & Information Science Program, Vanderbilt University (1973).

Wagner, R.A., and Brown, T.P., Order-*n* swap-extended correction of regular languages. Technical Report, Systems & Information Science Program, Vanderbilt University (1974).

Wagner, R.A., and Fischer, M.J., The string-to-string correction problem, *Journal of the Association for Computing Machinery* **21**, 168–173 (1974).

CHAPTER EIGHT

AN ANALYSIS OF THE GENERAL TREE-EDITING PROBLEM

Andrew S. Noetzel and Stanley M. Selkow

1. INTRODUCTION

Many processes can be described as transformations of ordered trees. Examples are found in the theory of evolution, hierarchical clustering, pattern recognition, and generation of sentences from a grammar. In the analysis of such processes, it is often desirable to compare two ordered trees by specifying the minimum-cost sequence of edit operations that will transform one tree to the other. For example, in the study of evolutionary processes, phylogenetic histories are represented by trees. The disparity between various proposed histories or the number of discrepancies between them is measured by the minimum-cost edit sequence that will transform the tree representation of one to that of the other. As another example, the minimum cost of the edit sequence that transforms one sentence structure to another is a measure of the difference of their generative processes.

Two specific cases of the tree-editing problem arise in these applications. In the first case, the operations of inserting and deleting nodes may be applied only at the leaves (nodes of the tree that have no descendants). In the second case, insertions and deletions may take place at any node within the tree. In this paper, we show for each case the analysis of edit operations on ordered trees that leads to algorithms for determining the minimum cost of transforming one tree to another.

Recursive versions of the algorithms for the general tree-edit problem are shown. For each case, the run time of the algorithm for determining the transformation having minimum cost will be a polynomial function of the sizes of the trees presented as input data.

2. DEFINITIONS

Trees

A *rooted, ordered tree* is a finite nonempty set of nodes **t** such that **t** has a distinguished node called the *root*, and the remaining nodes are partitioned into $m \geq 0$ sets, $t_1, t_2, \ldots t_m$, each of which is a tree.

The trees t_1, t_2, t_m are called *subtrees* of the root of t. The root of t_i, $1 \leq i \leq m$, is called the *i*th *son* of the root of **t**, and the root of **t** will be called the *father* of the root t_i.

In general, if a node y is a root of a subtree of any node x, then y is a son of x, and x is the father of y. Every node of a tree, except the root, has a father. The father of any node x will be denoted $f(x)$.

If, for any two nodes x and y there is a sequence of nodes $v_0, v_1, \ldots, v_n$, with $n > 0$, such that

$$v_0 = x, \qquad v_n = y, \qquad \text{and} \qquad v_i = f(v_{i+1}) \qquad \text{for } 0 \leq i < n,$$

then x is an *ancestor* of y and y is a *descendant* of x.

A node that has no descendants will be called a *leaf.*

Preorder Sequence

It will be convenient to describe the trees in their preorder sequence. The *preorder sequence* (or *preorder*) of a tree **t** is an ordering $x_1, x_2, \ldots, x_n$ of the nodes of **t** such that:

1. If $x_i = f(x_j)$, then $i < j$ (a father appears before its sons; hence, before all its descendants).
2. If $x_{j_1}, x_{j_2}, \ldots, x_{j_k}$ are the sons of x_i, in the order within the tree, then $j_1 < j_2 < \ldots < j_k$ (for nodes at the same level, the tree ordering holds),
3. If x_j is a descendant of x_i, then $x_{i+1}, x_{i+2}, \ldots, x_j$ are all descendants of x_i (the entire subtree of a node appears immediately after the node).

The preorder sequence is easily determined by applying the recursive rule:

1. List the root;
2. List each of the subtrees, from left to right, in preorder sequence.

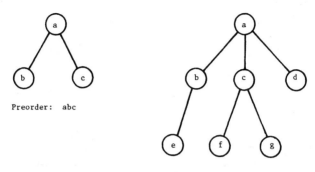

Preorder: abc

Preorder: abecfgd

Figure 1. Preorder sequence of ordered trees.

Examples are shown in Fig. 1.

Isomorphism

Trees **a** and **b** will be said to be *isomorphic* if **a** and **b** have the same number n of nodes, and for all $i = 1, \ldots, n$, the ith node in the preorder sequence of **a** has the same number of subtrees as the ith node of the preorder sequence of **b**.

If two trees are isomorphic, they are structurally identical.

Edit Operations

An *edit operation* on a tree **a** is either the *deletion* or *insertion* of a node, or the *replacement* of one node by another. Throughout the discussion, **a** will signify the tree to which the operation is applied, and **b** the transformed tree resulting from the edit operation or sequence of edit operations.

An edit operation involves only one node—either a new node, or a node of **a**. The remaining nodes of **a**, unchanged by the edit operation, will also be nodes of **b**. It is essential that each such node maintain its identity, even though, as a consequence of the operation, its position in tree **b** may differ from its position in **a**.

The particular edit operations are described as follows:

1. A *deletion* of node x of **a**, denoted $d(x)$. If node x has sons $x_1, x_2, \ldots, x_j$, and $f(x)$ has sons $y_1, y_2, \ldots, y_k$, where $y_i = x$, then in the tree **b** obtained by $D(x)$, the node $f(x)$ will have sons

$$y_1, \ldots, y_{i-1}, \quad x_1, \ldots, x_j, \quad y_{i+1}, \ldots, y_k.$$

The deletion operation is not valid for the root of **a**.

2. An *insertion* of node x into **a**, denoted $I(x)$. The node x will be inserted as the son of some node y of **a**. Node x may either be inserted with no sons or else take as sons any subsequence of the sons of y. If in **a**, y has sons $y_1, y_2, \ldots, y_k$, then, for some $0 \leq j$, $\ell \leq k$ with $j \leq \ell$, node y in **b** will have sons

$$y_1, \ldots ,y_j, \qquad x, \qquad y_{\ell+1}, \ldots, y_k,$$

and node x will either have no sons, if $j = \ell$, or else will have sons y_{j+1}, $y_{j+2}, \ldots, y_\ell$. Insertion of a new root of **a** is not a valid operation.

3. A *replacement* of node x of **a** by node y, denoted $R(x, y)$. The node y in tree **b** will have the same father and sons as node x in **a**.

The operation $R(x, y)$ following $R(y, x)$ will leave the tree in its original state. Likewise, $D(x)$ following $I(x)$ will restore the original tree. But $I(x)$ following $D(x)$ may restore the tree or not, depending on how the insertion is done. Examples are shown in Fig. 2.

Effect of edit operations on the preorder sequence

A replacement $R(x, y)$ operation changes the identity of a single node within a tree but leaves the structure of the tree unaltered. Therefore, the preorder of the nodes not involved in the replacement is not changed by the operation.

In addition, for both insertions and deletions, the preorder of the nodes not involved in the operation will not be affected by the operation. As can be seen from Fig. 2, $D(x)$ moves the subtrees of x higher in the tree, but does not change their order with respect to each other. Hence, $D(x)$ does not change the preorder of these subtrees. Neither does $D(x)$ change the subtrees' order with respect to the other subtrees of $f(x)$. The effect of $D(x)$ on the preorder sequence is only the deletion of node x. Since every insertion is the inverse of a deletion, its effect on the preorder sequence is only the insertion of the new node within the sequence.

Cost of edit operations

The cost of an edit operation will depend only on the type of operation and the nodes involved in the operation. Let $w_D(x)$ and $w_I(x)$ be the costs of deleting and inserting node x, respectively, and let $w_R(x, y)$ be the cost of replacing node x with node y. The costs are assumed to be nonnegative and finite, and for any nodes x, y, and z, the condition

$$w_R(x, z) \leq w_R(x, y) + w_R(y, z)$$

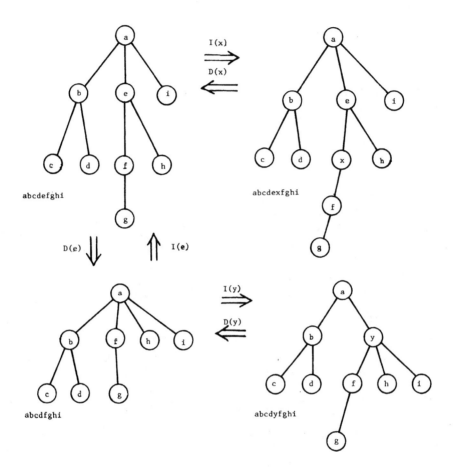

Figure 2. Insertion and deletion examples.

is assumed, so that no sequence of replacements can achieve at lower cost the same effect as a single replacement. There is no restriction, however, that a replacement must be of lower cost than a deletion and an insertion (though this may often be the case in practice).

The cost of a sequence of edit operations will be the sum of the costs of the operations.

For any node $x = x_0$, with descendants $x_1, x_2, \ldots, x_k$, for $k \geq 0$, let

$$W_D(x) = \sum_{i=0}^{k} w_D(x_i),$$

and

$$W_1(x) = \sum_{i=0}^{k} w_1(x_i).$$

$W_D(x)$ and $W_1(x)$ are the costs of edit sequences deleting and inserting, respectively, all nodes of the subtree with root x.

Distance between trees

The distance between any two trees **a** and **b** is the cost of the minimum-cost sequence of edit operations that transform **a** to **b**. Let $d(\mathbf{a}, \mathbf{b})$ be the distance between **a** and **b**. Then if a is the root of **a** and has sons $a_1, a_2, \ldots, a_k$, and b is the root of **b** and has sons $b_1, b_2, \ldots, b_\ell$, then $d(\mathbf{a}, \mathbf{b})$ can immediately be bounded by

$$d(\mathbf{a}, \mathbf{b}) \leq w_R(a, b) + \sum_{i=1}^{k} W_D(a_i) + \sum_{j=1}^{\ell} W_1(b_j),$$

by considering the sequence deleting all nodes of **a**, except the root a, replacing a by b, and inserting all nodes of **b**, except the root.

In general $d(\mathbf{a}, \mathbf{b}) \neq d(\mathbf{b}, \mathbf{a})$. Because any sequence that transforms **a** to **c**, and then transforms **c** to **b**, is a sequence that transforms **a** to **b**,

$$d(\mathbf{a}, \mathbf{b}) \leq d(\mathbf{a}, \mathbf{c}) + d(\mathbf{c}, \mathbf{b})$$

Trace

A *replacement* of node x will signify either x or any node resulting from one or more replacement operations beginning with node x.

For each edit sequence, the nodes of **a** may be partitioned into two classes: $\mathbf{a}_1$, the set of nodes whose replacements are deleted in the edit sequence, and $\mathbf{a}_2$, the set of nodes whose replacements appear in **b**. Likewise, the nodes of **b** are partitioned into $\mathbf{b}_1$, the set of nodes that are replacements of nodes inserted in the edit sequence, and $\mathbf{b}_2$, the set of nodes of **b** that are replacements of nodes of $\mathbf{a}_2$.

For any edit sequence transforming **a** to **b**, a *trace* is a one-to-one mapping from $\mathbf{a}_2$ to $\mathbf{b}_2$ that maps each node onto its replacement. Any node included in $\mathbf{a}_2$ or $\mathbf{b}_2$ will be said to be *touched* by the trace.

If each of the nodes of $\mathbf{a}_1$ is deleted, in any order, from tree **a**, the resulting tree will be called the *subtree of* **a** *induced by the trace*. Likewise, if the nodes of $\mathbf{b}_1$ are deleted from **b**, the result is the subtree of **b** induced by the trace.

Modifications of Edit Sequences

Any edit sequence that has a node x that is inserted and later has a replacement deleted may be rewritten, to effect the same transformation, in such a way that x is never inserted. Insertions and deletions of the sons of the temporary node x may be respecified to be insertions and deletions of sons of the father of x. It is also possible to rewrite any edit sequence such that all deletions take place before any insertions. If, in the original sequence, y is inserted as a son of x, and a replacement of x is later deleted, the edit sequence in which x is first deleted and y is inserted as a son of the father of x will have the same effect.

Associated with each edit sequence, therefore, is an equivalent edit sequence (transforming **a** to **b**, with the same trace) $E = E_D E_I$, consisting of a segment E_D, which has no insertions, and a segment E_I, which has no deletions. Let E_I^r be the sequence E_I written in reverse order, and with each $R(x, y)$ changed to $R(y, x)$ and each $I(x)$ replaced by $D(x)$. Let **a**′ be the tree obtained from **a** by E_D. Let **b**′ be the tree obtained from **b** by the edit sequence E_I^r. It is evident that **a**′ = **b**′.

Preorder form of trace

Unless otherwise stated, the nodes of a tree henceforth will be identified by their position in the preorder sequence. The preorder sequence of a tree with m nodes is $1, 2, \ldots, m$. The subsequence $r, r + 1, \ldots, i$, where i is a descendant of r, will be the preorder of the subtree consisting of the path from r to i, and all descendants of r to the left of that path. Descendants of i are not included.

If **a** and **b** have m and n nodes, respectively, let $T(1, m:1, n)$ represent the class of traces from **a** to **b**. If $T \in T(1, m:1, n)$ and $(i, j) \in T$, then T represents an edit sequence in which node j in the preorder of **b** is a replacement of node i in the preorder of **a**.

Necessarily $(1, 1) \in T$ for all $T \in T(1, m:1, n)$, since insertion or deletion at the root are not valid edit operations. Also, wherever $(i, j), (k, \ell) \in T$, if $i < k$, then $j < \ell$. If the trace is represented, as in Fig. 3, as a set of links between the nodes of **a** and **b**, in preorder, the above condition states that the lines of the trace do not cross. If **a** is transformed to **b** by replacement operations only, then $T = \{(i, i)\}$ for $i = 1, \ldots, m$. Since neither insertions nor deletions modify the preorder sequence of the nodes not involved in the operation, nor do these operations add to the trace, no sequence of valid operations will ever cause the lines of the trace to cross.

More generally, suppose node i is either r or a descendant of node r in **a**. Then $i \geq r$ and the set of nodes in the sequence $r, r + 1, r + 2, \ldots, i$, with the father–son relation taken from **a**, is a tree. Likewise, suppose node j is either s or a descendant of node s in **b**. Then $j \geq s$, and $s, s + 1, \ldots, j$, with the father–son relation taken from **b**, is a tree. In this case $T(r, i:s, j)$ will represent the set of

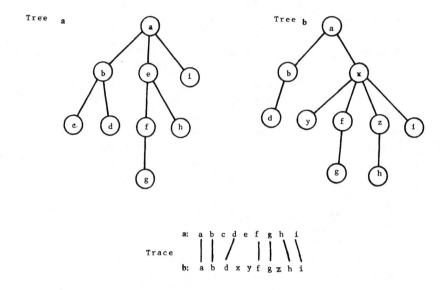

Figure 3. Preorder form of trace.

traces from tree $r, \ldots, i$ to tree $s, \ldots, j$. Necessarily, $(r, s) \in T$ for all $T \in T(r, i{:}s, j)$.

Following the above notation, when trees **a** and **b**, of m and n nodes, respectively, are described in preorder, we define

$$C(1, m{:}1, n) = d(\mathbf{a}, \mathbf{b})$$

to be the cost of the minimum-cost edit sequence transforming **a** to **b**.

3. INSERTIONS AND DELETIONS RESTRICTED TO LEAVES

We now consider the problem of finding the minimum-cost edit sequence when the edit operations of insertion and deletion are restricted to the leaves: Only leaves may be deleted, and a node may be inserted only as the son of a leaf.

Suppose the root of tree **a** had m subtrees. Let $i_1, i_2, \ldots, i_m$ be the last node of each subtree of the root, in the preorder sequence of **a**. Similarly, suppose the root of **b** has n subtrees, with $j_1, j_2, \ldots, j_n$ as the last node of each. Then if $(i, j) \in T$, for some $T \in T(1, i_m{:}1, i_n)$, then $(f(i), f(j))$ must also be in T and so must all pairs of ancestors of i and j, $(f^2(i), f^2(j)), \ldots, (1, 1)$. This is so because if no replacement of i in tree **a** was ever deleted, none of its ancestors could have been a leaf at any point in the edit sequence, and hence could not have been inserted or deleted. In particular, then, if nodes i and j are in subtrees p and q of the roots of **a** and **b**, respectively, then T must link the root of subtree

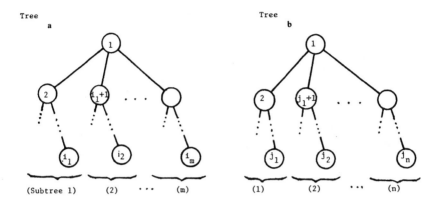

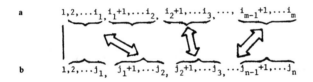

Figure 4. Form of trace when edit operations are restricted to leaves.

p to the root of subtree q. And therefore, if any other node in subtree p or q is touched by the trace, it must be linked to a node in subtree q or p, respectively.

Because the lines of the trace do not cross, if subtree p is linked to subtree q and subtree r is linked to subtree s, then $p < r$ implies $q < s$.

As a consequence, it must be possible to partition any trace into subtraces corresponding to each of the pairs of subtrees that are linked. A typical case is shown in Fig. 4.

Suppose $T \in T(1, i_k : 1, j_\ell)$ is the minimum-cost trace linking nodes up to subtrees k and ℓ. By definition, T has cost $C(1, i_k : 1, j_\ell)$. But T can be decomposed, depending on which of the following cases hold.

1. T touches no nodes of subtree ℓ. In this case

$$C(1, i_k : 1, j_\ell) = C(1, i_k : 1, j_{\ell-1}) + W_1(j_{\ell-1} + 1).$$

2. T touches no nodes of subtree k. In this case

$$C(1, i_k : 1, j_\ell) = C(1, i_{k-1} : 1, j_\ell) + W_D(i_{k-1} + 1).$$

3. T touches both subtrees k and ℓ. Then these subtrees must be linked to each other. Then T may be partitioned into $T_1 \in T(1, i_{k-1}:1, j_{\ell-1})$ and $T_2 \in T(i_{k-1} + 1, i_k:j_{\ell-1} + 1, j_\ell)$. T_1 and T_2 must be minimum-cost traces, and the cost of T is the sum of their costs:

$$C(1, i_k:1, j_\ell) = C(1, i_{k-1}:1, j_{\ell-1}) + C(i_{k-1} + 1, i_k:j_{\ell-1} + 1, j_\ell).$$

This formula also holds for the case where T touches neither subtree i_k nor j_ℓ.

The minimum cost of all traces is determined by whichever of the above cases yields the minimum cost,

$$C(1, i_k:1, j_\ell) = \min \begin{cases} C(1, i_k:1, j_{\ell-1}) + W_1(j_{\ell-1} + 1), \\ C(1, i_{k-1}:1, j_\ell) + W_D(i_{k-1} + 1), \\ C(1, i_{k-1}:1, j_{\ell-1}) + C(i_{k-1}, i_k:j_{\ell-1} + 1, j_\ell). \end{cases}$$
(1)

Since $T_2 \in T(i_{k-1} + 1, i_k:j_{\ell-1} + 1, j_\ell)$ is some trace from subtree k to subtree ℓ, it will be possible to determine $C(i_{k-1} + 1, i_k:j_{\ell-1} + 1, j_\ell)$ by the same rule as for $C(1, i_k:1, j_\ell)$. This forms the basis for a recursive algorithm for determining $C(1, i_m:1, j_n)$, the distance between trees **a** and **b**.

4. AN ALGORITHM FOR TREE DISTANCE WHEN INSERTIONS AND DELETIONS ARE RESTRICTED TO LEAVES

We define $E(r, i:s, j)$ to be the cost of the minimum cost $T(r, i:s, j)$, exclusive of the root replacement cost $w_R(r, s)$. Then

$$C(r, i:s, j) = E(r, i:s, j) + w_R(r, s).$$

Following the above, E may be written as a recursive function.

Function $E(r, i:s, j)$

Let $i_0 = r$, let $k \geq 0$ be the number of subtrees of r in $r, \ldots, i$, and let $i_1, i_2, \ldots, i_k$, with $i_k = i$, be the last node of each subtree.

Let $j_0 = s$, let $\ell \geq 0$ be the number of subtrees of s in $s, \ldots, j$, and let $j_1, j_2, \ldots, j_\ell$, with $j_\ell = j$, be the last node of each subtree.

begin

> if $r = i$ and $s = j$ then $E := 0$
> else if $s = j$ then $E := E(r, i_{k-1}{:}s, j) + W_D(i_{k-1} + 1)$
> else if $r = i$ then $E := E(r, i{:}s, j_{\ell-1}) + W_I(j_{\ell-1} + 1)$
> else $E := \min\{E(r, i_{k-1}{:}s, j) + W_D(i_{k-1} + 1),$
> $\qquad\qquad E(r, i{:}s, j_{\ell-1}) + W_I(j_{\ell-1} + 1),$
> $\qquad\qquad E(r, i_{k-1}{:}s, j_{\ell-1} + 1)$
> $\qquad\qquad\qquad + w_R({}^i{}_{k-1} + 1, j_{\ell-1} + 1)$
> $\qquad\qquad\qquad + E(i_{k-1} + 1, i_k{:}j_{\ell-1} + 1, j_\ell)\}$

end.

5. THE CASE OF GENERAL INSERTIONS AND DELETIONS

When insertions and deletions are allowed at any node of a tree (except the root), the minimum-cost trace will in general not be decomposable into minimum-cost subtraces. For example, consider the transformation of tree **a** to tree **b** of Fig. 5, with costs $w_D(x) = w_I(x) = 1$ and $w_R(x, y) = 0$ for any nodes x and y.

The edit sequence of minimum cost will have a cost of four, for the deletions of nodes 2 and 6, and their reinsertions as nodes 2 and 9. But any minimum-cost prefix trace $T(1, x{:}1, y)$, with $2 \le x, y \le 6$, will always include the element $(2, 2)$. This precludes the possibility of achieving a minimum-cost complete trace by extending minimum-cost prefix traces.

Suppose $T \in T(1, m{:}1, n)$ is the minimum-cost trace, and is known to include an element (i, j), $1 < i \le m$, $1 < j \le n$. Then T may be partitioned into traces $T_1 \in T(1, i - 1{:}1, j - 1)$ and $T_2 \in T(i, m{:}j, n)$. But the cost of T may not be assumed to be the sum of the minimum costs of traces of the form of T_1 and T_2 since not all $T(1, i - 1{:}1, j - 1)$ may be extended by the inclusion of (i, j).

A simple case is shown in Fig. 6. The trace $\{(1, 1), (2, 2)\} \in T(1, 2{:}1, 2)$ may not be extended by the addition of $(3, 3)$. The subtrees induced by the extended mapping are not isomorphic, so that mapping is not a trace.

A trace $T \in T(1, i - 1{:}1, j - 1)$ will be extendable by the inclusion of (i, j) if it contains the element $(f(i), f(j))$. For if the fathers of i and j appear in the isomorphic trees induced by T, then the induced subtrees will still be isomorphic when nodes i and j are appended as the rightmost sons of $f(i)$ and $f(j)$.

More generally, $T \in T(1, i - 1{:}1, j - 1)$ is extendable to $T' \in T(1, i{:}1, j)$ if it contains an element (r, s), where r is an ancestor of i, and s is an ancestor of j and no descendant of r on the path from r to i, nor any descendant of s on the path from s to j, is touched by T. In this case, since all intervening nodes do not appear, nodes r and s will play the roles of $f(i)$ and $f(j)$ in the isomorphic

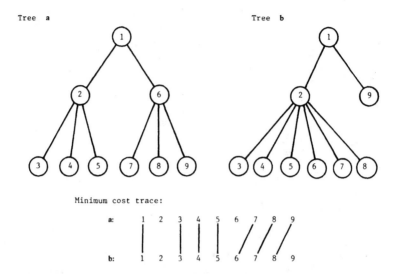

Figure 5. A minimum-cost trace which cannot be achieved by extending a minimum-cost prefix trace in $T(1, x; 1, y)$ with $2 \leq x, y \leq 6$.

subtrees obtained from T. Then i and j may again be appended as the rightmost descendants of r and s.

Suppose i is a uth descendant of r in tree **a**, for $u \geq 0$. Let $r_{i,0}, r_{i,1}, \ldots, r_{i,u}$ be the path from $r = r_{i,0}$ to $i = r_{i,u}$. Suppose j is a vth descendant of s in tree **b**, and $s_{j,0}, s_{j,1}, \ldots, s_{j,v}$ is the path from $s = s_{j,0}$ to $j = s_{j,v}$. For any $0 \leq k \leq u$ and any $0 \leq \ell \leq v$, let $T(r, r_{i,k}, i{:}s, s_{j,\ell}, j)$ be the class of all traces from $r, \ldots, i$ of **a** to $s, \ldots, j$ of **b** that contain (r, s) but do not touch any descendant of r in $r_{i,1}, r_{i,2}, \ldots, r_{i,k}$, nor any descendant of s in $s_{j,1}, s_{j,2}, \ldots, s_{j,\ell}$. The class $T(r, r_{i,k}, i : s, s_{j,\ell}, j)$ will be called a *restricted* class of traces, and its elements will be restricted traces.

Let $C(r, r_{i,k}, i{:}s, s_{j,\ell}, j)$ be the cost of the minimum-cost trace of the restricted traces $T(r, r_{i,k}, i{:}s, s_{j,\ell}, j)$. If trees **a** and **b** have m and n nodes, respectively, $T(1, 1, m{:}1, 1, n) = T(1, m{:}1, n)$, and $C(1, 1, m{:}1, 1, n) = C(1, m{:}1, n)$ is the minimum cost being sought.

A few special cases of the restricted trace are of interest. First, consider any trace between single-node subtrees. $T(r, r, r{:}s, s, s)$ has, by definition, minimum cost

$$C(r, r, r{:}s, s, s) = w_R(r, s). \qquad (2)$$

Next consider $T(r, i, i{:}s, s_{j,\ell}, j)$, with $r < i$. By definition of the restricted trace, it is seen that i may not be touched by any such trace. Hence, node i must be deleted in every edit sequence with this trace. The remainder of the trace, between $r, \ldots, i - 1$ and $s, \ldots, j$ is also restricted, since no descendant of r on

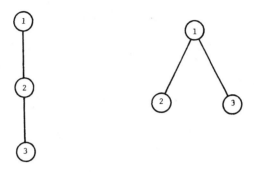

Figure 6. The trace $\{(1, 1), (2, 2)\}$ may not be extended by addition of $(3, 3)$.

the path from r to $f(i)$ may be touched by the trace. (If $f(i) = r$, this condition is vacuous.) And $f(i) \geq r$ is either $i - 1$ or an ancestor of $i - 1$, hence, is on the path from r to $i - 1$. Therefore, we may write

$$C(r, \; i, \; i{:}s, \; s_{j,\ell}, j) = C(r, \; f(i), \; i - 1{:}s, \; s_{j,\ell}, j) + w_{\mathrm{D}}(i). \qquad (3)$$

Next, consider restricted traces between a general subtree and a single-node subtree, $T(r, \; r_{i,k}, \; i{:}s, \; s, \; s)$ with $r < i$. Since $(r, \; s)$ must be in every such trace, i cannot be touched by any such trace. In fact, no node of $r + 1, \ldots, i$ may be touched by the trace. Hence, we may again write

$$C(r, \; r_{i,k}, \; i{:}s, \; s, \; s) = C(r, \; f(i), \; i - 1{:}s, \; s, \; s) + w_{\mathrm{D}}(i). \qquad (4)$$

By symmetrical arguments, we may write, for $s < j$,

$$C(r, \; r_{i,k}, \; i{:}s, \; j, \; j) = C(r, \; r_{i,k}, \; i{:}s, \; f(j), j - 1) + w_{\mathrm{I}}(j), \qquad (5)$$

and

$$C(r, \; r, \; r{:}s, \; s_{j,\ell}, j) = C(r, \; r, \; r{:}s, \; f(j), j - 1) + w_{\mathrm{I}}(j). \qquad (6)$$

For the general case, consider the minimum cost $T \in T(r, \; r_{i,k}, \; i{:}s, \; s_{j,\ell}, j)$, where $r \leq r_{i,k} < i$ and $s \leq s_{j,\ell} < j$. Then either T does not touch $r_{i,k+1}$, or it does not touch $s_{j,\ell+1}$, or else it contains the element $(r_{i,k+1}, s_{j,\ell+1})$.
If the first case holds

$$T \in T(r, \; r_{i,k+1}, \; i{:}s, \; s_{j,\ell}, j),$$

and if the second case holds

$$T \in T(r, \; r_{i,k}, \; i{:}s, \; s_{j,\ell+1}, j).$$

An example of the last case is shown in Fig. 7. Since $(r_{i,k+1}, s_{j,\ell+1}) \in T$, it is possible to partition T into traces

$$T_1 \in T(r, \; r_{i,k+1} - 1{:}s, \; s_{j,\ell+1} - 1)$$

and

$$T_2 \in T(r_{i,k+1}, \; i{:}s_{j,\ell+1}, j).$$

But T_1 is seen to be a member of a restricted class. The node $r_{i,k}$ is on the path from r to $r_{i,k+1} - 1$ as well as on the path to i, since $r_{i,k+1} - 1$ is either $r_{i,k}$ or the rightmost node of a subtree of $r_{i,k}$. Similarly, $s_{j,\ell}$ is on the path to $s_{j,\ell+1} - 1$. Therefore, one may write

$$T_1 \in T(r, \; r_{i,k}, \; r_{i,k+1} - 1{:}s, \; s_{j,\ell}, \; s_{j,\ell+1} - 1).$$

Furthermore, any member of the class $T(r, \; r_{i,k}, \; r_{i,k+1} - 1{:}s, \; s_{j,\ell}, \; s_{j,\ell+1} - 1)$ is extendable by the inclusion of $(r_{i,k+1}, s_{j,\ell+1})$. The only ancestors of $r_{i,k+1}$ and $s_{j,\ell+1}$ touched by any such trace are r and s, respectively, and these are linked.

The cost of T will be the sum of the costs of T_1 and T_2, and each of these will be the minimum-cost trace of its class.

Taking all three cases together, the minimum cost is expressed by

$$C(r, \; r_{i,k}, \; i{:}s, s_{j,\ell}, j) = \min \begin{cases} C(r, \; r_{i,k+1}, \; i{:}s, \; s_{j,\ell}, j), \\[2mm] C(r, \; r_{i,k}, \; i{:}s, \; s_{j,\ell+1}, j), \\[2mm] C(r, \; r_{i,k}, \; r_{i,k+1} - 1{:}s, \; s_{j,\ell}, \; s_{j,\ell+1} - 1) \\[2mm] + C(r_{i,k+1}, \; r_{i,k+1}, \; i{:}s_{j,\ell+1}, \; s_{j,\ell+1}, j). \end{cases}$$

$$(7)$$

6. AN ALGORITHM FOR TREE-EDIT DISTANCE IN THE CASE OF GENERAL INSERTIONS AND DELETIONS

A recursive algorithm for the computation of $C(1, m{:}1, n) = C(1, 1, m{:}1, 1, n)$ is straightforward. The above relations (2) through (7) are used. The algorithm is defined as a recursive function, for the distance between trees $r, \ldots, i$ and $s, \ldots, j$. The arguments x and y must be nodes on the paths from r to i and s to j, respectively.

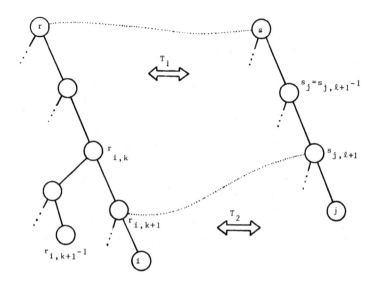

Figure 7

Function $C(r, x, i{:}s, y, j)$

Let $r_{i,0}, \ldots, r_{i,u}$, for $u \geq 0$, be the path from $r = r_{i,0}$ to $i = r_{i,u}$ in tree **a**, and let $x = r_{i,k}$.
    Let $s_{j,0}, \ldots, s_{j,v}$, for $v \geq 0$, be the path from $s = s_{j,0}$ to $j = s_{j,v}$ in tree **b**, and let $y = s_{j,\ell}$.

begin

$\quad$ if $r = i$ and $s = j$ then
$\qquad\qquad C := w_R(i, j)$
$\quad$ else if $x = i$ or $s = j$ then
$\qquad\qquad C := C(r, r_{i,k-1}, i-1{:}s, s_{j,\ell}, j) + w_D(i)$
$\quad$ else if $r = i$ or $y = j$ then
$\qquad\qquad C := C(r, r_{i,k}, i{:}s, s_{j,\ell-1}, j-1) + w_I(j)$
$\quad$ else

$\qquad\qquad C := \min\{ C(r, r_{i,k+1}, i{:}s, s_{j,\ell}, j),$
$\qquad\qquad\qquad C(r, r_{i,k}, i{:}s, s_{j,\ell+1}, j),$
$\qquad\qquad\qquad C(r, r_{i,k}, r_{i,k+1} - 1{:}s, s_{j,\ell}, s_{j,\ell+1} - 1)$
$\qquad\qquad\qquad + C(r_{i,k+1}, r_{i,k+1}, i{:}s_{j,\ell+1}, s_{j,\ell+1}, j)\}$

end.

7. CONCLUSIONS

A version of the tree-edit algorithm in which insertions and deletions are restricted to the leaves of a tree has been shown by Selkow (1977). That version of the algorithm combines iteration and recursion. It is also shown there that if n_1 and n_2 are the number of nodes of the two trees, and h_1 and h_2 are the depths of trees, with $h = \min(h_1, h_2)$, the algorithm will require $\mathcal{O}(n_1 n_2 h)$ steps to complete.

An iterative version of the tree-edit algorithm has been shown by Tai (1979). The iterative version computes the values of C as the (function) procedure of Sec. 6 does, but stores partial results in an array. The sequence in which the partial results are computed in order to bypass the recursiveness are shown in the first section of the algorithms given in that paper. (The solution shown here makes the latter sections unnecessary.) The analysis of the general tree-edit problem shows that $\mathcal{O}(n_1 n_2 h_1^2 h_2^2)$ steps are required for completion.

REFERENCES

Selkow, S.M., The tree-to-tree editing problem. *Information Processing Letters* **6**, 184–186 (1977).

Tai, K.C., The tree-to-tree correction problem. *Journal of the Association for Computing Machinery* **26** 422–433 (1979).

CHAPTER NINE

SIMULTANEOUS COMPARISON OF THREE OR MORE SEQUENCES RELATED BY A TREE

David Sankoff and Robert J. Cedergren

1. INTRODUCTION

In the analysis of molecular evolution, we generally consider N sequences at a time, where $N > 2$. The study of the relationships among N sequences is a large and difficult problem, and here we consider but one part of it: the case where we are given not only the sequences characteristic of N species, but also the phylogenetic tree representing the evolutionary history of these species. This tree is presented as in Fig. 1, where each of the N terminal nodes corresponds to a given, or observed, sequence, and each of the M interior nodes corresponds to an older sequence that can only be inferred. For reasons that will soon be clear, we assume that each interior node has at least three edges coming out from it. Somewhere along one of the edges of the tree is the point corresponding to the unknown ancestral sequence that gave rise to all the observed sequences. Evolutionary time is mirrored by movement along the edges, moving outward from the location of the ancestral sequence. The differences among the sequences are considered to be due to a number of substitutions, insertions, and deletions of sequence terms, each of which is presumed to have occurred at some specific point along some edge. Each interior node corresponds to an evolutionary point where one sequence gave rise to at least two different observed descendant sequences, due to separate evolution in different populations.

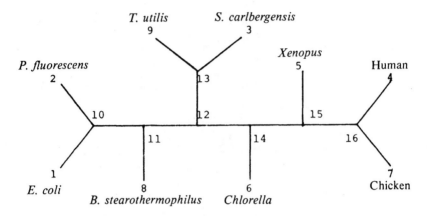

Figure 1. Phylogenetic relationship, or family tree diagram, of nine species. Numbered vertices represent hypothetical ancestral organisms. The "root" vertex, representing the first cellular organism, could probably be placed between vertices 11 and 12. Time is considered to elapse along the tree branches in a direction away from the root.

Given the tree and the observed sequences at the terminal nodes, we ask two questions:

1. What is the total amount of change that must have occurred in the tree?
2. To what extent can we reconstruct the unknown sequences corresponding to the interior nodes?

As a way of estimating the amount of change that occurred in the tree, we use the least possible amount of change consistent with the observed sequences. The reasons for choosing this *parsimony criterion* are two-fold: First, since changes are rare events, the fewer of them needed, the more plausible the reconstruction; and second, this choice leads to a tractable method. (The increase in N enriches the possibilities for alternative methods of estimation, but we do not pursue them here.)

This leads us to the problem of finding the smallest set (or, more generally, the lowest-cost set) of substitutions, insertions, and deletions that can explain the observed sequences. We have no way of estimating the order or location of changes *within* an edge, so we ask no more than to associate the changes with edges of the tree. (This is why each interior node is assumed to have at least three edges.) Such association is feasible to a considerable extent, and gives a substantial reply to the second question above, since if we know what change occurred on what edges we can also reconstruct the sequences as they existed at the nodes of the tree.

If there are N observed sequences and M interior nodes, and if all the observed sequences have length n, then the algorithm described here (due to

Sankoff (1975); Sankoff *et al.* (1973, 1976)) requires time approximately equal to $(2n)^N Mt_1$, where t_1 is the time required to carry out a fairly simple procedure that will be described later. For $N = 3$, this is quite feasible for (say) RNA sequences of length about 120, but the exponential growth with N leads to prohibitive computing cost for $N = 5$ or more. However, large trees, particularly those where each interior node has exactly three edges coming out of it, can be decomposed into sets of overlapping subtrees where the $N = 3$ version of the algorithm may be applied. By repeated application of this version to the subtrees in an appropriate order, an iterative approximation procedure based on local optimization can give the correct, or almost correct solution, to the problem in reasonable time.

2. TREE ALIGNMENTS

One way to formalize these ideas is in terms of *tree alignments*. Suppose we are given a tree, like the one in Fig. 2(a), together with a sequence, from some alphabet, at every node, both terminal and interior. (Ordinarily, only the sequences at terminal nodes are known, or given, while sequences at interior nodes are to be reconstructed.) Suppose we pad all the sequences out to the same length by inserting nulls (ϕ's) at selected positions, and then put the sequences together into a matrix A, using each sequence as one row [see Fig. 2(b)]. Each row corresponds to one node of the tree. The combination of

   i) matrix,
  ii) tree, and
 iii) correspondence between rows and nodes

is called a *tree alignment*. To avoid meaningless padding, we impose the condition that no column of the matrix consist entirely of ϕ's. For convenience, we shall place the rows containing observed sequences in the upper part of the matrix, and the rows containing reconstructed sequences in the lower part of the matrix.

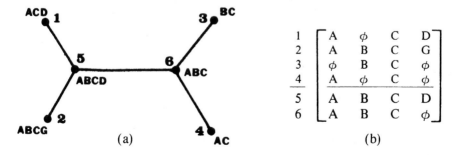

Figure 2. Tree of sequences and tree alignment matrix.

In effect, a tree alignment is just a coordinated collection of ordinary alignments (i.e., alignments of just two sequences), one for each edge of the tree. Given an edge e of the tree, we take the two rows corresponding to the nodes at the ends of e, and put one above the other to form an ordinary two-row alignment, $a(e)$. This alignment implicitly displays what changes take place along the edge. A column in $a(e)$ of the form $[\begin{smallmatrix} x \\ x \end{smallmatrix}]$ indicates that no change has taken place along e in the corresponding element of the sequence; a column of the form $[\begin{smallmatrix} x \\ y \end{smallmatrix}]$ shows that x has been substituted by y along e, or vice versa (depending on which end of e is nearer to the location of the ancestral sequence); and a column of the form $[\begin{smallmatrix} x \\ \phi \end{smallmatrix}]$ or $[\begin{smallmatrix} \phi \\ x \end{smallmatrix}]$ indicates that x has been deleted or inserted along e.

Let the *extended alphabet* be the given alphabet together with the null character ϕ. Suppose that weights $w(x, y)$ are given for all x and y in the extended alphabet, and suppose that

   i) $w(x, x) = 0$ for all x,
   ii) $w(x, y) > 0$ if $x \neq y$, and
   iii) $w(x, y) = w(y, x)$

Then the *length* of any ordinary alignment

$$a(e) = \begin{bmatrix} x_1 \ldots x_p \\ y_1 \ldots y_p \end{bmatrix}$$

is defined as

$$\sum_{i=1}^{p} w(x_i, y_i).$$

(For fuller discussion, see the section on Levenshtein and other distances in Chapter 1). The *length* of a tree alignment is defined as the sum of the lengths of all the ordinary alignments of which it is composed:

$$\text{length of tree alignment} = \sum_{e \text{ in tree}} \text{length}(a(e)).$$

Given a tree and observed sequences at its terminal nodes, we define the *tree distance* among the sequences to be the length of the shortest possible tree alignment connecting them. Note that minimization of tree length involves choosing

   1. a reconstructed sequence at each interior node of the tree, and
   2. positions in all the sequences (both observed and reconstructed) at which ϕ's are to be inserted.

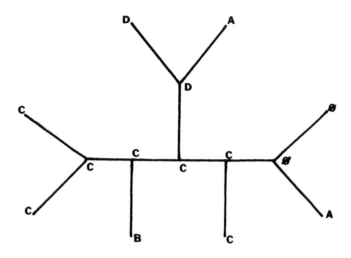

Figure 3. Diagram of one column of a tree-alignment matrix.

An algorithm to perform this minimization is the subject of this chapter.

Because all sequences are padded to the same length, the tree alignment provides an explicit correspondence between all the sequences involved. A single column of the matrix shows which elements or gaps (ϕ's) correspond to each other across all the sequences at once, and corresponds to a diagram like that shown in Fig. 3. For each edge of the diagram, if the same element is shown at both ends, then no change is presumed to have occurred along that edge, while if different nonnull elements occur at the ends, a single substitution is presumed. If one end has a null and the other end a nonnull element, an insertion or deletion is presumed. A tree alignment can be thought of as a series of such diagrams, one for each column of the matrix (all based on the same tree, of course).

We define the length of a column c of A in the obvious way, namely, as the sum of the weights of all the corresponding columns from the ordinary alignments $a(e)$:

$$\text{length of } c = \sum_{e \text{ in tree}} \text{weight of column } c \text{ in } a(e).$$

Then the length of the tree alignment is simply the sum of the lengths of the columns of A:

length of tree alignment $= \sum\limits_{e \text{ in tree}} \text{length } (a(e))$

$$= \sum_{e} \left[\sum_{c \text{ in } A} \text{weight of column } c \text{ in } a(e) \right]$$

$$= \sum_{c} \text{length of } c.$$

Finally, note that the length of c is just the length of a tree like that in Fig. 3.

3. AN ALGORITHM TO CALCULATE TREE DISTANCE

The distance between two sequences **a** and **b** is defined to be the shortest length of any alignment of **a** and **b**. The basic dynamic-programming algorithm to find this distance can be described in terms of the values

$$d_{ij} = \text{distance between } \mathbf{a}^i \text{ and } \mathbf{b}^j,$$

where $\mathbf{a}^i$ and $\mathbf{b}^j$ are the initial segments of **a** and **b** of length i and j. The dynamic-programming recurrence can then be written

$$d_{ij} = \min \begin{cases} d_{i-1,j} & + w(a_i, \phi), \\ d_{i-1,j-1} + w(a_i, b_j), \\ d_{i,j-1} & + w(\phi, b_j). \end{cases}$$

Note that each d on the right-hand side has the form

$$d_{i-\Delta_1, j-\Delta_2},$$

where:

Each Δ is in $\{0,1\}$ and $\Delta_1 + \Delta_2 \neq 0$.

In addition, note that each w on the right-hand side has the form

$$w(\Delta_1 a_i, \Delta_2 b_j),$$

where

$$\Delta x = \begin{cases} \phi & \text{if } \Delta = 0, \\ x & \text{if } \Delta = 1. \end{cases}$$

Using the new notation, the recurrence can be written

$$d_{ij} = \min_{\Delta_1 + \Delta_2 \neq 0} [d_{i-\Delta_1, j-\Delta_2} + w(\Delta_1 a_i, \Delta_2 b_j)].$$

The minimizing $\Delta_1 a_i$ and $\Delta_2 b_j$ then form the final column

$$\begin{vmatrix} \Delta_1 a_i \\ \Delta_2 b_j \end{vmatrix}$$

of some minimal-length alignment of a^i and b^j.

Suppose that the observed sequences at the N terminal nodes of the tree are **a, b, . . . , c**. A dynamic-programming algorithm can be described in terms of the N-dimensional array of quantities

$$d_{ij...k} = \text{tree distance among } a^i, b^j, \ldots, c^k.$$

To develop the recurrence equation, consider a minimum-length tree alignment, and suppose its matrix is A. Recall that the upper part of A corresponds to the N observed sequences, and the lower part corresponds to the M reconstructed sequences at interior nodes of the tree.

Let the final column of A be c. Then the upper part of c must be one of $2^N - 1$ alternatives:

$$\begin{vmatrix} \Delta_1 a_i \\ \Delta_2 b_j \\ \cdot \\ \cdot \\ \cdot \\ \Delta_N c_k \end{vmatrix}, \qquad \text{where } \Delta_1 + \Delta_2 + \ldots + \Delta_N \neq 0.$$

(The only aspect of this statement that does not follow immediately from the definition of a tree alignment is the restriction that $\Delta_1 + \ldots + \Delta_N \neq 0$. To justify this restriction, note that if the upper part of c is entirely null, then deleting c from A yields a new tree-alignment matrix. By positivity of the weights and the restriction that no column can be entirely null, the new tree-alignment matrix has smaller length than A, which contradicts the definition of A.)

At this point, we face a problem that did not exist in the basic algorithm, namely, the need to assign letters (from the extended alphabet) in the lower part of c. Given the upper part of c, we need to find entries $x_1, \ldots, x_M$ in the lower part that minimize the length of c. In other words, given an assignment of a letter from the extended alphabet to each terminal node of the tree, we need to assign a letter from the extended alphabet to each interior node so as to minimize the length of the tree. Then the recurrence equation can be written in this form:

$$d_{ij\ldots k} = \min_{\Delta_1 + \ldots + \Delta_N \neq 0} [d_{i-\Delta_1, j-\Delta_2, \ldots, k-\Delta_N} + \min_{x_1, \ldots, x_M} (\text{length of} \begin{vmatrix} \Delta_1 a_i \\ \cdot \\ \cdot \\ \Delta_N c_k \\ x_1 \\ \cdot \\ \cdot \\ x_M \end{vmatrix})].$$

4. A DYNAMIC-PROGRAMMING ALGORITHM
FOR THE INNER MINIMIZATION

Note that the inner minimization in the recurrence must be carried out $2^N - 1$ times for each $d_{ij\ldots k}$ in the N-dimensional array of values. Thus it is fortunate that this minimization need not be done by exhaustive search. A rapid dynamic-programming algorithm has been proposed on several occasions (Fitch, 1971; Hartigan, 1973; Sankoff and Rousseau, 1975). The algorithm has computing time proportional to M, since it consists primarily of certain operations carried out at each interior node.

In one special case, the number of such operations that are necessary can be drastically reduced, and the algorithm takes on an abbreviated form. This case is the one where every substitution, insertion, and deletion has weight 1. We describe the special algorithm first, and then the general case.

Before either version of the algorithm, there is a preliminary step that needs to be done only once for the phylogenetic tree. Note that due to the symmetry of the weight function w, it makes no difference in the calculation of the length of an edge alignment $a(e)$ which of the two endpoints of e is considered to represent an earlier point in time. Thus a solution to the minimization problem as we have stated it will not depend on the location of the ancestral sequence in the tree.

Nevertheless, the algorithm requires that one interior node of the tree be selected as the root; we make this choice arbitrarily. Consider every edge of the tree as directed away from the root (i.e., each edge is given a direction that can be indicated by an arrowhead on the edges, and these directions are such that every path that starts from the root follows the edges in the direction in which they point.) For each interior node v, we call the nodes that can be reached (in one step) by following an edge in the proper direction the *children* of v (there are at least two such nodes), and the single node that can be reached by following an edge in the wrong direction the *parent* of v. (The root is the only node without a parent.) A node v, together with its children, their children, and so on, forms a subtree that is said to *hang* fom v.

To complete the preliminary step, select an ordering of the nodes for use in the recursive procedure. We want the order to be such that, before any node v is reached, the children of v have already been reached. We start by selecting any interior node whose children are all terminal nodes. At each subsequent step, we select a node whose children are either terminal nodes or interior nodes that were previously selected. When this procedure is complete, the root is necessarily the final node.

4.1 The special case algorithm

Now we describe the algorithm itself, for the special case where every operation has weight 1. Recall that the input is a tree with a letter (from the extended alphabet) assigned to each terminal node. The output is an assignment of letters to the interior nodes so as to minimize the number of operations, i.e., edges containing a substitution, insertion, or deletion.

The algorithm has three stages: initialization of the terminal nodes; a formal recursion through the interior nodes (in the order selected above) to find the minimum possible number of operations; and a backward recursion ("backtracking," in reverse order) to find the minimizing assignment itself. For each node v, the algorithm makes use of a *tree length* $L(v)$, a *primary set* $P(v)$, and a *secondary set* $S(v)$. The primary and secondary sets are nonoverlapping subsets of the extended alphabet, and indicate letters that may, under appropriate circumstances, be used in an optimum assignment. The primary set is never empty, but the secondary set may be. The tree length $L(v)$ is the shortest possible length of the subtree that hangs from v, using any assignment of letters from the extended alphabet to interior nodes in the subtree, and the given assignment of letters to terminal nodes. Thus $L(\text{root})$ is the minimum length we are looking for. The secondary sets $S(v)$ are used only during the backtracking, and are not needed if we are interested in finding only the minimal length possible for a tree, and not the actual assignments at the internal nodes.

To initialize the recursive process, we consider each terminal node v, and set $L(v) = 0$, $P(v)$ to consist of the one letter that was assigned to v, and $S(v)$ to be empty.

The recursive step at each interior node v is as follows. For each letter a in the extended alphabet, define

$n(a) = $ the number of children v' of v for which $P(v')$ contains a.

Define

$$m = \max_a n(a).$$

Then the recurrence equation for tree length is given by

$$L(v) = \sum_{\substack{v' \text{ child} \\ \text{of } v}} (1 + L(v')) - m.$$

Define the primary set and secondary sets by

$$P(v) = \text{the set of all } a \text{ such that } n(a) = m,$$
$$S(v) = \text{the set of all } a \text{ such that } n(a) = m - 1.$$

When the forward recursion is complete, $L(\text{root})$ gives the minimum possible length for any assignment of letters to the interior nodes, where the terminal nodes have the assignment that was given during initialization. If desired, an optimum assignment of letters to the interior nodes can be obtained by backtracking. (Every possible optimum assignment can be obtained by this procedure.) First, assign to the root any element of $P(\text{root})$. Then proceed recursively along the nodes in reverse order. To assign a letter at node v, consider what letter a has been selected at the parent of v. If a belongs to $P(v)$, select a. If a belongs to $S(v)$, select either a or any element of $P(v)$. If a belongs to neither $P(v)$ nor $S(v)$, select any element of $P(v)$.

4.2 The general-case algorithm

The algorithm for the general case of arbitrary weights w follows the same general approach, but the structure attached to each node v is more elaborate. For each interior node v, we calculate a function $f_v(a)$, which has numerical values, where a ranges over the extended alphabet. If an optimum assignment is to be found by backtracking, then a subset of the extended alphabet, $M_v(a)$, is also needed at v.

Consider a node v and the subtree that hangs from it. We define $f_v(a)$ to be the minimum possible length of the subtree under any assignment of letters such that a is assigned to v (consonant, of course, with the given assignment of letters to terminal nodes). Thus $\min_a f_{\text{root}}(a)$ is the minimum length we are looking for.

Now we describe the general recursive process. To initialize, consider each terminal node v, and suppose that $c(v)$ is the letter assigned to v. Define

$$f_v(a) = \begin{cases} 0 & \text{if } a = c(v), \\ \infty & \text{otherwise.} \end{cases}$$

Then the interior nodes are run through in the order described above, and the following calculations are made at each node v. First, calculate

$$f_v(a) = \sum_{\substack{v' \text{ child} \\ \text{of } v}} \min_b \left[f_{v'}(b) + w(a, b) \right] \qquad \text{for every } a.$$

Then, to prepare for backtracking, let

$M_{v'}(a)$ = the set of letters b for which $f_{v'}(b) + w(a, b)$ is minimal.

After the recursion is complete, the desired minimum length is $\min_a f_{\text{root}}(a)$. If an optimum assignment of letters is desired, it may be found by backtracking as follows. (Every possible optimum assignment can be found in this way.) First, assign to the root any letter b for which $f_{\text{root}}(b)$ achieves its minimum. Then proceed through the interior nodes in the reverse order, excluding the root. At node v, consider what letter a has previously been assigned to the parent of v, and assign to v any element of $M_v(a)$.

What is the connection between the special-case algorithm and the general one? Setting $w(x, y) \equiv 1$ for $x \neq y$ in the extended alphabet, we can show that

$$\left.\begin{aligned} f_v(a) &= L(v) \\ M_v(a) &= \{a\} \end{aligned}\right\} \quad \text{for } a \text{ in } P(v),$$

$$\left.\begin{aligned} f_v(a) &= L(v) + 1 \\ M_v(a) &= P(v) \cup \{a\} \end{aligned}\right\} \quad \text{for } a \text{ in } S(v)$$

for each nonterminal node v in the tree, and that all solutions to the minimization problem can be found without taking into account $f_v(b)$ or $M_v(b)$ for any b not in $P(v)$ or $S(v)$.

REFERENCES

Fitch, W.M., Towards defining the course of evolution: minimum change for a specific tree topology. *Systematic Zoology* **20**, 406–416 (1971).

Hartigan, J.A., Minimum mutation fits to a given tree. *Biometrics* **29**, 53–65 (1973).

Sankoff, D., Minimal mutation trees of sequences. *SIAM Journal on Applied Mathematics* **78**, 35–42 (1975).

Sankoff, D., Cedergren, R.J., and Lapalme, G., Frequency of insertion–deletion, transversion, and transition in the evolution of 5S ribosomal RNA. *Journal of Molecular Evolution* **7**, 133–149 (1976).

Sankoff, D., Morel, C., and Cedergren, R.J., Evolution of 5S RNA and the nonrandomness of base replacement. *Nature New Biology* **245**, 232–234 (1973).

Sankoff, D., and Rousseau, P., Locating the vertices of a Steiner tree in an arbitrary metric space. *Mathematical Programming* **9**, 240–246 (1975).

CHAPTER TEN

AN ANTHOLOGY OF ALGORITHMS AND CONCEPTS FOR SEQUENCE COMPARISON

Joseph B. Kruskal and David Sankoff

Basic to the literature on sequence comparison is the concept of distance from one sequence to another and algorithms for calculating the distance. Distance means the shortest way of converting one sequence into the other using substitutions and *indels* (insertions and deletions). The basic algorithms are time-quadratic dynamic-programming algorithms, like the one described fully in Chapter 1 and summarized briefly here in Figs. 1(a) and 1(b). This chapter is devoted to variations on the basic algorithm that solve modified problems occurring in various fields, to practical modifications of the algorithm intended to speed up computation, and to some mathematical discussion of distance concepts.

1. DIRECTED NETWORKS

Machine processing of human speech, in connection with speech-controlled machines and speech research, frequently makes use of machine-produced phonetic transcriptions (though the phonetic categories used may differ greatly from those ordinarily used by phoneticians). Since it is notoriously difficult to achieve accurate machine transcription, it is desirable to have the machine produce several alternative transcriptions, in the hope that one of them will be correct. For example, Fig. 2 shows several examples in which the material between a_i and a_{i+1} is given two alternative transcriptions. Short uncertainties

GIVEN $\mathbf{a} = a_1 \ldots a_m$ $\mathbf{b} = b_1 \ldots b_n$

Costs (or weights): substitution $w(x,y)$, deletion $w(x,\phi)$, insertion $w(\phi,y)$

DEFINITIONS   ● $\begin{bmatrix} x_1 \ldots x_p \\ y_1 \ldots y_p \end{bmatrix}$ is an alignment if each x_i and y_i is a letter of the alphabet or ϕ,

but no column is $\begin{bmatrix} \phi \\ \phi \end{bmatrix}$.

● It is an alignment between $\mathbf{a}$ and $\mathbf{b}$ if ϕ's can be inserted in $\mathbf{a}$ to make x and in $\mathbf{b}$ to make

y.

● Its length is $\Sigma\, w(x_i, y_i)$.

● $d(\mathbf{a},\mathbf{b}) =$ minimum possible length of any alignment

PROBLEM Find $d(\mathbf{a},\mathbf{b})$ and the minimum length alignment

AUXILIARY CONCEPTS $\mathbf{a}^i = a_1 \ldots a_i$, $\mathbf{b}^j = b_1 \ldots b_j$, $d_{ij} = d(\mathbf{a}^i, \mathbf{b}^j)$

INITIALIZATION $d_{00} = 0$, and $d_{ij} = \infty$ if either i or j negative.

RECURRENCE

$$d_{ij} = \min \begin{cases} d_{i-1,j} + w(a_i,\phi), \\ d_{i-1,j-1} + w(a_i,b_j), \\ d_{i,j-1} + w(\phi,b_j), \end{cases} \quad \text{pointer}(i,j) = \begin{cases} (i-1,j) & \text{or} \\ (i-1,j-1) & \text{or} \\ (i,j-1) & \end{cases} \begin{array}{l} \text{if corresponding} \\ \text{term is} \\ \text{minimum} \end{array}$$

SOLUTION $d(\mathbf{a},\mathbf{b}) = d_{mn}$, and the optimum alignment can be found by using the pointers to "backtrack" from (m,n).

WORKSPACE

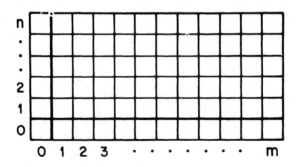

Figure 1a. A basic dynamic programming algorithm.

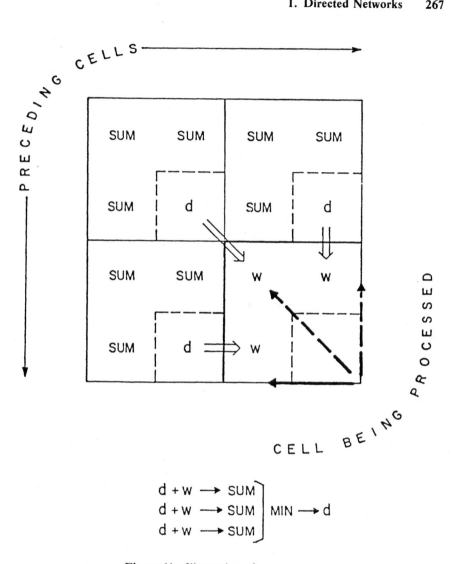

Figure 1b. Illustration of recurrence.

of this kind are very common, and even a single sentence may contain several of them. Furthermore, the situation is sometimes more complicated, as illustrated in Fig. 3. Any such collection of alternative transcriptions will be referred to as a "directed network." (A formal definition is given later).

Mark Liberman (personal communication) has described a potential application in which an ordinary sequence transcription accurately made by a phonetician must be compared with a network transcription made by machine, to find the sequence within the network that most closely approximates the accurate transcription. One can also visualize applications where two networks

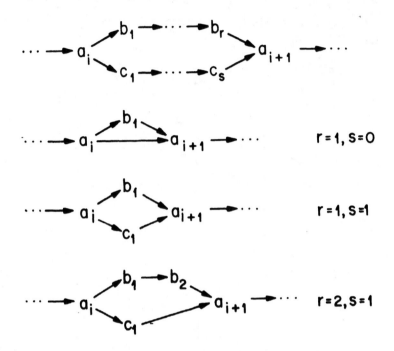

Figure 2. Alternative transcriptions.

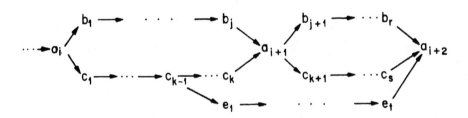

Figure 3. Complicated network of alternative transcriptions.

are to be compared, so as to find an ordinary sequence in each of them which are closest to one another.

Of course, one method for solving these problems is to list all possible alternative sequences in the network, and compare them one by one, subsequently choosing the best match. This may require a large number of comparisons, however, since if two alternatives exist at k different spots, there are 2^k different sequences in the network. If both items are networks, every sequence in one must be compared with every sequence in the other, and the inefficiency is still greater. Fortunately, a much more elegant and efficient method exists, in which the alternative transcriptions contribute additively instead of multiplicatively to the computing time.

Informally, the idea is to generalize the array (like that at the bottom of Fig. 1(a)) on which the recursion is performed, and to extend the recurrence (like that in Fig. 1(b)) in a suitable manner. For example, Fig. 4 shows the array that would be used to compare an ordinary sequence $a_1 \ldots a_7$ with the network shown there. In mathematical terms, the array can be described as the Cartesian product of one sequence or network with the other sequence or network. For "ordinary" cells, the procedure for finding d in each cell is the same as in Fig. 1(b): add the d values from the three predecessor cells to the three w values associated with this cell, and take the minimum. The procedure used with cells like those on the right of Fig. 5 is also the same: the difference from Fig. 1(b) is that these cells share predecessors. The procedure with some other special cells, which have five predecessors each instead of three, is illustrated in Fig. 6. Where a single predecessor cell has been replaced by two cells, such as cells a_5c_6 and a_5e_4, the minimum of the two d values is taken before addition to the weight w and minimization.

More formally (see Fig. 7), we consider a sequence $a_1 \ldots a_m$ as a straight-line directed graph, with the letters a_i as nodes. We define a *directed network* over an alphabet to be any directed graph, containing no directed cycles, whose nodes consist of letters from the alphabet. An example is shown in Fig. 8. Of course, the same letter from the alphabet may be used for more than one node. A *successor* to a node is any other node that can be reached in one step by following an arrow, and a *predecessor* is any other node that can be reached in one step by travelling backwards along an arrow. A *source* is a letter without a predecessor, and a *sink* is a letter without a successor. A *full sequence* means any path from a source to a sink; an *initial sequence* means any path starting at a source. The distance between two networks means the smallest distance between any full sequence in one network and any full sequence in the other. Thus, network distance is defined in terms of sequence distance, and different types of sequence distance yield different types of network distance.

The following algorithm, which is a formal version of the method described informally above, yields the distance between two networks based on simple alignment distance between sequences. (It is not difficult to generalize the

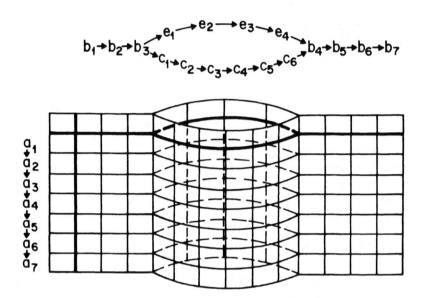

Figure 4. Workspace or array to compare a network with a sequence.

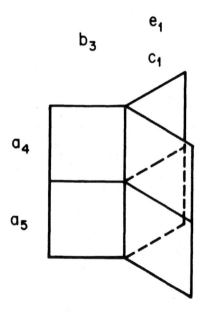

Figure 5. Some cells from the array.

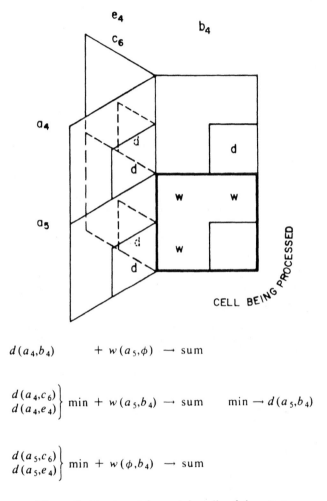

$$d(a_4,b_4) \qquad + w(a_5,\phi) \;\longrightarrow\; \text{sum}$$

$$\left.\begin{array}{l} d(a_4,c_6) \\ d(a_4,e_4) \end{array}\right\} \min + w(a_5,b_4) \;\longrightarrow\; \text{sum} \qquad \min \longrightarrow d(a_5,b_4)$$

$$\left.\begin{array}{l} d(a_5,c_6) \\ d(a_5,e_4) \end{array}\right\} \min + w(\phi,b_4) \;\longrightarrow\; \text{sum}$$

Figure 6. Treatment for certain cells of the array.

$$a_1 \longrightarrow a_2 \longrightarrow a_3 \longrightarrow \cdots \longrightarrow a_{m-1} \longrightarrow a_m$$

Figure 7. A sequence represented as directed network.

validity proof from Chapter 1 to cover this situation, but we do not give the proof here.) We refer to the nodes in one network as $a_1 \ldots a_m$ and those in the other as $b_1 \ldots b_n$.

First, adjoin a *supersource*, called a_0 or b_0 respectively, to each network. This means a new node, which uses a special new letter not drawn from the alphabet, that is the predecessor of every source. After adding the supersource, every other node has at least one predecessor (and if the network was a sequence, every other node has exactly one predecessor). Also, each full and each initial sequence starts with a_0 or b_0, respectively. We require every alignment to match a_0 with b_0, and this match is assigned weight 0. (It is easy to see that these changes are merely a formality, and do not make any substantive change in the sequences and alignments.) Then all pairs $a_i b_j$ may be considered a general abstract formal equivalent of the array illustrated in Fig. 4. The pairs involving a supersource correspond to the special initial row and column.

Let d_{ij} be the smallest distance between any initial sequence in one network terminating at a_i and any initial sequence in the other network terminating at b_j. The algorithm finds d_{ij} for all i and j by recursion. To start with, $d_{00} = 0$. To state the recurrence equation, we let a_{i*} indicate all the predecessor(s) of a_i, and b_{j*} indicate all the predecessor(s) of b_j. Then the recurrence is given by

$$d_{ij} = \min \begin{cases} w(a_i, \phi) + \min_{i*} d_{i*j} \\ w(a_i, b_j) + \min_{i*} \min_{j*} d_{i*j*} \\ w(\phi, b_j) + \min_{j*} d_{ij*}. \end{cases}$$

To find the distance between the networks, let a_i and b_j run through all the sinks in the two networks, and use

$$d(\mathbf{a}, \mathbf{b}) = \min_i \min_j d_{ij}.$$

If a pointer is set at each minimization to record which term(s) provide the minimum value, then the optimum sequence(s) in each network and their optimum alignment(s) can be found very easily by tracing backward.

The algorithm can also be carried out on an array like the one at the bottom of Fig. 1(a), as illustrated in Fig. 9. It is convenient to assign numbers to the nodes so that every node is assigned a larger number than any of its predecessor nodes, as illustrated in Figs. 7 and 8. (It is a mathematical property of directed networks that this is always possible.) With this convention, the cells used in calculating d_{ij} always lie to the left of or above cell ij.

If u is the maximum number of predecessors of any node in one network, and v the corresponding number for the other, then asymptotically for long

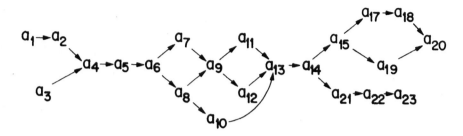

Figure 8. A directed network.

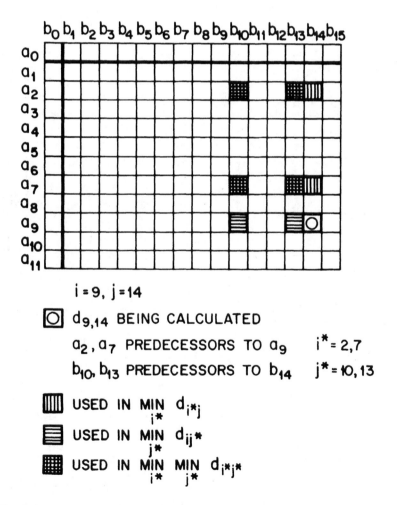

i = 9, j = 14

d_{9,14} BEING CALCULATED

a_2, a_7 PREDECESSORS TO a_9   i* = 2,7

b_{10}, b_{13} PREDECESSORS TO b_{14}   j* = 10, 13

USED IN MIN d_{i*j}
       i*

USED IN MIN d_{ij*}
       j*

USED IN MIN MIN d_{i*j*}
       i*   j*

Figure 9. A workspace or computational array to compare two networks.

networks the dominant part of the computing time is $(uv + u + v)mnt$, where t is the basic time for an appropriate simple operation.

2. LEVEL-BUILDING IN RECOGNITION OF CONTINUOUS SPEECH

Suppose utterances are formed by combining words from a fixed dictionary having D entries. We do not consider questions of syntax or semantics here, so any sequence of words is considered legitimate. Each word is made up of short units of sound, which we will refer to as *letters*. In an observed utterance $\mathbf{b} = b_1 \ldots b_n$, however, we do not expect to encounter the words in their precise dictionary form: some deformation will have occurred. Then one version of the continuous speech-recognition problem is to interpret $\mathbf{b}$ as a sequence of dictionary words in an optimal way. A major difficulty is that $\mathbf{b}$ contains no explicit indication of the boundaries between words.

To give this problem precise meaning, suppose that a distance function d between sequences is given, where d is similar to the distance functions used elsewhere in this volume. Then we ask for the sequence $\mathbf{x}^1, \ldots, \mathbf{x}^k$ of dictionary words whose concatenation $\mathbf{a} = \mathbf{x}^1 \cdot \ldots \cdot \mathbf{x}^k$ minimizes $d(\mathbf{a}, \mathbf{b})$ over all sequences (of all lengths) of dictionary words. The direct approach to solving this problem is to compare every possible sequence of k dictionary words with $\mathbf{b}$, for some reasonable range of values of k, to find the one that minimizes $d(\mathbf{a}, \mathbf{b})$. This is prohibitively costly, however, since for each k there are D^k possibilities to compare with $\mathbf{b}$. Even to recognize one telephone number from among all ten-digit possibilities involves 10^{10} comparisons. Since the sequences $\mathbf{a}$ will typically have roughly n letters (where n is the length of $\mathbf{b}$), each comparison requires n^2 steps, and D^k comparisons require $D^k n^2$ steps.

A much faster method, which Myers and Rabiner (1981) call *level-building,* is sometimes used in speech-recognition work. Though level-building, as presented in the literature, appears to be very complex, it can be understood much more easily as a special case of the directed-networks algorithm presented in the previous section. One network is simply the observed sequence $\mathbf{b}$. We let this be the factor that runs vertically in our computational array (like the one in Fig. 4).

To describe the other network, we use the D words in the dictionary, which are

$$\mathbf{w}^1 = \mathbf{w}_1^1 \ldots w_{m_1}^1, \qquad \ldots, \qquad \mathbf{w}^D = w_1^D \ldots w_{m_D}^D.$$

Also, it is convenient to set

$$m = \min(m_1, \ldots, m_D) \quad \text{and} \quad M = \max(m_1, \ldots, m_D)$$

for the minimum and maximum length of any word in the dictionary. Then Fig. 10 describes the other network: Except for the empty sequence at the bottom of

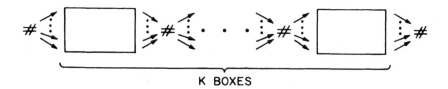

Figure 10a. A directed network for level-building algorithm.

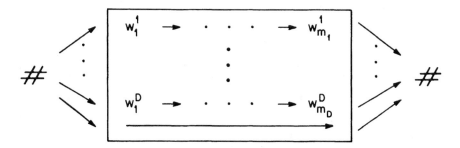

Figure 10b. Meaning of the box.

the box in Fig. 10(b), each sequence in the box is one of the dictionary words. The element # between boxes is a dummy element, which can be thought of as a word separator. It is not to be matched but to be deleted. This fact can be built into the minimization algorithm, or can be guaranteed by using deletion and insertion weights of 0 for #, while substitution weights are ∞. The number K of boxes to be used in the network should be some reasonable upper limit, such as n/m, for the number of words making up **b**. The empty sequence at the bottom of the box effectively permits any box to be skipped, so that less than K words may be matched to **b**.

Each box together with the following # contains $\leqq DM + 2$ letters, and the whole network contains $\leqq K(DM + 2) + 1$ letters. Using $K = n/m$, this is of order $(M/m)Dn$. Multiplying by the length n of **b**, we see that the directed-network algorithm may require computing time (and memory space) proportional to Dn^2 in order to solve the stated problem. This can be costly if D is large, but it is far better than the exponential cost of the direct approach.

In addition, various space-saving techniques and time-saving heuristics are available. For example, a very elementary technique to save memory space in comparing two ordinary sequences is to fill the array one column (say) at a time. Working through the array in this order, it is easy to get along with no more than one column of memory for the minimum distances of the partial sequences. Essentially the same technique can be used here. For a given letter in the network, call it a, we call the n cells $(a, b_1), \ldots, (a, b_n)$ of the Cartesian product the column that corresponds to a. Only D columns containing Dn memory spaces are needed to store distances at any one time.

However, if backtracking is to be used to find the optimal alignment, pointers need to be saved also, and this elementary technique cannot be used for pointer memory. Myers and Rabiner, however, have a way of saving some pointer memory in their level-building algorithm. Straightforward use of the algorithm from the previous section would require one pointer to be stored for every cell of the Cartesian-product array, i.e., on the order of $KDMn$ pointers. By their technique, the factor D can be reduced to $\min(n, D)$. Since n is often much smaller than D, this may be a substantial saving.

To describe their space-saving technique in our terms, consider one box of the directed network, and consider the section of the Cartesian-product array that corresponds to that box, i.e., the union of all columns that correspond to elements in that box. Let the ith *sheet* (for that box) be the subsection of the Cartesian product that corresponds to word $\mathbf{w}^i$ in the box: corresponding to any box there are D sheets. In terms of columns, the ith sheet is the union of columns that correspond to the elements in $\mathbf{w}^i$. In terms of elements, the ith sheet consists of all elements of the form (w_p^i, b_j) for all $p(1 \leq p \leq m_i)$ and all j $(1 \leq j \leq n)$, where w_p^i is in the given box.

To use the Myers–Rabiner space-saving technique (which assumes $n < D$), we fill the Cartesian-product array with pointer values, one column at a time. Whenever we have finished filling a column that corresponds to a #, however, we can reduce the pointer memory, corresponding to the preceding box, from $D(Mn)$ to $n(Mn) = Mn^2$. For each b_j $(1 \leq j \leq n)$, the pointer in cell $(\#, b_j)$ must point to some cell of the form $(w_{m_i}^i, b_j)$. For each such pointer, we retain the sheet containing the cell it points to. All sheets not so selected we discard. Thus the memory is reduced from D sheets to at most n sheets. Discarding in this way is acceptable, because during backtracking we never need sheets that have been discarded.

3. CUTTING CORNERS TO IMPROVE CALCULATION TIME

Suppose sequence $\mathbf{a}$ has length m and sequence $\mathbf{b}$ has length n. Then common basic algorithms (like the one in Fig. 1) to find the simple alignment distance between $\mathbf{a}$ and $\mathbf{b}$ require computing time whose dominant term is $3mnt$, where t is the time required to carry out a simple basic task (three additions and one trichotomous minimization).

Because the calculation time reduces to $3n^2t$ if $m = n$, which is quadratic in n, the algorithm is referred to as *time-quadratic*. In some application areas, the possibility of finding the minimum in calculation time that is merely quadratic in sequence length was considered a major advance, since methods previously contemplated (based, e.g., on the idea of listing distance) require time that is exponential in sequence length. In other fields, however, the need to spend calculation time that is substantially more than linear has been a heavy burden.

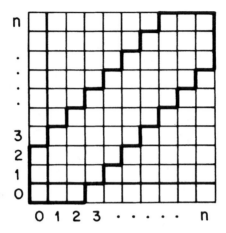

Figure 11. Cutting corners when $m = n$.

In this section we briefly discuss two approaches to reducing the calculation time while possibly *improving* the realism of the distance function between sequences.

One way to reduce the calculation time of the basic algorithm is to reduce the number of cells in the array that must be processed during the calculation, typically by eliminating cells in the remote corners of the array (i.e., the corners that are remote from the beginning and the ending of the path). Of course, this means that certain alignments are being excluded from consideration. In many cases, however, the excluded alignments are considered relatively implausible, so excluding them is no great loss. After all, a path that passes near a remote corner of the array corresponds to an alignment in which most of one sequence is deleted and most of the other inserted. In fact, the excluded alignments may be so unrealistic that their exclusion is actively desired, because it makes more sense to define the distance as the minimum amount of realistically achievable change.

While this idea can save significant time when comparing two sequences, it can achieve really dramatic savings in a related problem involving the simultaneous comparison of *three* sequences. We now explore the savings in both the two-sequence and three-sequence cases.

The first approach to cutting the corners off the array is to require that the alignment never deviate too far from a uniformly spaced alignment. To start with, consider the simple case $m = n$, and consider the constraint that every match of some a_i with some b_j in the alignment must satisfy $|i - j| \leq K$, as illustrated in Fig. 11. If $K = 0$, this permits only the trivial alignment in which a_i is matched with b_i for all i. On the other hand, if K is as large as n, the constraint has no force at all. Intermediate values of K are meaningful, and interest would

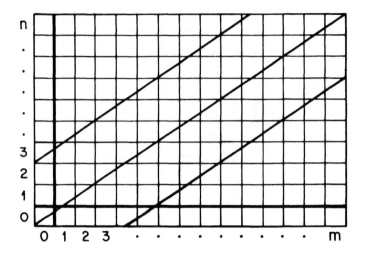

Figure 12. Cutting corners when $m \neq n$.

center on fairly small values. The number of cells that need to be processed is reduced

$$\text{from} \quad (n+1)^2 \quad \text{to} \quad (2K+1)(n+1) - (K+1)K,$$

with a proportional reduction in the dominant part of the computing time. For K small, the reduction ratio is approximately $(2K+1)/(n+1)$: for $K = 5$ and $n = 100$, this is 0.11.

Now consider the same approach when the lengths of **a** and **b**, which we denote by m and n, may be different. It is not entirely obvious how to generalize the previous constraint, but one possiblity is based on the distance from the center of the (i, j) cell to the diagonal line connecting the outer corners of the array. (Of course, each cell is taken to be 1 by 1.) The previous constraint restricts attention to cells for which this distance $\leq K/\sqrt{2}$. Using this as the constraint in the general case, as illustrated in Fig. 12, and doing some algebra, the constraint is that

$$\left| \frac{i}{m+1} - \frac{j}{n+1} \right| \leq K \frac{1}{\sqrt{2}} \frac{\left[\dfrac{m+1}{n+1} + \dfrac{n+1}{m+1} \right]}{\sqrt{(m+1)^2 + (n+1)^2}}$$

Approximate calculation shows that the reduction factor from $(m+1)(n+1)$ is

$$K\lambda[2 - K\lambda]$$

where

$$\lambda = \frac{1}{\sqrt{2}} \left[\frac{1}{(m+1)^2} + \frac{1}{(n+1)^2} \right]^{1/2}$$

For K small, this is approximately

$$\begin{cases} 2K/[\min(m, n) + 1] & \text{if } m \text{ and } n \text{ are approximately equal,} \\ \sqrt{2}\, K/[\min(m, n) + 1] & \text{if the ratio of } m \text{ to } n \text{ is very far from 1.} \end{cases}$$

For $K = 5$ and $m = n = 100$, this becomes 0.10, and for $K = 5$, $m = 100$ and $n = 1000$, it becomes 0.07.

Sankoff, Cedergren, and Lapalme (1976) consider the simultaneous comparison of three sequences, **a**, **b**, and **c**, to find a single "three-sequence alignment." This is an important special case of the method described in Chapter 9 by Sankoff and Cedergren. It involves processing all the cells in a three-dimensional array, which is $n + 1$ by $n + 1$ by $n + 1$ if the sequences all have length n. Cutting corners in a three-dimensional array yields much greater savings than cutting corners in a two-dimensional array, for it reduces the volume to be processed from the cube to a thin cylinder of hexagonal cross-section surrounding the main diagonal of the cube. If the constraint is $|i - j| \leq K$, $|j - k| \leq K$, $|k - i| \leq K$, then the reduction ratio is approximately $(\frac{3}{4})(2K + 1)/(n + 1)^2$. For $K = 5$ and $n = 100$, this is 0.0008.

Another constraint that has been used to reduce computation time has greater intuitive appeal, namely, that no more than F consecutive deletions and no more than G consecutive insertions are permitted (ordinarily, F and G are taken to be equal, and typical values are 1, 2, or 3). One approach to implementing this constraint is "cutting corners" in the manner used above, i.e., using the basic algorithm of Fig. 1 but ignoring the cells outside an appropriate region. It is often overlooked, however, that this provides only an *approximate* value for the minimum alignment distance under the stated constraint, as explained below. A method to find the true minimum alignment distance under this constraint is described in Sec. 4.2, but it may require more computation time than the basic algorithm of Fig. 1. The simple expedient of following the basic algorithm, but in a constrained region, is often used, however, and sometimes the users do not appear to realize its limitation.

The appropriate region is shown in Fig. 13. Any cell in this region can be reached by an alignment satisfying the constraints, and no cell outside it can be reached. The reason why the simple time-saving algorithm using this region is not exact is that many alignments not satisfying the constraint *also* lie entirely within the region. For example, an alignment that starts with F deletions, immediately followed by $F + G$ insertions, stays within the region thus far, and may easily stay entirely within the region, even though the constraint is violated by the $F + G$ consecutive insertions.

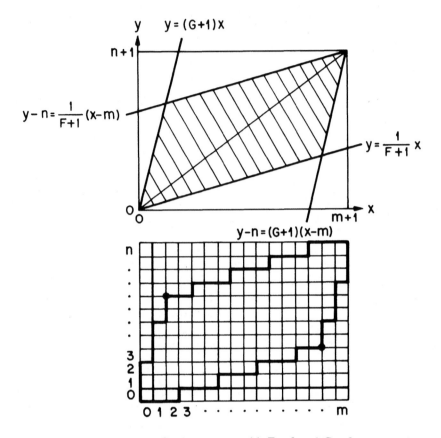

Figure 13. Cutting corners with $F = 2$ and $G = 2$.

To an approximation, the simple algorithm based on this region saves computation time by a factor equal to (area of region)/(area of rectangle). Some algebraic manipulation shows that this factor is

$$\frac{\left[1 - \dfrac{m+1}{n+1}\left(\dfrac{1}{F+1} - \dfrac{1}{m+1}\right)\right]\left[1 - \dfrac{n+1}{m+1}\left(\dfrac{1}{G+1} - \dfrac{1}{n+1}\right)\right]}{\left[1 - \dfrac{1}{(F+1)(G+1)}\right]}.$$

If an unlimited number of deletions or insertions is possible, use of $F = m$ or $G = n$ yields an appropriate value. If $m = n$ and $F = G$, the time-saving factor reduces to

$$\frac{\left[1 - \left(\dfrac{1}{F+1} - \dfrac{1}{n+1}\right)\right]^2}{\left[1 - \dfrac{1}{(F+1)^2}\right]}$$

$$= \frac{F}{F+2} + \frac{2(F+1)}{F+2} \frac{1}{n+1} + \frac{(F+1)^2}{(F+1)^2 - 1} \frac{1}{(n+1)^2} \, ,$$

which yields the following several values:

F	Factor for general n	Factor for $n = 100$
1	$\dfrac{1}{3} + \dfrac{4}{3} \dfrac{1}{n+1} + \dfrac{4}{3} \dfrac{1}{(n+1)^2}$	0.347
2	$\dfrac{1}{2} + \dfrac{3}{2} \dfrac{1}{n+1} + \dfrac{9}{8} \dfrac{1}{(n+1)^2}$	0.515
3	$\dfrac{3}{5} + \dfrac{8}{5} \dfrac{1}{n+1} + \dfrac{16}{15} \dfrac{1}{(n+1)^2}$	0.616

4. DELETION AND INSERTION CONSTRAINTS

In some situations, it is appropriate to limit the insertions and deletions in various ways. In this section, we describe algorithms, based on the algorithm of Figure 1 and closely related to each other, that can find the true minimum alignment distance subject to any one of the following constraints:

a) No more than K indels are permitted;

b) No more than F *consecutive deletions* and no more than G *consecutive insertions* are permitted (this constraint was discussed in Sec. 3);

c) No more than K *strings* of indels are permitted, where each string consists either of consecutive insertions or of consecutive deletions.

The same techniques can also be used to handle many combinations, variations, and generalizations of these constraints. Incidentally, there is nothing unique about the algorithms shown below. For each problem, many alternative versions exist which will produce the correct answer, though we doubt the existence of practical alternatives with dramatically improved computing time.

4.1 Limit on the total number of indels

We are interested in the minimum alignment distance between **a** and **b** subject to the constraint that the alignment contain no more than K indels. One algorithm to find this is based on the following definitions, with $i \geq 0$, $j \geq 0$, and $0 \leq k \leq K$ (where $\mathbf{a}^i$ and $\mathbf{b}^j$ are the initial segments of **a** and **b** of length i and j):

A_{ij}^k is the set of all alignments between $\mathbf{a}^i$ and $\mathbf{b}^j$ using no more than k indels;
d_{ij}^k is the minimum length of any alignment in A_{ij}^k (and is ∞ if that set is empty).

If **a** and **b** have length m and n, respectively, then d_{mn}^K is the desired answer.

Note that A_{ij}^k will be empty if $k < |i - j|$ because it is impossible to align two sequences of different length without using enough indels. Also note that ∞ is the natural value to use in mathematics and computer science for a minimum taken over no values at all, and that in practical computation any sufficiently large number can be used.

The values of d_{ij}^k may be calculated by the algorithm shown in Fig. 14 and further illustrated in Fig. 15. To understand the recurrence equation, consider some shortest alignment in the named set A_{ij}^k, and ask whether it has k indels or $<k$ indels. If it has k indels, break the alignment up into its final operation (a deletion, a matching, or an insertion) and the shorter alignment that precedes the final operation; if it has $<k$ indels, simply take the entire alignment. Depending on which of the four possible cases occurred, the resulting alignment has the following properties and is in an earlier named set as indicated:

Deletion: it aligns $\mathbf{a}^{i-1}$ and $\mathbf{b}^j$, has $k-1$ indels, is in $A_{i-1,j}^{k-1}$,
Matching: it aligns $\mathbf{a}^{i-1}$ and $\mathbf{b}^{j-1}$, has k indels, is in $A_{i-1,j-1}^k$,
Insertion: It aligns $\mathbf{a}^i$ and $\mathbf{b}^{j-1}$, has $k-1$ indels, is in $A_{i,j-1}^{k-1}$,
Has $<k$ indels: it aligns $\mathbf{a}^i$ and $\mathbf{b}^j$, has $\leq k-1$ indels, is in A_{ij}^{k-1}.

Each of these four cases yields one of the four possibilities in the recurrence minimization.

To actually justify the validity of the recurrence, it is necessary to show that if we synthesize an alignment by reversing the steps above, using any shortest alignment from the earlier named set, then we obtain an alignment in A_{ij}^k. In this situation (as in many others), a slightly stronger statement is true: If we synthesize using *any* alignment from the earlier named set, we obtain an alignment in A_{ij}^k. The key to finding algorithms in the spirit of those illustrated here is to find named sets such that if we analyze and then resynthesize, we stay within the original named set.

INITIALIZATION

$d_{00}^0 = 0$, $d_{i0}^0 = d_{0j}^0 = \infty$ for $i = 1$ to m, $j = 1$ to n. Also $d_{ij}^k = \infty$ if i, j, or k is negative.

RECURRENCE

Optimum alignment uses k indels and ends with:

$$d_{ij}^k = \min \begin{cases} w(a_i,\phi) + d_{i-1,j}^{k-1}, & \text{deletion} \\[2em] w(a_i,b_j) + d_{i-1,j-1}^k, & \text{match} \\[2em] w(\phi,b_j) + d_{i,j-1}^{k-1}, & \text{insertion} \\[2em] d_{i,j}^{k-1}, & \text{Optimum alignment uses } < k \text{ indels} \end{cases}$$

WORKSPACE

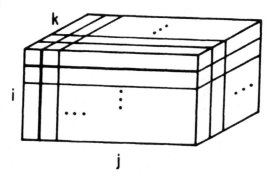

Figure 14. Algorithm when the total number of indels is limited.

Computation can be visualized in an array like that shown in Fig. 1, where the (i, j)-cell contains the $K + 1$ values $d_{ij}^0, \ldots, d_{ij}^K$ at different "levels" of k. Alternatively, computation can be visualized and actually carried out in a three-dimensional array as shown in Fig. 14, where each level of k corresponds to a two-dimensional array. While the values can be evaluated in many different orders, one natural order is to calculate the $k = 0$ level first, then the $k = 1$ level, and so on, where each level can be evaluated with the same freedom available in Fig. 1.

Computation time may be saved by restricting attention to cells that have meaningful entries. Since k indels permit an alignment of $\mathbf{a}^i$ with $\mathbf{b}^j$ only when $|i - j| \le k$, level k of the array has meaningful entries only in the main diagonal

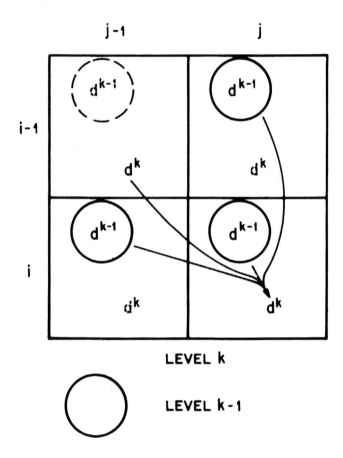

Figure 15. Recurrence when the total number of indels is limited.

and in k diagonals to either side of it, i.e., for $k = 0$ the main diagonal only, for $k = 1$ in three diagonals only, for $k = 2$ five diagonals only, etc. If $m = n$ and $K \ll n$, then the number of three-dimensional cells to be processed is about $n(K + 1)^2$ since

$$1 + 3 + \ldots + (2K + 1) = (K + 1)^2.$$

Each cell requires three additions and one four-way minimization, so the dominant part of the computation time is $n(K + 1)^2 t_1$ where t_1 is the time required for these steps. By comparison, the time for the basic algorithm in Fig. 1 is about $n^2 t_1$ (with a similar t_1), so this computation is actually faster than the basic algorithm if $(K + 1)^2 < n$.

A slightly different version of this algorithm would be a little faster. Define B_{ij}^k to be the set of all alignments between $\mathbf{a}^i$ and $\mathbf{b}^j$ using *exactly* k indels, and define d_{ij}^k as the minimum over B_{ij}^k instead of A_{ij}^k. With this approach, the recurrence equation is the same, except that the fourth term of the minimization is omitted. Dropping this term provides the time-saving. In this algorithm, the desired answer is no longer d_{mn}^K as before but $\min_{0 \leq k \leq K} d_{mn}^k$ instead.

4.2 Limit on the consecutive deletions and insertions

We are interested in the minimum alignment distance between $\mathbf{a}$ and $\mathbf{b}$ subject to the constraint previously discussed in Sec. 3, namely, that the alignment should nowhere contain more than F deletions *consecutively* or more than G insertions *consecutively* (though it may contain an unlimited number of deletions and insertions in total). One algorithm to find this is based on the following definitions:

Definitions. The *tail* of an alignment is a specific terminal part of it: namely, if the final operation is a match, then the tail consists of that single operation; if the final operation is a deletion, then the tail consists of the longest possible final string of consecutive deletions; and if the final operation is an insertion, the tail consists of the longest possible final string of insertions.

A_{ij} (for $i = 0$ to m, $j = 0$ to n) is the set of all alignments between $\mathbf{a}^i$ and $\mathbf{b}^j$.
A_{ij}^0 is the set of alignments in A_{ij} whose tail consists of a match;
A_{ij}^{-f} (for $f = 1$ to F) is the set of alignments in A_{ij} whose tail consists of f deletions.
A_{ij}^{+g} (for $g = 1$ to G) is the set of alignments in A_{ij} whose tail consists of g insertions.
d_{ij}^0 (or d_{ij}^{-f} or d_{ij}^{+g}) is the minimum length of any alignment in the corresponding set (and is ∞ if the corresponding set is empty).

Note that A_{ij}^{-f} will be empty if $\mathbf{a}^i$ is not long enough to have f terms deleted from it (that is, if $i < f$), and similarly A_{ij}^{+g} is empty if $\mathbf{b}^j$ is not long enough to have had g terms inserted into it (that is, if $j < g$).

The desired answer is then

$$
d_{mn} = \min \begin{cases} d_{mn}^0 \\ \min_f d_{mn}^{-f} \\ \min_g d_{mn}^{+g} \end{cases}.
$$

INITIALIZATION

$d_{00}^0 = 0, \quad d_{00}^k = \infty \text{ for } k \neq 0$

RECURRENCE

Next to last operation is:

Tail is a match

$$d_{ij}^0 = w(a_i, b_j) + \min \begin{cases} d_{i-1,j-1}^0 & \text{match} \\ \min_f d_{i-1,j-1}^{-f} & \text{deletion} \\ \min_g d_{i-1,j-1}^{+g} & \text{insertion} \end{cases}$$

Tail is exactly one deletion

$$d_{ij}^{-1} = w(a_i, \phi) + \min \begin{cases} d_{i-1,j}^0 & \text{match} \\ \\ \min_g d_{i-1,j}^{+g} & \text{insertion} \end{cases}$$

Tail is exactly f deletions ($f = 2$ to F)

$$d_{ij}^{-f} = w(a_i, \phi) + d_{i-1,j}^{-(f-1)} \qquad\qquad \text{deletion}$$

Tail is exactly one insertion

$$d_{ij}^{+1} = w(\phi, b_j) + \min \begin{cases} d_{i,j-1}^0 & \text{match} \\ \\ \min_f d_{i,j-1}^{-f} & \text{deletion} \end{cases}$$

Tail is exactly g insertions ($g = 2$ to G)

$$d_{ij}^{+g} = w(\phi, b_j) + d_{i,j-1}^{+(g-1)} \qquad\qquad \text{insertion}$$

Figure 16. Algorithm when the number of consecutive deletions is limited to F and the number of consecutive insertions is limited to G.

The quantities involved may be calculated by the algorithm shown in Fig. 16 and further illustrated in Fig. 17. (The verbal material associated with the recurrence is not part of the recurrence itself, but is merely explanatory.) To understand any one of the recurrence equations, choose some shortest alignment from the corresponding named set, and break it up into the final operation and the shorter alignment that precedes the final operation. Consider what properties the shorter alignment must have, and which of several earlier named sets it must belong to. Each such set corresponds to one term in the minimization. (To justify the recurrence equation, it is necessary to show that resynthesis stays within the original named set, as discussed earlier.)

Computation can be visualized in an array like that shown in Fig. 1, where each cell contains $1 + F + G$ values. Alternatively, computation can be visualized and actually carried out in a three-dimensional array that is $m + 1$ by $n + 1$ by $1 + F + G$. While the values can be evaluated in many different orders, one natural order is to sweep through the cells of the two-dimensional array in any order available in Fig. 1, evaluating all the $1 + F + G$ values in that cell when the cell is reached. Computation time can be calculated as $(m + 1)(n + 1)t_2$, where t_2 is the time for $1 + F + G$ additions and three minimizations, one over $1 + F$ elements, one over $1 + G$ elements, and one over $1 + F + G$ elements. Interest would ordinarily rest in small values of F and G, for which computation would not much exceed that for the basic algorithm of Fig. 1. Specifically, compare the following computations:

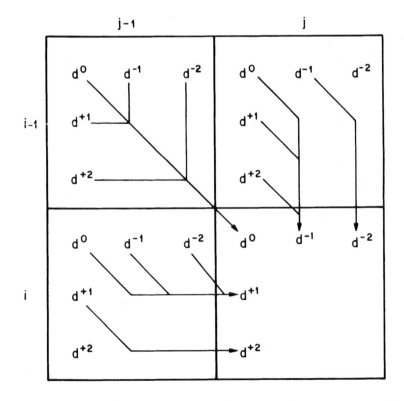

Figure 17. Recurrence when the number of consecutive deletions is limited to F and the number of consecutive insertions is limited to G.

Basic algorithm: $(m + 1)(n + 1)$ $\begin{cases} 3 \text{ additions,} \\ 1 \text{ minimization over 3 elements;} \end{cases}$

$F = G = 1$: $(m + 1)(n + 1)$ $\begin{cases} 3 \text{ additions,} \\ 2 \text{ minimizations over 2 elements each,} \\ 1 \text{ minimization over 3 elements;} \end{cases}$

$F = G = 2$: $(m + 1)(n + 1)$ $\begin{cases} 5 \text{ additions,} \\ 2 \text{ minimizations over 3 elements each,} \\ 1 \text{ minimization over 5 elements.} \end{cases}$

Furthermore, we can make an additional saving while retaining the rigorous correctness of the algorithm by considering only cells that can be reached. As discussed in Sec. 3, in the case $F = G = 2$ this will yield a saving of about 50% if $m = n$ and more if $m \neq n$.

4.3 Limit on the number of strings of deletions and insertions

We are interested in the minimum alignment distance between **a** and **b**, subject to the constraint that the indels in the alignment are all contained within at most K strings (of arbitrary length), where each string consists entirely of consecutive deletions or entirely of consecutive insertions. One algorithm to find this is based on the following definitions:

A_{ij}^k (for $k = 0$ to K, $i = 0$ to m, $j = 0$ to n) is the set of all alignments between $\mathbf{a}^i$ and $\mathbf{b}^j$ in which the indels are all contained within at most k strings, where each string consists either of consecutive deletions or of consecutive insertions;

A_{ij}^{k-} is the set of alignments in A_{ij}^k in which the final operation is a deletion (necessarily, of a_i);

A_{ij}^{k+} is the set of alignments in A_{ij}^k in which the final operation is an insertion (necessarily, of b_j);

A_{ij}^{k0} is the remaining alignments in A_{ij}^k;

d_{ij}^{k0} (or d_{ij}^{k-} or d_{ij}^{k+}) is the minimum length of any alignment in the corresponding set (and is ∞ if the corresponding set is empty).

Note that ordinarily A_{ij}^{k0} consists of alignments whose final operation is a match (necessarily, of a_i with b_j). The only peculiarities occur when A_{ij}^{k0} is empty (for example, $k = 0$ and $i \neq j$), though this is not truly an exception, or when $i = j = 0$ (any value of k), since A_{00}^{k0} contains just one alignment, which is the null alignment and does not contain any final operation. Note that A_{ij}^{0-} and A_{ij}^{0+} are all empty, since if there are *no* strings of indels, there are no indels, and hence there are no alignments that end with a deletion or an insertion.

INITIALIZATION

$d_{00}^{00} = 0$. Also $d_{ij}^{0-} = d_{ij}^{0+} = \infty$, and if either i or j is negative, d is taken to be ∞.

RECURRENCE

	Preceding operation:	Preceding number of strings:

Final operation is a deletion
(use for $k \geq 1$)

$$d_{ij}^{k-} = w(a_i,\phi) + \min \begin{cases} d_{i-1,j}^{k-} & \text{deletion} & k \\ d_{i-1,j}^{k-1,0} & \text{match} & k-1 \\ d_{i-1,j}^{k-1,+} & \text{insertion} & k-1 \end{cases}$$

Final operation is a match
(use for $k \geq 0$)

$$d_{ij}^{k0} = w(a_i,b_j) + \min \begin{cases} d_{i-1,j-1}^{k-} & \text{deletion} & k \\ d_{i-1,j-1}^{k0} & \text{match} & k \\ d_{i-1,j-1}^{k+} & \text{insertion} & k \end{cases}$$

Final operation is an insertion
(use for $k \geq 1$)

$$d_{ij}^{k+} = w(\phi,b_j) + \min \begin{cases} d_{i,j-1}^{k-1,-} & \text{deletion} & k-1 \\ d_{i,j-1}^{k-1,0} & \text{match} & k-1 \\ d_{i,j-1}^{k+} & \text{insertion} & k \end{cases}$$

Figure 18. Algorithm when the number of strings of pure deletions and pure insertions is limited to K.

The desired answer is then

$$d_{mn}^K = \min(d_{mn}^{K0}, d_{mn}^{K-}, d_{mn}^{K+}).$$

The quantities involved may be calculated by the algorithm shown in Fig. 18 and further illustrated in Fig. 19. The recurrence may be understood and justified as in the preceding algorithm.

Computation can be visualized in a three-dimensional array, which is made of a series of levels, or layers, one for each value of k. Each level is a two-dimensional array like that in Fig. 1. The (i, j)-cell of level k contains three

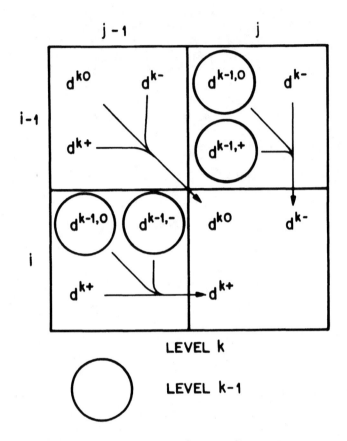

Figure 19. Recurrence when the number of strings of pure deletions and pure insertions is limited to K.

values, d_{ij}^{k-}, d_{ij}^{k0}, and d_{ij}^{k+} While the values can be calculated in many orders, one natural order would be to calculate the levels in order, first level 0, then 1, etc. Within each level, the cells would be calculated in any order that was available in Fig. 1. Whenever a cell is reached, all three values in it would be calculated.

5. A BIOLOGICAL APPLICATION OF INSERTION AND DELETION CONSTRAINTS: 5S RNA FROM HUMANS AND *E. COLI*

Sometimes it is difficult to find a sensible basis for determining the relative weights of different operations for various reasons, including a lack of data. One common way to handle this situation is to do a series of comparisons, using different weights, and attempt to determine which weights seem most appro-

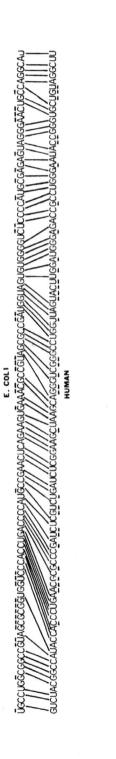

Figure 20. Two traces connecting 5S RNA from *E. Coli* and humans. Horizontal bars mark deleted and inserted letters. Trace lines are shown only for identities, not for substitutions. Though upper trace has more identities, lower trace is more plausible because the deletions and insertions occur in fewer strings.

priate, with the aid of the results. Another approach, which is illustrated here, is to make use of a series of constraints. This early comparison of human and *E. coli* 5S RNA, which is drawn from Sankoff and Cedergren (1973), permitted a systematic statistical approach to determining *ex post facto* which constraint level was most appropriate. At the time, nothing precise was known about the relative weights of the different operations, but there was a biological consensus that indels might occur far less often than substitutions and in a limited number of sites, so a series of constraints were used to limit the number of strings of insertions and deletions (in the sense of Sec. 4.3).

The study was not explicitly phrased in terms of alignment distance, but in terms of another closely related measure,

M = the length of the longest common subsequence of
    a and **b**
  = the maximum number of identities in any
    alignment of **a** and **b**

It is well known and not hard to show that if **a** has length m and **b** has length n, then

$$M = \frac{1}{2}(m + n - d),$$

where d is based on weights for which w(substitution) ≥ 2 and w(indel) $= 1$.

The 5S RNA sequences **a** from humans and **b** from *E. coli* bacteria each have length 120. A maximum of k strings of indels was permitted, with $k = 0$, 1, ... The second meaning of M extends naturally to the constrained situation, namely,

M_k = the maximum number of identities in any alignment
     of **a** and **b** that uses no more than k strings of indels.

Using essentially the same method described in the preceding section, M_k can be directly calculated. The unconstrained value of M will be denoted by M_∞, since it corresponds to permitting an arbitrarily large number of strings of indels in the alignment. In fact, $M_\infty = 81$ and it actually used 37 strings of indels.

In order to judge which value of k is appropriate to use, 100 pairs of sequences of length 120 were randomly generated with the same statistical composition as **a** and **b**. The values $M_{k+1} - M_k$ from the random pairs were used to see whether the extra freedom of one more string of indels is significant in the real pair. In particular, the average value and the standard deviation of $M_{k+1} - M_k$ was calculated for the random pairs of sequences. Based on these values, $M_{k+1} - M_k$ for **a** and **b** was found to be larger than the average of the random pairs by a statistically significant amount for $k = 0, 1, 2, 3, 4$ and 5, but not for $k = 6$ and up. Thus the alignment using six strings of indels was selected. Figure 20 shows traces of the alignments for $k = \infty$ and $k = 6$.

The interpretation behind this choice is that there is a "true" alignment that has some number k_0 of strings of indels. For $k \geq k_0$, we would not expect $M_{k+1} - M_k$ to be significant, since the increase of M_{k+1} over M_k simply results from the optimization procedure taking advantage of chance features of the pair. On the other hand, for $k < k_0$ the increase of M_{k+1} over M_k should be significant, since it permits incorporation of features of the "true" alignment of the pair.

This method has also been used by Cedergren and Sankoff (1976) to prove that no evolutionary homology exists between 5S RNA and another ribosomal RNA, namely, 5.8S RNA. It has also been studied in the context of the theory of partially ordered sets by Sankoff and Sellers (1972).

6. FINDING SIMILAR PORTIONS OF TWO SEQUENCES

If often happens in molecular biology that two sequences will contain *portions* that are similar (i.e., homologous) to each other, even though the entire sequences are not. In speech processing, it may be necessary to compare two utterances that contain a word or phrase in common, even though they differ elsewhere. Thus methods for seeking such similar portions and the alignment between them are of considerable interest. The phrase *local alignment* will be used to indicate an alignment between portions of two sequences. Smith and Waterman (1981) suggest an algorithm for finding local alignments, which we present here in modified form. Sellers (1980), which we discuss further below, also deals with this problem. A closely related problem, finding a portion of one sequence that is similar to an entire other sequence, is treated in Chapter 2, and is called "word-spotting" in speech-processing.

At first glance, it might seem reasonable to find portions of a and b with minimum distance, as in the basic algorithm. Unfortunately, in the absence of any constraint to keep the portions long, the minimum distance of zero will always be possible with a meaningless alignment of very short portions, such as a single element from one sequence with a single element from the other. This happens because the weights (whose sum forms the length by which we evaluate a homology) act only as penalties (for indels and substitutions), and do not reward matches. To solve this, we replace the weights by *scores* that are both positive and negative, such as

$$s(x, x) = +1 \qquad \text{for any match,}$$
$$s(x, y) = -\tfrac{1}{3} \qquad \text{for any substitution,}$$
$$s(x, \phi) = s(\phi, y) = -\tfrac{4}{3} \qquad \text{for any indel,}$$

and call the sum of scores the *quality* (instead of length). Note that we are using opposite algebraic sign for scores than we did for weights, because it is easier to think about contributions of both signs if positive values are to be preferred and negative values to be avoided. Thus quality is to be maximized, as opposed to length, which is minimized. The score values shown above are those proposed by Smith and Waterman (1981) (though they actually use more general weights permitting block deletion and insertion, resembling the approach described in the following section). The choice of values will have to be guided by biological intuition and practical experience, but other values worth considering might be

$$s(x, x) = +1 \qquad\qquad \text{for a match,}$$
$$s(x, y) = -1 \qquad\qquad \text{for a substitution,}$$
$$s(x, \phi) = s(\phi, y) = -1 \text{ or } -2 \text{ or } -3 \quad \text{for an indel.}$$

The scores are required to be positive or negative as appropriate, namely,

$$s(x, x) > 0 \qquad \text{for every } x,$$
$$s(x, y) < 0 \qquad \text{whenever } x \neq y \text{ (including indels),}$$

so that matches are encouraged and anything else is discouraged.

To find the maximum-quality local alignment between $\mathbf{a} = a_1 \ldots a_m$ and $\mathbf{b} = b_1 \ldots b_n$, we proceed in a manner that closely resembles the basic algorithm. Define $\mathbf{a}^i = a_1 \ldots a_i$ and $\mathbf{b}^j = b_1 \ldots b_j$ to be the initial portions of $\mathbf{a}$ and $\mathbf{b}$. Let q_{ij} be the maximum quality of any local alignment between $\mathbf{a}^i$ and $\mathbf{b}^j$ that includes a_i and b_j. (Note that these alignments need only include portions of $\mathbf{a}^i$ and $\mathbf{b}^j$, that is, portions of portions.) However, as a vital special case, if all such alignments have negative quality, then let $q_{ij} = 0$ instead (because it is better to start fresh, forming a new local homology as we proceed from this point on). Since every local alignment between $\mathbf{a}$ and $\mathbf{b}$ contains some final a_i and final b_j, and hence is considered in connection with one or another of the q_{ij}, the maximum quality available must be

$$\max_{1 \leq i \leq m,\ 1 \leq j \leq n} q_{ij}.$$

(A mathematically more sophisticated and elegant definition of q_{ij} is the maximum quality of any local alignment between portions *ending at* a_i and b_j. This eliminates the vital special case, since its role is taken by the local alignment ending at (i, j) that has 0 columns.)

To find the values of q_{ij}, the following recurrence may be used:

$$q_{ij} = \max \begin{cases} q_{i-1,j} + s(a_i, \phi), & \text{(for alignment ending with deletion),} \\ q_{i-1,j-1} + s(a_i, b_j), & \text{(for alignment ending with substitution),} \\ q_{i,j-1} + s(\phi, b_j), & \text{(for alignment ending with insertion),} \\ 0, & \text{(if no positive quality alignment exists).} \end{cases}$$

Initial values are $q_{i0} = 0$ and $q_{0j} = 0$ for all i and j. To obtain the best local alignment itself, we also need to keep track of pointers for backtracking. For this purpose, we may accompany each step of the above recurrence by the step below:

$$\text{pointer}(i,\ j) = \left\{ \begin{array}{ll} (i-1, j) & \\ (i-1, j-1) & \text{or} \\ (i, j-1) & \text{or} \\ \phi & \text{or} \end{array} \right\},$$

depending on which line is maximum above. (If several lines are all maximum, then the pointer value is the corresponding set of alternatives.) The initial values are pointer$(i,\ j) = \phi$ if $i = 0$ or $j = 0$. Backtracking starts from the pair $(i,\ j)$ for which q_{ij} attains its maximum value, and proceeds just as in the basic algorithm. If a pointer value of ϕ is reached, backtracking stops. If a set of alternative values is reached during backtracking, any one of them may be used. To obtain all the maximum-quality local alignments, all possible choices must be used whenever choice is possible.

This algorithm finds one portion in each sequence that is similar to that in the other. In some applications, however, there may be several portions in each sequence that are similar to corresponding portions in the other, and we may wish to find them all. To do this, one's first thought might be to look for other large values of q_{ij} in the table and backtrack from them. While this works reasonably well when applied with common sense to small examples, it does not yield very satisfactory results for examples of realistic size, or if applied literally by a computer. The difficulties will be illustrated and explained in a moment. Smith and Waterman (1981) suggest a simple modification, but it does not adequately solve the problem. A satisfactory practical method is suggested below.

Suppose the maximum quality is $q_{67,\ 70} = 100$, resulting from the local alignment

$$\begin{bmatrix} a_{11} & a_{12} & - & a_{13} & a_{14} & a_{15} & a_{16} & \cdots & a_{67} \\ b_{17} & b_{18} & b_{19} & b_{20} & b_{21} & b_{22} & a_{23} & \cdots & b_{70} \end{bmatrix}$$

Suppose there is also a second local alignment,

$$\begin{bmatrix} a_{105} & a_{106} & a_{107} & \cdots & a_{135} \\ b_{207} & b_{208} & b_{209} & \cdots & b_{240} \end{bmatrix},$$

which yields $q_{135,240} = 50$. Associated with the first local alignment will be a vast number of minor variations whose quality is almost 100, but which tell us nothing new. For example, we can remove a few columns from either end of the local alignment, or add a few. We can make minor changes in the middle, e.g.,

$$\begin{bmatrix} \cdots & a_{14} & a_{15} & - & a_{16} & \cdots \\ \cdots & b_{21} & - & b_{22} & b_{23} & \cdots \end{bmatrix}.$$

We can use combinations of these changes. As a result, all the cells near (67, 70) will have values not much less than 100, and all the cells within a considerable region of it will have values larger than 50.

One difficulty is that we don't want to look at the local alignments associated with all these cells. They tell us nothing new, once we have seen the maximum-quality alignment from which they are derived. An even more serious problem, however, is that they conceal the existence of the other local alignment of quality 50. How do we know that we ought to look at the value of 50 and ignore all the values of 99, 98, 97, etc., associated with the first alignment?

Scanning the table visually, while keeping in mind the location at which the values are found, would be one possibility, but this is burdensome with a large table and tricky to automate. Smith and Waterman suggest avoiding cells on the backtracking path of any previously identified local alignment. This is a step in the right direction, but far from sufficient, since there are many other cells near the high-quality cell at the beginning of the backtracking path. Fortunately, there is a method for suppressing all the minor variations of selected local alignments.

Mark every cell (i, j) in the backtracking path of each selected alignment as "used." (If a cell is marked more than once, this does no harm.) Then carry out the algorithm described above once again, but with one difference: For every cell that is "used," q_{ij} is set to 0, regardless of what it would otherwise be. After the calculation is complete, the q_{ij} values will ignore all minor variations on the previously selected local alignments. By repeated use of this algorithm, adding one or more selected alignments each time, we can find all similar portions of two sequences.

Sellers (1980) takes quite a different approach to finding partial alignments, which is an extension of the method of Erickson and Sellers described in Chapter 2. He introduces a definition of "local optimality" for partial alignments, and describes a test for whether a given alignment has this property. Unfortunately, it appears that if the sequences have length n, then the number of local optimal partial alignments may be of order n^4, so that local optimality may be impractical as a criterion for selecting interesting partial alignments.

7. WEIGHTS FOR INSERTION AND DELETION OF CONSECUTIVE STRINGS

In some biological contexts, there are mechanisms that can result in the insertion or deletion of an entire consecutive string into a macromolecule by a single action. In such contexts, an alignment where ten elements are inserted consecutively at one place and five elements deleted consecutively at another place might seem far more plausible than an alignment consisting of five isolated insertions of one element each. When this is true, ordinary Levenshtein distance does not correspond to implausibility very well.

As illustrated in the preceding section, one way of handling such a context is to place an upper bound on the number of strings of indels, where each string consists either of consecutive insertions or of consecutive deletions. A computational method to cover this constraint is given in Sec. 4.3.

Another way to handle this context is described in this section. Our method is similar to but more general than a method sometimes used in comparing macromolecular sequences under the assumption known as "linear gap weights" (Gotoh (1982)). Our method is based on generalizing Levenshtein distance to permit weights to be associated with *strings* of consecutive insertions or consecutive deletions, in addition to the ordinary weights associated with individual indels and substitutions. We present a simple generalization of the algorithm in Fig. 1, which is suitable for finding the minimum-cost distance in this sense.

Specifically, let $w(x, y)$, $w_-(x)$, and $w_+(y)$ be used to indicate the ordinary substitution, deletion, and insertion weights. In addition, use *beginning weights* and *ending weights*,

$$w_{b-}(x), \qquad w_{e-}(x), \qquad w_{b+}(y), \qquad w_{e+}(y),$$

to indicate the weight associated with

> beginning (b), or
> ending (e),
> a consecutive string of deletions ($-$) with deletion of x, or
> a consecutive string of insertions ($+$) with insertion of y.

For any alignment, the total length is formed by summing the ordinary terms for substitutions and indels, and in addition two new terms for each string of consecutive insertions or consecutive deletions. (Even a string of length 1 has two associated terms: the beginning and ending weights are associated with the same element.)

To illustrate how these weights can achieve the desired effect, suppose the beginning weights are all set to some large constant value, say 10, and the ending weights are all set to 0. Suppose all the ordinary indel weights are equal and small, say 1. Then the total weight associated with any consecutive string of insertions or deletions is 10 + (length of string). Thus in the example discussed above, a string of ten consecutive insertions and a string of five consecutive deletions would cost $20 + 15 = 35$, while five isolated insertions would cost $5 \times 11 = 55$. This rates the former as considerably more plausible than the latter, as desired.

Many further variations, beyond the limits of the above system, are practical without making the algorithm much more expensive. For example, if it were desirable in applications, the same principle that was applied to insertions and deletions could be applied to substitutions; i.e., beginning and ending weights could be used for strings of consecutive substitutions. Also, weights could be associated not only with the beginning and ending elements but also with the next-to-beginning and next-to-ending elements of the string, and so on.

One method to calculate the minimum alignment distance between **a** and **b**, using the above definition of distance, is based on the following definitions:

A_{ij} is the set of all alignments between $\mathbf{a}^i$ and $\mathbf{b}^j$;
A_{ij}^0 is the set of alignments in A_{ij} that
   end with a match (necessarily, between a_i and b_j);
A_{ij}^- is the set of alignments in A_{ij} that
   end with a deletion (necessarily, of a_i);
A_{ij}^+ is the set of alignments in A_{ij} that
   end with an insertion (necessarily, of b_j);
d_{ij}^0 (or d_{ij}^- or d_{ij}^+) is the minimum alignment distance
   over all alignments in the corresponding set.

A slight modification of the definitions of d_{ij}^- and d_{ij}^+, however, reduces the computational cost of the algorithm somewhat, at the cost of making the algorithm a little harder to understand:

d_{ij}^- (or d_{ij}^+) is the minimum of (alignment distance minus final
   ending weight) over all alignments in the corresponding set.

Then the desired distance is

$$d_{mn} = \min(d_{mn}^0, \quad d_{mn}^+ + w_{e+}(b_n), \quad d_{mn}^- + w_{e-}(a_m)).$$

The quantities involved may be calculated by the algorithm shown in Fig. 21 and further illustrated in Fig. 22.

We remark that for each cell, the basic algorithm of Fig. 1 requires three additions and one trichotomous minimization, while this algorithm requires 15 additions (apparently 17, but 2 can be saved) and three trichotomous minimizations. Thus the dominant term in the computation time is about four times as much here as there, about 12 mnt_1 instead of 3 mnt_1.

8. GENERALIZED SUBSTITUTIONS: QUARTIC-, CUBIC-, AND QUADRATIC-TIME ALGORITHMS

In most of this volume, the elementary operations are either substitutions, insertions, or deletions, which can be thought of as follows:

INITIALIZATION

$$d_{00}^0 = 0, \quad d_{00}^- = d_{00}^+ = \infty.$$

RECURRENCE

$$d_{ij}^- = \min \begin{cases} d_{i-1,j}^- + w_-(a_i) \,, \\ d_{i-1,j}^0 + w_-(a_i) + w_{b-}(a_i) \,, \\ d_{i-1,j}^+ + w_-(a_i) + w_{b-}(a_i) + w_{e+}(b_j) \,, \end{cases}$$

$$d_{ij}^0 = \min \begin{cases} d_{i-1,j-1}^- + w(a_i,b_j) + w_{e-}(a_{i-1}) \,, \\ d_{i-1,j-1}^0 + w(a_i,b_j) \,, \\ d_{i-1,j-1}^+ + w(a_i,b_j) + w_{e+}(b_{j-1}) \,, \end{cases}$$

$$d_{ij}^+ = \min \begin{cases} d_{i,j-1}^- + w_+(b_j) + w_{b+}(b_j) + w_{e-}(a_i) \,, \\ d_{i,j-1}^0 + w_+(b_j) + w_{b+}(b_j) \,, \\ d_{i,j-1}^+ + w_+(b_j) \,. \end{cases}$$

Figure 21. Algorithm when there are weights for each consecutive string of insertions and deletions.

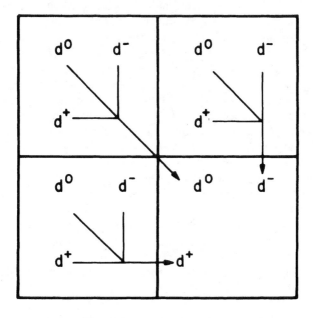

Figure 22. Recurrence when there are weights for each consecutive string of insertions and deletions.

substituting a sequence of length 1 by a sequence of length 1;
substituting a sequence of length 0 by a sequence of length 1;
substituting a sequence of length 1 by a sequence of length 0.

Chapter 7 by Wagner also includes transpositions (swaps), which can be thought of as

substituting a sequence of length 2 by a sequence of length 2.

In a discussion above, we referred to biological contexts in which a whole string is inserted or deleted as a single action:

substituting a sequence of length 0 by a sequence of length k;
substituting a sequence of length k by a sequence of length 0.

In time-warping of sequences derived from continuous signals, as explained in Chapter 4 by Kruskal and Liberman, expansion and compression are used:

substituting a sequence of length 1 by a sequence of length k;
substituting a sequence of length k by a sequence of length 1.

This leads to the idea of generalized substitutions, which include ordinary substitutions, indels, transpositions, and all the other examples above as special cases. In particular, let $x \rightarrow y$ denote the generalized substitution of x by y, where these sequences may have any lengths, and suppose that some fixed set G of acceptable generalized substitutions has been specified, along with associated weights $w(x, y)$. We assume, consistent with our approach to ordinary substitutions, that if $x \rightarrow y$ is in G, then $x \neq y$. The elements of G are called generalized elementary operations. Then the concepts of trace, alignment, and listing (as described in Sec. 3 of Chapter 1) all generalize in a straightforward way, as illustrated in Fig. 23. Similar straightforward generalization occurs for the concepts of trace length, alignment length, and listing length (each is the sum of weights of the elementary operations used in the analysis) and for the corresponding concepts of trace distance, alignment distance, and listing distance (each is the minimum possible length of any analysis of the corresponding kind).

Since all these concepts are based on the set G of generalized elementary operations, we will indicate the concepts when necessary as *G-based*. This distinguishes them from the corresponding simple concepts, and also shows which set of generalized elementary operations is being used. Of course, if E is the usual set of elementary operations, then E-based distance is the same as simple distance.

A basic property that we want for G-based traces (and alignments, and listings) is that it should be possible to compare any two sequences. It is easy to

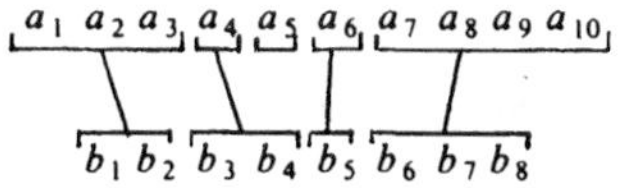

$$a_1\ a_2\ a_3\ a_4\ a_5\ a_6\ a_7\ a_8\ a_9\ a_{10}$$

TRACE BASED ON GENERALIZED SUBSTITUTIONS

$$\begin{bmatrix} a_1 a_2 a_3 & a_4 & a_5 & a_6 & \phi & a_7 a_8 a_9 a_{10} \\ b_1 b_2 & b_3 b_4 & \phi & b_5 & b_6 & b_7 b_8 \end{bmatrix}$$

ALIGNMENT BASED ON GENERALIZED SUBSTITUTIONS

$$a_1\ a_2\ a_3\ a_4\ a_5\ a_6\ a_7\ a_8\ a_9\ a_{10}$$
$$a_4 \to b_3\ b_4$$
$$a_1\ a_2\ a_3\ b_3\ b_4\ a_5\ a_6\ a_7\ a_8\ a_9\ a_{10}$$
$$a_5 \to \phi$$
$$a_1\ a_2\ a_3\ b_3\ b_4\ a_6\ a_7\ a_8\ a_9\ a_{10}$$
$$a_7\ a_8\ a_9\ a_{10} \to b_7\ b_8$$
$$a_1\ a_2\ a_3\ b_3\ b_4\ a_6\ b_7\ b_8$$
$$a_1\ a_2\ a_3 \to b_1\ b_2$$
$$b_1\ b_2\ b_3\ b_4\ a_6\ b_7\ b_8$$
$$a_6 \to b_5$$
$$b_1\ b_2\ b_3\ b_4\ b_5\ b_7\ b_8$$
$$\phi \to b_6$$
$$b_1\ b_2\ b_3\ b_4\ b_5\ b_6\ b_7\ b_8$$

LISTINGS BASED ON GENERALIZED SUBSTITUTIONS

Figure 23. Trace, alignment, and listing based on generalized substitutions.

see that a necessary and sufficient condition for this is that G should contain all ordinary indels. (This holds whether we consider traces, alignments, or listings.) Thus we shall henceforth assume that G contains all ordinary indels.

The relationships among the three kinds of G-based distance, however, are in general by no means the same as among their simple counterparts. While it is true that G-based trace distance is a special case of G-based alignment distance, and G-based alignment distance is a special case of G-based listing distance, the extra generality of G-based listing distance over the other two G-based distances may be of great significance in practical applications, even for the nicest weights. This is in strong contradistinction to the fact, explained elsewhere in detail, that simple listing distance is equal to simple alignment distance using the same weights, if the weights satisfy a mild condition (specifically, the triangle inequality). Equality between listing and alignment distance is important, since in many contexts the conceptual starting point is listing distance, and alignment distance is introduced only for the great computational saving it permits.

8.1 Relationship to Wagner's work

As an illustration, consider Chapter 7 by Wagner, in which the set of generalized elementary operations consists of transpositions (swaps) together with ordinary indels and ordinary substitutions. We shall call that set W. Wagner's definition of distance is equivalent to W-based listing distance. In practice, however, he finds it advantageous to work with traces instead of listings, but his traces are quite different from our present definition of W-based traces. He is able to define a distance on his traces that is equivalent to W-based listing distance, but his trace distance is, of course, quite different from W-based trace distance.

To get some feeling for the difference between W-based versions of listing distance d_L and alignment distance d_A, we compare the sequences cddd and dddc, assuming that w is nonnegative everywhere and 0 only for identical sequences. First we calculate d_L. Every listing from cddd to dddc has cost $\geq$ the cost of one of the following three listings:

cddd		cddd		cddd	
ddd	Delete c	dddd	Substitute c by d	dcdd	Transpose c and d
dddc	Insert c	dddc	Substitute d by c	ddcd	Transpose c and d ,
				dddc	Transpose c and d

so

$$d_L(\text{cddd, dddc}) = \min \begin{cases} w(\text{c}, \phi) + w(\phi, \text{d}), \\ w(\text{c, d}) + w(\text{d, c}), \\ 3w(\text{cd, dc}). \end{cases}$$

Next we calculate d_A. Here we find that there is no helpful way to make use of transpositions in an alignment from cddd to dddc. Every such alignment has cost $\geq$ the cost of one of the following two:

$$\begin{bmatrix} \text{c} & \text{d} & \text{d} & \text{d} & \phi \\ \phi & \text{d} & \text{d} & \text{d} & \text{c} \end{bmatrix} \qquad \begin{bmatrix} \text{c} & \text{d} & \text{d} & \text{d} \\ \text{d} & \text{d} & \text{d} & \text{c} \end{bmatrix},$$

so

$$d_A(\text{cddd, dddc}) = \min \begin{cases} w(\text{c}, \phi) + w(\phi, \text{d}), \\ w(\text{c, d}) + w(\text{d, c}). \end{cases}$$

These alignments correspond to the first two listings, but there is no alignment that corresponds to the third listing. If $w(\text{cd, dc})$ is small enough compared to the other costs, then the value of d_L will be different from that of d_A.

It might be possible to describe Wagner's traces and his trace distance in the present framework, but to do so would require an extremely large set G of generalized elementary operations and a complicated weight function w. Quite possibly, G might have to consist of all possible pairs of unequal sequences.

Working with trace or alignment distances in our sense is far simpler than working with listing distances. (In fact, Lowrance and Wagner (1975) incorporate transpositions in just such a manner.) It is precisely because alignment distances are used that the methods described in many other parts of the book have the simplicity they do. It is precisely because this simplicity is not available that Wagner's extended string-to-string correction problem is so difficult, and requires such powerful techniques to solve. In some applications involving generalized substitutions, it might be appropriate to ask whether trace or alignment distances in the present sense are acceptable (even though they are not mathematically equivalent to listing distances), in view of the simplicity they provide. In applications like those of Wagner, for instance, W-based trace or alignment distance might be quite reasonable, when one considers the human slip-of-the-finger mechanism that gives rise to the transpositions in the first place. In fact, it is conceivable that W-based trace distance, like that of Lowrance and Wagner (1975), might measure implausibility of change even better than does the listing distance.

8.2 G-based alignment distance: quartic and quadratic algorithms.

It is easy to work with alignment distance in the sense described here. To generalize the recurrence equation from Fig. 1, we introduce some notation: Let $\mathbf{x} \cdot \mathbf{y}$ indicate the sequence formed by concatenating $\mathbf{x}$ with $\mathbf{y}$, and note that $\mathbf{a^*} \cdot \mathbf{x} = \mathbf{a}^i$ just means that $\mathbf{a}^i$ can be split up into two parts, with the former being $\mathbf{a^*}$ and the latter being $\mathbf{x}$. Then the recurrence is

$$d(\mathbf{a}^i, \mathbf{b}^j) = \min_{\substack{\mathbf{x} \to \mathbf{y} \text{ in } G \\ \mathbf{a^*} \cdot \mathbf{x} = \mathbf{a}^i \\ \mathbf{b^*} \cdot \mathbf{y} = \mathbf{b}^j}} \{d(\mathbf{a^*}, \mathbf{b^*}) + w(\mathbf{x}, \mathbf{y})\}.$$

This recurrence may be evaluated computationally by at least two different approaches. In its simplest form, Approach 1 assumes that the list of generalized substitutions in G is finite and not too long: it runs through this list one by one. For each element $\mathbf{x} \to \mathbf{y}$ of G, it is easy and rapid to check whether $\mathbf{x}$ agrees with some terminal segment of $\mathbf{a}^i$ and whether $\mathbf{y}$ agrees with some terminal segment of $\mathbf{b}^j$. If they do, a term is formed and included in the minimization. The computation time $t_1(G)$ for one evaluation of the recurrence equation is dominated by the number of elements in G, but also reflects the time to compute w and also the lengths of the elements of G (because it takes longer to check whether **x** and **y** agree with terminal segments if **x** and **y** are long). The computation time for the entire algorithm is then

$$(m + 1)(n + 1)t_1(G) \approx mnt_1(G).$$

Approach 2 runs through every combination of a terminal segment of $\mathbf{a}^j$ and a terminal segment of $\mathbf{b}^j$, and checks whether the substitution of one by the other is in G. More specifically, let h and k be the lengths of the terminal segments. In its simplest form, Approach 2 runs through all pairs (h, k) with $0 \leq h \leq i$ and $0 \leq k \leq j$, excluding $(h, k) = (0, 0)$, and assumes that there is a simple method not requiring too much time, to check whether

(terminal segment of length h of $\mathbf{a}^i$) $\rightarrow$ (terminal segment of length k of $\mathbf{b}^j$)

is in G. If it is, a term is formed and included in the minimization. The computation time for the evaluation of this recurrence is dominated by

$$[(i + 1)(j + 1) - 1]t_2(G) = (ij + i + j)t_2(G),$$

where $t_2(G)$ includes the time required to check whether some given $\mathbf{x} \rightarrow \mathbf{y}$ is in G. Then the computation time for the entire algorithm is dominated by the sum of these times over $0 \leq i \leq m$ and $0 \leq j \leq n$, and the dominant part of the sum is $\frac{1}{4}m^2n^2t_2(G)$.

Comparing the computation times of the two approaches, we see that Approach 1 is quadratic in sequence length (taking $m = n$, it is proportional to n^2), while Approach 2 is quartic in sequence length. Approach 1 relies on the small number of elements in G to achieve more rapid computing. However, a more sophisticated version of Approach 1 would be necessary in most practical cases. Consider, for example, the elementary operations used by Wagner. If the alphabet has r letters in it, then there are

$r^2 - r$	ordinary substitutions,
r	deletions,
r	insertions, and
$r^2 - r$	transpositions (swaps),

for a total of $2r^2$ generalized substitutions. In applications, r might be in the range of 50 to several hundred, so $2r^2$ is hardly small.

To rescue Approach 1 in such a situation, it may be possible to divide the generalized substitutions into a small number of types, such as the four types just shown, and run through the *types* rather than running through the individual elements of G. This version of Approach 1 assumes that the list of types is not too long, and that for each *type* it is easy and rapid to check whether the type matches some combination of a terminal segment of $\mathbf{a}^i$ and a terminal segment of $\mathbf{b}^j$. If they do, a term is formed and included in the minimization. It is necessary to recognize that in some situations (though not here), a single type might match *several* combinations of terminal segments. (For example, suppose that the type consists of reversal of a sequence of any length, that is, $x_1 \ldots x_h \to x_h \ldots x_1$ for any h. If $\mathbf{a}^i$ and $\mathbf{b}^j$ are

$$c \; d \; c \; d \; c \; d \; c \; d$$

and

$$d \; c \; d \; c \; d \; c \; d \; c,$$

then a match would occur for $(h, k) = (2, 2), (4, 4), (6, 6)$ and $(8, 8)$.) Division into types can also be applied to Approach 2, though the advantages are not as clearcut as for Approach 1. We shall not, however, elaborate these ideas further.

8.3 Block deletions and insertions: a cubic-time algorithm

When sequence comparison arises from discretization of a continuous signal, as in speech research, there are situations where it may be desirable to permit a deletion or insertion of a consecutive block of length k to cost less than k separate deletions or insertions. Several ways to approach this situation have been discussed previously. Another is to consider deletion of any consecutive block of length k to be a single type of generalized substitution, with weight depending only on k, and similarly for insertion of a block of length k. Thus we let G consist of ordinary substitutions and all possible block insertions and deletions of all lengths. Ordinarily we would expect the weight to be an increasing but concave function of k, that is, if $w_k = w(a_1 \ldots a_k \to \phi)$, then

$$w_{k+1} - w_k < w_k - w_{k-1}.$$

However, in the present approach we permit w_k to be arbitrary.

In principle, the blocks can be arbitrarily long, so there are an infinite number of types of generalized substitutions, and Approach 1 is not helpful. (In practice, of course, we might limit the size of k.) It might appear that we could rescue Approach 1 by defining the types in a different manner, but that simply shifts the difficulty rather than resolving it.

Consider Approach 2 with the present set G. The only possible (h, k) for which

(terminal segment of length h of $\mathbf{a}^i$) $\to$ (terminal segment of length k of $\mathbf{b}^j$)

can be in G are

$$(h,\ k) = \begin{cases} (1,\ 1) & \text{ordinary substitution} \\ h = 0,\ k \text{ anything} & \text{block insertion,} \\ k = 0,\ h \text{ anything} & \text{block deletion,} \end{cases}$$

so it is necessary to run through only these $i + j + 1$ cases. The dominant part of the computation time for this iteration is $(i + j + 1)t_2(G)$, and for the entire algorithm the dominant term is $\frac{1}{2}(m + n)mnt_2(G)$, so in this case Approach 2 is cubic in sequence length.

9. CONSISTENCY AND METRIC PROPERTIES: SOME CONDITIONS

In one popular approach to defining the distance function d (not used in this chapter), d itself is given for the elementary operations, and the definition can be viewed as extending d from these special cases to the general case. This is called the "extension of d from elementary operations" approach, or the "extension" approach for short, and it gives rise to the question of consistency: namely, does d as defined agree with d as given for the elementary operations?

Another approach to defining d (which is used throughout this chapter), bases the definition on a separate weight function w which is defined over the elementary operations. This is called the "definition of distances from weights" approach, or the "weights" approach for short. Essentially the same question can be asked here: do d and w agree on the elementary operations? In this form, of course, the question is less urgent and may be avoided, which is the chief reason for choosing this approach above.

In this section we examine some conditions under which various closely related properties hold for distances:

metric properties;
agreement of d with w in the "weights" approach, which also covers consistency of d in the "extension" approach;
equality of different types of distance.

We deal with listing distance, alignment distance, and trace distance. We first consider the simple distances (in the sense of Chapter 1), and then deal in less detail with G-based distances.

In the mathematical literature, the word "distance" is ordinarily used to indicate a function d which satisfies the *metric properties*.

1. *Nonnegative Property:* $d(\mathbf{a},\ \mathbf{b}) \geq 0$ for all $\mathbf{a}$ and $\mathbf{b}$.
2. *Zero Property:* $d(\mathbf{a},\ \mathbf{b}) = 0$ if and only if $\mathbf{a} = \mathbf{b}$.
3. *Symmetry:* $d(\mathbf{a},\ \mathbf{b}) = d(\mathbf{b},\ \mathbf{a})$ for all $\mathbf{a}$ and $\mathbf{b}$.
4. *Triangle Inequality:* $d(\mathbf{a},\ \mathbf{b}) + d(\mathbf{b},\ \mathbf{c}) \geq d(\mathbf{a},\ \mathbf{c})$ for all $\mathbf{a},\ \mathbf{b},\ \mathbf{c}$.

Although we do not restrict the word "distance" in this way, we consider it desirable for d to have these important properties. We shall start by considering what properties of w are necessary and/or sufficient to achieve these properties for d. It is not surprising that the same properties, as applied to w (instead of d), play a prominent role.

We assume throughout this section that w has the nonnegative property, in order to avoid complications. When dealing with a set G of generalized elementary operations, we assume that G includes all ordinary indels. We assume that w is defined for all elementary operations but not for anything else, and we write $w(\mathbf{x}, \mathbf{y})$ to indicate the weight of the elementary operation $\mathbf{x} \rightarrow \mathbf{y}$. We shall use ϕ for the null sequence (i.e., sequence of length 0), and we shall consider an element as a sequence of length 1 whenever that is convenient. By this means, the familiar notation $w(x, y)$, $w(x, \phi)$, and $w(\phi, y)$ for ordinary substitutions, deletions, and insertions falls out from the systematic notation applicable to generalized elementary operations.

Simple listing distance. Obviously d has the nonnegative property (because w is assumed to have it), and d is symmetric if w is symmetric. It is easy to see that $d(\mathbf{a}, \mathbf{a}) = 0$ always holds, and that $d(\mathbf{a}, \mathbf{b}) = 0$ is possible with $\mathbf{a} \neq \mathbf{b}$ if and only if w is 0 somewhere (recall that $w(x, x)$ is not defined). Therefore, d has the zero property if and only if w has it. It is interesting to note, however, that d satisfies the triangle inequality whether or not w satisfies it. This is because a listing of length $d(\mathbf{a}, \mathbf{b})$ from $\mathbf{a}$ to $\mathbf{b}$ can be joined with a listing of length $d(\mathbf{b}, \mathbf{c})$ from $\mathbf{b}$ to $\mathbf{c}$ to obtain a listing of length $d(\mathbf{a}, \mathbf{b}) + d(\mathbf{b}, \mathbf{c})$ from $\mathbf{a}$ to $\mathbf{c}$, which shows that $d(\mathbf{a}, \mathbf{c}) \leqq d(\mathbf{a}, \mathbf{b}) + d(\mathbf{b}, \mathbf{c})$ as desired.

For agreement of d with w (so that d is an extension of w), it is necessary and sufficient that w satisfy the triangle inequality. If w does not satisfy the triangle inequality, however, we can define weights w^* very simply as the restriction of d to sequences of length $\leqq 1$, that is,

$$w^*(x, y) =_{\text{def}} d(x, y)$$

for any x and y which are in the alphabet or are ϕ. Then the distance function derived from w^* is again d (and w^* satisfies all the metric axioms which d satisfies, including the triangle inequality and any of the axioms that w satisfies). Thus any d obtained from weights can also be obtained by the extension approach.

Simple alignment distance and simple trace distance. It is not hard to see that every alignment has the same length as its corresponding trace. (For a length function introduced above, which incorporates block insertion and deletion weights, the corresponding statement is not true.) Thus simple alignment distance is always the same as simple trace distance, so this discussion covers the properties of both, though we refer only to the former.

As before, d must be nonnegative (because w is assumed nonnegative), d has the zero property if and only if w has the zero property, and d is symmetric if w is symmetric. In contrast to the preceding situation, however, d need not satisfy the triangle inequality in general; but if w satisfies the triangle inequality then d does also. The reason that alignment distance differs in this way from listing distance is that when two successive alignments are combined into a single alignment, two successive elementary operations affecting the same element cannot in general be left as separate operations (as they can for listings), but must be combined into a single elementary operation (two substitutions combine into a substitution, a substitution followed by a deletion combine into a deletion, etc.). Thus lack of the triangle inequality in w can be transmitted to d. There is one exception, however, whose effect will show up below. When a deletion, say $[^x_\phi]$ is followed by an insertion, say $[^\phi_y]$, they need not be combined into a single substitution $[^x_y]$ but may alternatively be left as two elementary operations $[^{x\phi}_{\phi y}]$. As a result, d satisfies a small part of the triangle inequality even if w doesn't.

For agreement of d with w (so that d is an extension of w), part of the triangle inequality on w is necessary and sufficient, namely, the assumption that

$$w(x, \phi) + w(\phi, y) \geqq w(x, y) \qquad \text{for all } x \text{ and } y.$$

This condition is necessary, because

$$d(x, \phi) + d(\phi, y) \geqq d(x, y)$$

always holds (see above).

In case w does not satisfy the partial triangle inequality above, new weights w^* can be defined by restriction of d, as they were for simple listing distance, and then d is obtained from w^* and agrees with it. Thus any d obtained from weights can also be obtained by the extension approach.

Equality of distances. We have already seen that simple alignment distance is the same as simple trace distance, based on the same weights. Simple listing distance, however, may be different from other distances based on the same weights, since it must satisfy the triangle inequality, and the others need not. If w satisfies the triangle inequality, however, then simple listing distance is the same as the other distances based on the same weights. (To see this, note that in the presence of the triangle inequality, combining two operations in one thread of a listing cannot make the listing longer. Therefore, there is a minimum-length listing which has only one operation in each thread. The simple length of such a listing is the same as the simple length of the corresponding alignment.)

If the weights may be changed, then a stronger result holds. Simple listing distance based on some weights w is always the same as simple alignment distance based on some modified weights w^*. (The modified weights may be obtained as shown above.)

Modification of w. In practice, modification of some initial w to achieve the triangle inequality or other metric properties is quite common. While a variety of reasons for doing this may be stated, it is often possible to see the influence of one or both of the following ideas, though the connection may be hard to make due to differences in viewpoint and notation:

i) to insure that the alignment distance, which can be computed by a rapid dynamic-programming algorithm, is the same as the listing distance, which is the conceptual starting point;

ii) to insure that the listing distance has some desired properties.

G-based distances. Some of the facts here closely follow those for the simple case. Obviously G-based listing distance d has the nonnegative property (because w is assumed to have it), and d is symmetric if w is symmetric. It is easy to see that $d(\mathbf{a}, \mathbf{a}) = 0$ always holds, and that $d(\mathbf{a}, \mathbf{b}) = 0$ is possible with $\mathbf{a} \neq \mathbf{b}$ if and only if w is 0 somewhere (recall that $w(\mathbf{x}, \mathbf{x})$ is not defined). Therefore d has the zero property if and only if w has it. Finally, d satisfies the triangle property whether or not w satisfies it. This, however, is as far as we have been able to go.

G-based alignment distance and G-based trace distance are equal, so the following statement covers both. The distance d must be nonnegative (because w is assumed nonnegative), d is symmetric if w is symmetric, and d has the zero property if and only if w has the zero property. On the other hand, we have already noted that G-based alignment distances and G-based listing distances are most certainly not the same in general.

The triangle inequality for G-based distances has an interesting and significant consequence, namely,

$$d(\mathbf{a}, \mathbf{b}) \leq c(m + n),$$

where m and n are the length of $\mathbf{a}$ and $\mathbf{b}$, and c is the maximum deletion or insertion cost for a single letter,

$$c = \max_x [d(x, \phi), d(\phi, x)].$$

Thus while the cost of replacing one subsequence by another (or of deleting or inserting a subsequence) may increase with the number of letters involved, this increase cannot be more than linear.

REFERENCES

Cedergren, R. J., and Sankoff, D., Evolutionary origin of 5.8S ribosomal RNA. *Nature* **259**:74–76 (1976).

Gotoh, O., An improved algorithm for matching biological sequences. *Journal of Molecular Biology.* **162**:705–708 (1982).

Lowrance, R., and Wagner, R. A., An extension of the string-to-string correction problem. *Journal of the ACM* **22**(2): 177–183 (1975).

Myers, C. S., and Rabiner, L. R., A level-building dynamic time-warping algorithm for connected-word recognition. *IEEE Transactions on Acoustics, Speech, and Signal Processing* ASSP-**29**(2): 284–297 (1981).

Sankoff, D., and Sellers, P. H., Shortcuts, diversions, and maximal chains in partially ordered sets. *Discrete Mathematics* **4**: 287–293 (1972).

Sankoff, D., and Cedergren, R. J., A test for nucleotide-sequence homology. *Journal of Molecular Biology* **77**: 159–164 (1973).

Sankoff, D., Cedergren, R. J., and LaPalme, G., Frequency of insertion–deletion, transversion, and transition in the evolution of 5S ribosomal RNA. *Journal of Molecular Evolution* **7**: 133–149 (1976).

Sellers, P. H., The theory and computation of evolutionary distances: Pattern recognition. *Journal of Algorithms* **1**: 359–373 (1980).

Smith, T. F., and Waterman, M. S., Identification of common molecular subsequences. *Journal of Molecular Biology.* **147**: 195–197 (1981).

CHAPTER ELEVEN

DISSIMILARITY MEASURES FOR CLUSTERING STRINGS

James M. Coggins

1. BACKGROUND

The need to quantify the dissimilarities between strings arises in several fields, one of which is grammatical inference. In the grammatical inference problem, we are given a set of sample strings from some language, and our task is to infer a generative grammar for those strings that in some sense "explains" the structure of the strings. Applications of grammatical inference appear in the areas of programming-language design (Crespi-Reghizzi *et al.*, 1973) and in syntactic pattern recognition (Lu and Fu, 1977a,b,c).

A general scheme for grammatical inference by a constructive method involves two steps: First, a canonical grammar is constructed for each string in the given sample of strings. Second, the canonical grammars are merged to produce a candidate grammar for the language containing the sample. The generation of canonical grammars for individual strings is not difficult in either the finite-state or the context-free case. The merging of the grammars, however, is complicated by the fact that the order in which the productions of the canonical grammars are merged has a significant impact on the form of the inferred candidate grammar.

A clustering of the sample strings by structural similarity can be used to guide and constrain the order in which the productions of the canonical grammars are merged. The canonical grammars for strings within a cluster are merged first; then the grammars for different clusters are merged later. The effect of the clustering is to reduce the arbitrariness of the merging order by providing constraints based on the structure of the strings in the sample. Placing constraints on the merging order simultaneously constrains the set of grammars that can be inferred.

The use of a clustering by structural similarity to constrain the order in which productions are merged involves the intuitively reasonable assumption that strings with similar structure are generated by similar sequences of productions. We can, of course, construct grammars that violate this assumption, but in the absence of contradictory information, this assumption may be adopted as a working hypothesis. There is a basis for this assumption involving acceptance criteria for candidate grammars. The ideal grammatical inference procedure would infer a grammar with few productions that "explains" the structure of the sample strings. Given a set of grammars that generate the sample, we tend to favor the grammar with the fewest productions. To minimize the number of productions in the candidate grammar, we ensure that strings with similar structure are produced by similar sequences of productions.

2. STRING DISSIMILARITY MEASURES

The effectiveness of the clustering depends on the measure of structural similarity (or dissimilarity) used to generate it. The most common string dissimilarity measure used in the literature is the Levenshtein distance d. This is defined as the minimum cost of a sequence of edit operations (change, delete, insert) that change one string into another. Each change operation costs 2. Delete and insert operations cost 1. Wagner and Fischer (1974) note another form for d which is the basis of the results in this paper and which will be given here as the definition of d:

$$d(\mathbf{a}, \mathbf{b}) = \text{length}(\mathbf{a}) + \text{length}(\mathbf{b}) - 2q(\mathbf{a}, \mathbf{b}), \qquad (1)$$

where length $(\mathbf{x})$ denotes the length of string $\mathbf{x}$ and $q(\mathbf{x}, \mathbf{y})$ denotes the length of the longest common subsequence between strings $\mathbf{x}$ and $\mathbf{y}$. Wagner and Fischer (1974) mention the formula (1) in noting that the length of the longest common subsequence, $q(\mathbf{x}, \mathbf{y})$, can be computed from $d(\mathbf{x}, \mathbf{y})$. At that time, faster algorithms existed for computing $d(\mathbf{x}, \mathbf{y})$ than for $q(\mathbf{x}, \mathbf{y})$.

Lu and Fu (1977a, b, c) use a modified form of d to cluster syntactic patterns before applying a grammatical-inference procedure. They modify d in two ways. First, different weights are used for insertion in the middle of a string and insertion at the end of a string. Second, the specific problem they investigate enables them to take advantage of detailed information about the structure of the alphabet, which is, for them, a set of encodings of pattern primitives. The cost of an edit operation depends on the letters (pattern primitives) involved in the operation.

The availability of such detailed information distinguishes their work from the work of Liou (1977) and Liou and Dubes (1977). Liou examines the case where detailed information about the structure of the alphabet is not available, so the costs of edit operations cannot depend on the letters involved. In this case, the dissimilarity between the strings must be determined solely from the structure of the sample strings. In this chapter we are concerned with the same problem as Liou.

Liou observes two problems in using d alone as the basis for a clustering procedure. First, d is insensitive to important structural features of strings such as common substrings and common symbols. (A substring is a sequence of consecutive elements of a string.) Substring relationships are very important for grammatical-inference procedures because such relationships can point out recursive structure that should be captured in the inferred grammar.

The second problem Liou observes is that d often assigns the same dissimilarity values to different pairs of strings. That is, d produces large numbers of ties. Ties pose a serious problem in this context because they force the clustering algorithm to make arbitrary decisions. This implies a more serious problem for the use of clustering in grammatical inference. The clustering procedure is used to reduce the arbitrariness in the order in which productions of canonical grammars are merged. The results of the clustering will be used to eliminate certain merging orders, and thereby to eliminate certain grammars from the set of grammars that can be inferred. If the clustering is itself arbitrary, we could be eliminating many accepable grammars from consideration, and our results could be less useful than they were before clustering was attempted.

Liou attempts to solve these problems in two ways. First, he attempts to fine-tune d by weighting edit operations according to their positions in the minimum-cost sequence of edit operations. For example, the first few operations would be weighted more heavily than later operations, or vice versa. Second, he develops a heuristic clustering procedure that resolves ties on the basis of substring and character information. The modified clustering procedure has several problems. The clustering algorithm is based on a minimum spanning tree (MST) that is constructed using d alone, before additional structural features are incorporated. Thus, the order in which the strings are presented to the MST construction algorithm has a great effect on the structure of the MST. Thus, the clustering algorithm works from a largely arbitrary MST, and the problem of arbitrary clustering is not solved. The problem can again be traced to the tendency of d to produce large numbers of ties. In this case, the ties result in problems for the MST algorithm.

One common method for eliminating ties when using d is to make the cost of each edit operation depend on the characters involved in the operation. The resulting string dissimilarity measure is called the weighted Levenshtein distance. However, assigning those weights in a meaningful, nonarbitrary way requires detailed information about the alphabet and the way it is used. We are interested here in the situation where such detailed information is not available. The approach we will use is to construct new string dissimilarity measures that are computed solely from structural properties of strings, not from detailed information about the alphabet. We begin by explaining the source of the difficulty when we use d alone.

When $d(\mathbf{x}, \mathbf{y})$ is defined as in formula (1), the reasons for the problems Liou observed become obvious. The insensitivity of d to important structural properties, especially the length of the longest common substring, is obvious. From (1), the value of d depends only on the lengths of the strings and on the length of their longest common subsequence.

Let $q(\mathbf{x}, \mathbf{y})$ denote the length of the longest common subsequence, as before, and let $s(\mathbf{x}, \mathbf{y})$ denote the length of the longest common substring between $\mathbf{x}$ and $\mathbf{y}$. The values of $q(\mathbf{x}, \mathbf{y})$ and $s(\mathbf{x}, \mathbf{y})$ are not independent of each other since $q(\mathbf{x}, \mathbf{y}) \geq s(\mathbf{x}, \mathbf{y})$ for any strings $\mathbf{x}$ and $\mathbf{y}$. Nevertheless, the actual substring and subsequence can involve completely different and nonoverlapping parts of the strings being measured. For example, let $\mathbf{x} = $ fadgcdeafabc and let $\mathbf{y} = $ abcfdgdef. Now, $q(\mathbf{x}, \mathbf{y}) = 6$ and $s(\mathbf{x}, \mathbf{y}) = 3$. The longest common subsequence is fdgdef and the longest common substring is abc. The longest common subsequence and the longest common substring have no characters in common and they do not overlap in either string. Thus, no subsequence measure (or weighted subsequence measure) can provide substring information. Nevertheless, we want to consider substring information when computing string dissimilarities.

The second problem Liou observed was that d produces large numbers of ties. The nature of this problem is also obvious from (1). The largest value that can be assumed by d is equal to the sum of the lengths of the strings. Since an even number is subtracted from that sum, the parity of the d value will always be the same as the parity of the sum of the lengths of the strings. Thus, the number of values that d can assume is severely limited. Consider a sample S of N strings, with no string longer than r characters. We would like to apply a clustering algorithm to the strings based on d. We observe from (1) that the largest value that d can assume for any pair of strings in S is $2r$. Let us assume that S is constructed so that all integral values between 1 and $2r$ appear in the dissimilarity matrix. We observe that there are $N(N - 1)/2$ entries in the dissimilarity matrix to fill and that d may assume any of $2r$ values. In order to preclude ties in the dissimilarity matrix, we require that $2r \geq N(N - 1)/2$. Table 1 shows the minimum string lengths required to enable the possibility of zero ties in the dissimilarity matrix for some reasonable values of N. It is interesting that Lu and Fu use a set of 51 syntactic patterns, the longest of which has $r = 17$ characters. Except for the extra information they inferred about the structure of the alphabet, they would have had severe problems with ties.

3. IMPROVED STRING DISSIMILARITY MEASURES

We see that d cannot assume enough values to prevent large numbers of ties, and the problem increases rapidly with N. The problem is too severe to be corrected by weighting functions like those of Liou. Even the detailed weighting functions used by Lu and Fu are not completely satisfactory since the weights

Table 1. Minimum r for some values of N for a dissimilarity matrix with no ties

N	r	N	r	N	r
4	3	9	18	24	138
5	10	13	39	28	189
8	14	16	60	40	390

are still largely arbitrary. We need to be able to incorporate more structural features from the strings in the dissimilarity measure.

In incorporating further structural information, we do not wish to override the ordering of the d distances (at least, not without good reason). By incorporating further structural information in the dissimilarity measure, we hope to refine the ordering of d distances. We would like to maintain the order of the string distances given by d, but we want to impose a reasonable order among string distances that are equal under d.

One attraction of d is that it is metric. A dissimilarity measure, $D(\mathbf{x}, \mathbf{y})$, is metric if the following conditions are satisfied:

a) $D(\mathbf{x}, \mathbf{y}) \geq 0$ for all $\mathbf{x}$ and $\mathbf{y}$;
b) $D(\mathbf{x}, \mathbf{y}) = 0$ if and only if $\mathbf{x} = \mathbf{y}$;
c) $D(\mathbf{x}, \mathbf{y}) = D(\mathbf{y}, \mathbf{x})$ for all $\mathbf{x}$ and $\mathbf{y}$;
d) $D(\mathbf{x}, \mathbf{y}) \leq D(\mathbf{x}, \mathbf{z}) + D(\mathbf{z}, \mathbf{y})$ for all $\mathbf{x}, \mathbf{y}, \mathbf{z}$.

We will maintain these properties in any modification of d we construct.

We observe that the form of (1) can be described as a conversion of a similarity measure, $q(\mathbf{x}, \mathbf{y})$, into a dissimilarity measure. The form is based on a similarity measure, S, and a reversal function, R, and is

$$D(\mathbf{x}, \mathbf{y}) = R(\mathbf{x}, \mathbf{y}) - 2S(\mathbf{x}, \mathbf{y}), \qquad (2)$$

where the dissimilarity measure D is defined by subtracting twice the value of S from a maximum dissimilarity given by the reversal function. For d, the reversal function is the sum of the string lengths. The nature of the reversal function depends on what the similarity measure is measuring—length, number of different symbols, etc. Its definition must be such that $D(\mathbf{x}, \mathbf{y})$ is metric.

We can immediately define other dissimilarity measures of the same form as (2) that are based on different structural features of strings. For example,

$$D_s(\mathbf{a}, \mathbf{b}) = \text{length}(\mathbf{a}) + \text{length}(\mathbf{b}) - 2s(\mathbf{a}, \mathbf{b}), \qquad (3)$$

where $s(\mathbf{x}, \mathbf{y})$ is the length of the longest common substring between $\mathbf{x}$ and $\mathbf{y}$.

Table 2: Liou's data set with the distance of each string from "(a)".

	Computed by:				Computed by:		
String	d	D_2	D_3	String	d	D_2	D_3
(a)	0	0	0	a + a + b	6	12	40
				a + b + a	6	12	40
a + (a)	2	4	5	a + b + b	6	12	40
a	2	4	6	b + a + a	6	12	40
(a) + b	2	4	6	b + a + b	6	12	40
((a))	2	6	8	b + b	6	12	41
(a + a)	2	6	9				
(b)	2	6	10				
(a + b)	2	6	10				
(b + a)	2	6	10	b + b + b	8	16	69
a + a	4	8	19				
b	4	8	20				
a + b	4	8	20				
a + (b)	4	10	26				
(b) + a	4	10	26				
((b))	4	10	26				
b + (b)	4	10	27				

Another example is

$$D_c(\mathbf{a}, \ \mathbf{b}) = \text{char}(\mathbf{a}) + \text{char}(\mathbf{b}) - 2c(\mathbf{a}, \ \mathbf{b}), \tag{4}$$

where char($\mathbf{x}$) gives the number of different characters from the alphabet used in $\mathbf{x}$ and $c(\mathbf{x}, \mathbf{y})$ gives the number of different characters from the alphabet that $\mathbf{x}$ and $\mathbf{y}$ have in common. We now see that d is only one of a class of string dissimilarity measures having the form of (2). The metric properties of these measures are easily shown.

Composite string-dissimilarity measures can be formed from those already constructed. For example, we can construct a new measure as

$$D_2(\mathbf{a}, \ \mathbf{b}) = D_s(\mathbf{a}, \ \mathbf{b}) + d(\mathbf{a}, \ \mathbf{b}), \tag{5}$$

which can be rewritten to reflect the form of (2) by substituting (1) and (3) into (5), giving

$$D_2(\mathbf{a}, \ \mathbf{b}) = 2(\text{length}(\mathbf{a}) + \text{length}(\mathbf{b})) - 2(q(\mathbf{a}, \ \mathbf{b}) + s(\mathbf{a}, \ \mathbf{b})).$$

Other weightings and combinations are possible. For example, let

$$D_3(\mathbf{a}, \mathbf{b}) = d(\mathbf{a}, \mathbf{b})D_s(\mathbf{a}, \mathbf{b}) + D_c(\mathbf{a}, \mathbf{b}). \tag{6}$$

This metric uses d and D_s as its primary measures and then adds D_c to break any remaining ties.

4. EXAMPLE

We will now apply dissimilarity measures d, D_2, and D_3 to a set of strings from Liou (1977). The strings are listed in Table 2 along with the distance of each from the length-3 string "(a)" as computed by d, D_2, and D_3. It should be noted that, since all the string lengths have the same parity (they are all odd), the problem of ties for d is twice as severe. The sum of the string lengths will always be even, so the values of d and D_2 will always be even, as noted before. Thus, d can assume only half of the possible values in its range, 0 to $2r$. From Table 2, we observe that the number of ties decreases markedly as more structural information is incorporated into the dissimilarity measure. Note that the refinement becomes coarse as the distance from the given string increases. This is because the value of $s(\mathbf{x}, \mathbf{y})$ reaches zero or one quickly, so further refinement is made only on the basis of character information in D_3. The values of d dominate the composite measures. We also observe that the groups of ties produced by d are refined by the additional structural information with no

Table 3. Distributions of the values of d, D_2, and D_3 in the entire dissimilarity matrix.

d		D_2		D_3			
Value	Freq.	Value	Freq.	Value	Freq.	Value	Freq.
2	52	4	27	4	7	27	13
4	125	6	20	5	11	28	2
6	58	8	72	6	9	32	2
8	16	10	55	8	4	34	2
10	1	12	47	9	6	36	1
		14	15	10	10	38	8
		16	16	12	4	39	13
		20	1	13	1	40	16
				14	1	41	5
				16	7	49	2
				17	10	50	6
				18	19	51	6
				19	18	52	1
				20	12	67	1
				24	4	68	12
				25	12	69	3
				26	24	105	1

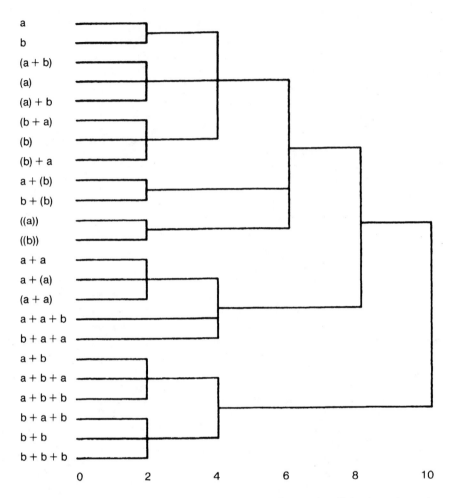

Figure 1. Dendrogram resulting from application of complete-link clustering using distance d.

crossovers in this case. But this table shows distances from only one string in the sample.

Table 3 shows summary statistics from the entire dissimilarity matrix. While the problem of ties is not totally eliminated, much more precise guidance is now available for clustering algorithms applied to this set of strings. Figures 1 through 3 show the dendrograms resulting from the application of a complete-link clustering algorithm to the same sample strings using d, D_2, and D_3 as the dissimilarity measures. The ordering imposed by the addition of substring and character information results in more refined and useful clusterings than occurred using d alone.

5. CONCLUSION

We have presented, within the context of the grammatical-inference and clustering applications, a generalization of Levenshtein distance that allows more structural information from the strings to be incorporated in the dissimilarity measure. The general form of a family of metric dissimilarity measures was presented, and some particular string dissimilarity measures from that family were constructed and demonstrated.

The effects of these new string dissimilarity measures in complete grammatical inference systems and in other applications such as spelling

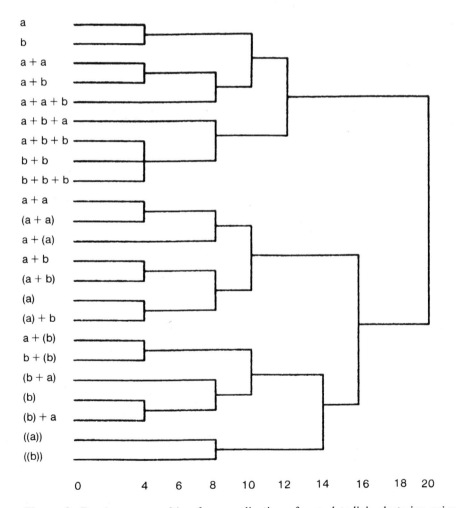

Figure 2. Dendrogram resulting from application of complete-link clustering using distance D_2.

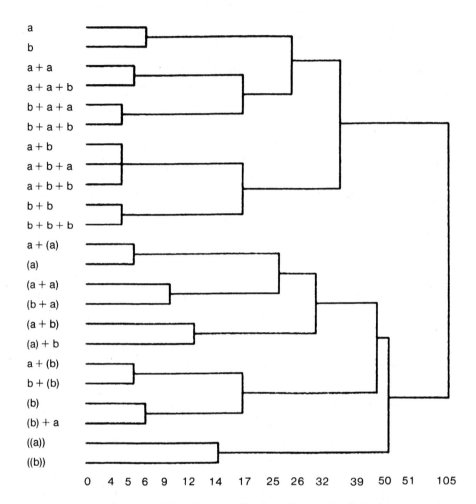

Figure 3. Dendrogram resulting from application of complete-link clustering using distance D_3.

correction (Okuda *et al.,* 1976) and syntactic pattern recognition will be subjects for further research. The construction of composite measures other than those presented here and the development of other basic metrics may also prove fruitful.

REFERENCES

Crespi-Reghizzi, S., *et al.* The use of grammatical inference for designing programming languages, *Communications of The Association for Computing Machinery* **16**, Feb. 1973, p. 83.

Hunt, J. W., and Szymanski, T. G., A fast algorithm for computing longest common subsequences, *Communications of the Association for Computing Machinery* **20**, May 1977, p. 350.

Liou, J., Grammatical Inference by Constructive Method. Ph.D. Thesis, Department of Computer Science, Michigan State University, 1977.

Liou, J., and Dubes, R. C., A Constructive Method for Grammatical Inference Based on Clustering. Technical Report TR 77–01. Department of Computer Science, Michigan State University, 1977.

Lu, S. Y., and Fu, K. S., Error-correcting Parsing for Syntactic-Pattern Recognition. Ph.D. Thesis, School of Electrical Engineering. Purdue University, 1977a.

Lu, S. Y., and Fu, K. S., A clustering procedure for syntactic patterns, *IEEE Transactions on Systems, Man, and Cybernetics,* Oct. 1977b, p. 734.

Lu, S. Y., and Fu, K. S., A sentence-to-sentence clustering procedure for pattern analysis, *Proceedings of IEEE Conference on Image Processing,* 1977c.

Okuda, T., Tanaka, E., and Kasai, T., A method for correction of garbled words based on the Levenshtein metric, *IEEE Transactions on Computers* **C25**, Feb. 1976, p. 172.

Wagner, R. A., and Fischer, M. J., The string-to-string correction problem, *Journal of the Association for Computing Machinery* **21**, Jan. 1974, p. 168.

PART FOUR

COMPUTATIONAL COMPLEXITY

1. INTRODUCTION

The chapters in Part IV focus largely on the computational complexity of sequence comparison problems. The latter parts of Chapter 7 (in Part III), by Wagner, are also concerned with this topic.

Consider the well-studied comparison problem in which each substitution is weighted at least twice as heavily as an insertion or deletion. In this case, substitutions may be ignored, so finding the minimum weight matching of two sequences is essentially the same as finding their longest common subsequence. Chapter 12, by Hirschberg, reviews what is known about the computational complexity of this and three other closely related problems: finding the shortest common supersequence, finding the longest common substring (i.e., consecutive subsequence), and finding the shortest common superstring. It also surveys more general versions of these problems, in which N sequences are compared instead of two, and the relationship of all these problems to other well-known problems of computer science.

Chapter 14, by Masek and Paterson, sets an upper bound on the complexity of the basic sequence-comparison problem through a detailed presentation and analysis of an algorithm which, asymptotically for long sequences, is the fastest currently available. (This algorithm may be used whenever the alphabet is finite and the given weights are rational.) The running time of the algorithm is proportional to $n^2/\log n$ for two sequences of length n, and to $mn/\min(m, \log n)$ for sequences of lengths m and n with $m \leq n$.

Sometimes a sequence is compared not with another sequence but with a large class, possibly infinite, of sequences defined by a "grammar." The problem is to find the grammatical sequence that is closest to the given sequence and/or to find the distance involved. Chapter 13, by Wagner, surveys what is known about the complexity of string-to-grammar correction when the grammar is regular, or context-free, or context-sensitive, and sketches a proof that the problem is NP-complete for a context-sensitive grammar. Chapter 10 (of Part III) discusses two related problems under the headings of Directed Networks and Level-Building.

CHAPTER TWELVE

RECENT RESULTS ON THE COMPLEXITY OF COMMON-SUBSEQUENCE PROBLEMS

Daniel S. Hirschberg

1. INTRODUCTION

We present an overview of recent results in the solution of a variety of common-sequence problems.

Consider two strings $\mathbf{a} = a_1 a_2 \ldots a_m$ and $\mathbf{b} = b_1 b_2 \ldots b_n$ whose elements are members of an alphabet (for illustrative purposes, we shall take it to be the English alphabet). If there is a mapping $F: \{1, \ldots, m\} \to \{1, \ldots, n\}$ such that $F(i) = k$ only if $a_i = b_k$ and F is a monotone strictly increasing function (that is, $F(i) = u$, $F(j) = v$ and $i < j$ imply that $u < v$), then we say that string $\mathbf{a}$ is a *subsequence* of string $\mathbf{b}$ and that string $\mathbf{b}$ is a *supersequence* of string $\mathbf{a}$. String $\mathbf{a}$ could be obtained by deleting zero or more symbols from string $\mathbf{b}$, while string $\mathbf{b}$ could be obtained by inserting zero or more symbols into string $\mathbf{a}$. For example, "course" is a subsequence of "computer science."

If string $\mathbf{a}$ can be obtained from string $\mathbf{b}$ by deleting a (possibly null) prefix and/or suffix of $\mathbf{b}$, then we say that string $\mathbf{a}$ is a *substring* of string $\mathbf{b}$ and that string $\mathbf{b}$ is a *superstring* of string $\mathbf{a}$. For example, "our" is a substring of "course."

A string $\mathbf{c}$ is a *common subsequence* of strings $\mathbf{a}$ and $\mathbf{b}$ if it is a subsequence of both. Similar definitions hold for common supersequence, common substring, and common superstring. Here $\mathbf{c}$ is a *longest common subsequence* (LCS) if it is a common subsequence and is as long as any other common subsequence of $\mathbf{a}$ and $\mathbf{b}$. Similar definitions hold for shortest common supersequence

(SCS), longest common substring (LCG), and shortest common superstring (SCG). For example, "cled" and "coed" are the LCSs of "schooled" and "encyclopedia."

We define the N–LCS problem as: Find a string **c** that is an LCS of (a given set of) N strings. That is, find a string of maximum length that is a subsequence of each of the N strings that are given as input. We similarly define the N–SCS, N–LCG and N–SCG problems.

These problems naturally occur in a number of computing and data-processing applications. For example, we can consider a file to be a string in which each line of the file is a single symbol. An LCS of two files can then be used to determine the discrepancies between the two files.

A genetic application arises in the study of the evolution of long molecules such as proteins; there, a 2–LCS is used to measure the correlation between two such molecules (Needleman and Wunsch, 1970; Sankoff, 1972).

2. NP-COMPLETENESS RESULTS

Richard Karp (1972) has introduced the notion of NP-completeness. Essentially, there is a class of "hardest" problems that can be solved in polynomial time on a nondeterministic Turing machine. However, there is, as yet, no known polynomial time solution for any of these problems (on a deterministic machine). In addition, should a polynomial solution be devised for any one of the problems in this class, then there will be an effective polynomial solution for all of the NP-complete problems. Conversely, if it can be shown that any one of the NP-complete problems requires time not bounded by any polynomial, then this will be proof that all the problems in that class have no polynomial solution.

The travelling salesman problem, knapsack problem, and maximal clique problem are examples of problems that are known to belong to this class. It is a common conjecture that there is no polynomial solution for any of the NP-complete problems.

David Maier has shown (1978 and personal communication) that if N is arbitrary and is part of the input specifications, then the N–LCS problem (for alphabet size ≥ 2), the N–SCS problem (for alphabet size ≥ 5), and the N–SCG problem (for alphabet size ≥ 2) are all NP-complete problems. Hence, we must fix N if we hope for time-efficient algorithms.

3. POLYNOMIAL ALGORITHMS

The 2–LCS problem has received considerable attention in the literature. Wagner and Fischer (1974) demonstrated an algorithm for the 2–LCS problem that required quadratic time and space. The algorithm uses the dynamic-programming approach. Let $d(i, j)$ be the length of the 2–LCS of the i-prefix and j-prefix of the two input strings. A simple recurrence relation exists between $d(i, j)$ and $d(i - 1, j)$, $d(i, j - 1)$, and $d(i - 1, j - 1)$. A d-matrix can be filled in with the correct values of the function d in quadratic time, since each value depends on considering at most three previously computed values. (Initially, $d(i, 0)$ and $d(0, j)$ are set to zero for all i and j.) The value of $d(i, j)$ will be equal to the maximum of the three other values, possibly plus one, depending on whether or not $a_i = b_j$. If, in addition, we store pointers indicating which of the three values $d(i, j)$ was dependent on, we can recover an LCS by tracing these threads from $d(m, n)$ back to $d(0, 0)$. This method requires quadratic sqace (to hold the d-matrix).

Hirschberg (1975) reduced the space complexity of the 2–LCS problem to linear. The basic idea is that, since each row of the d-matrix can be evaluated from knowing the values of just the previous row, the entire d-matrix need not be stored. Recovery of an LCS, however, is not as simple as before.

The "curve" that recovers an LCS was obtained by following the threads through the d-matrix after the entire matrix was computed. As mentioned earlier, any single row of the matrix can be computed in linear space. We can compute the middle row of d and also the middle row of the matrix d' that solves the LCS problem for the reverse strings. It can be shown that their sum is maximized at points where an LCS curve intersects. By applying this procedure recursively, the quartile intersections can be recovered, etc. Each iteration uses linear space, and it can be shown that the total time used is still quadratic (though with a slightly higher coefficient).

The N–LCS and N–SCS problems can similarly be solved using the dynamic-programming approach and requiring time and space $O(n^N)$. It is possible that the space complexity may be reduced, using a method similar to that for the 2–LCS problem.

Morris and Pratt (1970) and Weiner (1973) showed how to solve the 2–LCG problem in linear time (and linear space) using the bi-tree data structure. Several related problems (such as determination of the longest substring **c** of **a** that occurs in k places within **a**) can also be solved in linear time. McCreight (1976) developed another algorithm with similar characteristics that has smaller space requirements.

The 2–SCG problem can be solved in linear time. The solution will consist of one of the two input strings if it is a superstring of the other or it will consist of one string followed by the suffix of the other where the "missing prefix" of the second string is a suffix of the first string. Determining whether **a** is a substring of **b** and finding a maximum length "prefix–suffix" can both be done in linear time (see, for example, Aho, Hopcroft and Ullman, 1974, pp. 329–335). This technique can be extended so as to get a solution to the N–SCG problem in time $O(n \cdot N!)$. Simply enumerate all orderings of the N strings and iterate, merging together the first two strings into a 2–SCG.

4. LOWER BOUNDS

Aho, Hirschberg, and Ullman (1976) have shown that algorithms for the 2–LCS problem that are restricted to making "equal–not equal" comparisons must make $O(n^2)$ comparisons for alphabets of unrestricted size and $O(ns)$ comparisons for alphabets of size restricted to s.

It has also been shown (Hirschberg, 1978) that $O(n \log n)$ is a lower bound on the number of "less than–equal to–greater than" comparisons required to solve the 2–LCS problem.

See also Wong and Chandra (1976).

5. OTHER ALGORITHMS FOR THE 2–LCS PROBLEM

Hirschberg (1977) presents algorithms that, depending on the nature of the input strings, may not require quadratic time to recover an LCS. One of the algorithms finds an LCS in time $O(pn + n \log n)$, where p is the length of the LCS. Thus, this algorithm may be preferred for applications where the expected length of an LCS is small relative to the lengths of the input strings. The second algorithm presented in that paper requires time bounded by $O((m + 1 - p)p \log n)$. In the common special case where p is close to m (the length of the shorter of the two input strings), this algorithm takes time much less than n^2. See also Hunt and Szymanski (1977).

The only known algorithm for the LCS problem with worst-case behavior less than quadratic is due to Paterson (1974) and applies to the case of a finite alphabet of size s. The algorithm has complexity $O(n^2 \log \log n/\log n)$ in its basic form. It uses a "Four Russians" approach (see Arlazarov et al. (1970) or Aho, Hopcroft and Ullman (1974)). Essentially, instead of calculating the matrix d, the matrix is broken up into boxes of some appropriate size, k. The "high" sides of a box (the $2k - 1$ elements of d on the edges of the box with largest indices) are computed from d-values known for boxes adjacent to it on the "low" side and from the relevant symbols of **a** and **b** by using a lookup table that was precomputed.

There are $2k + 1$ elements of d adjacent to a box on the "low" side. Two adjacent d-elements can differ by either zero or one. There are thus 2^{2k} possibilities in this respect. The **a**- and **b**-values range over an alphabet of size s for each of the $2k$ elements, yielding a multiplicative factor of s^{2k} and the total number of boxes to be precomputed is therefore $2^{2k(1+\log s)}$. Each such box can be precomputed in time $O(k^2)$ for a total precomputing time of $O(k^2 2^{2k(1+\log s)})$.

There are $(n/k)^2$ boxes to be looked up, each of which will require $O(k \log k)$ time to be read (since the numbers being read are in the range 0 to k) for a total time of $O(n^2 \log k/k)$.

The total execution time will therefore be

$$O(k^2\ 2^{2k(1+\log s)} + n^2\ \log k/k).$$

If we let $k = \log n/2$ $(1 + \log s)$, we see that the total execution time will be $O(n^2 \log \log n/\log n)$, which is better than quadratic.

Masek and Paterson (1976, and Chapter 14 in this volume) and Szymanski (1976) have independently observed that the sides of a box can be stored as "steps" consisting of 0s and 1s indicating whether adjacent elements of the side differ by 0 or 1. A box can therefore be looked up in time $O(2k)$, reducing that component of the execution cost from $O(n^2 \log k/k)$ to $O(n^2/k)$. Choosing the same value for k as before, the total execution time will be $O(n^2/\log n)$.

Szymanski has also demonstrated (1976) that it is possible to modify this algorithm for the case when the alphabet is of unrestricted size, with the resulting time complexity of $O(n^2 (\log \log n)^2/\log n)$.

If we consider the LCS problem with the restriction that no symbol appears more than once within either input string, then this problem can be solved in time $O(n \log n)$ (Szymanski, 1975).

If, in addition, one of the input strings is the string of integers 1 through n, this problem is equivalent to finding the longest ascending subsequence in a string of distinct integers. If we assume that a comparison between integers can be done in unit time, this problem can be solved in time $O(n \log \log n)$ using the techniques of van Emde Boas (1974).

We note that the 2–LCS problem is equivalent to the string-to-string correction problem. That problem is as follows: Given strings **a** and **b**, find the minimum-length sequence of insert-symbol and/or delete-symbol instructions that will transform string **a** into string **b**. Lowrance and Wagner (1975) have studied an extension of this problem that allows character interchange as an instruction (that is, $\ldots ab \ldots \rightarrow \ldots ba \ldots$). They showed that this problem can also be solved in quadratic time. (See also Wagner (1975) and Chapter 7 in this volume).

6. AN OPEN PROBLEM

The 2–LCS problem has a general lower bound of $n \log n$ and a general upper bound of $n^2 (\log \log n)^2/\log n$. For restricted alphabets, say $\{0, 1\}$, a lower bound of n and an upper bound of $n^2/\log n$ are the best known. Can any of these four bounds be improved upon?

REFERENCES

Aho, A. V., Hirschberg, D. S., and Ullman, J. D., Bounds on the complexity of the longest common subsequence problem, *Journal of the Association for Computing Machinery* **23**:1, 1–12 (1976).

Aho, A. V., Hopcroft, J. E., and Ullman, J. D., *The Design and Analysis of Computer Algorithms.* Reading, Mass.: Addison-Wesley, 1974.

Arlazarov, V. L., Dinic, E. A., Kronrod, M. A., and Faradzev, I. A., On economic construction of the transitive closure of a directed graph. *Dokl. Akad. Nauk.*

SSSR **194**, 487–488 (in Russian). English translation in *Soviet Math. Dokl.* **11**:5, 1209–1210 (1970).

Hirschberg, D. S., A linear-space algorithm for computing maximal common subsequences. *Communications of the Association for Computing Machinery* **18**:6, 341–343 (1975).

Hirschberg, D. S., Algorithms for the longest-common-subsequence problem. *Journal of the Association for Computing Machinery* **24**:3, 664–675 (1977).

Hirschberg, D. S., An information-theoretic lower bound for the longest-common-subsequence problem. *Info. Proc. Letters* **7**:1, 40–41 (1978).

Hunt, H. B., and Szymanski. T. G., A fast algorithm for computing longest common subsequences. *Communications of the Association for Computing Machinery* **20**:5, 350–353 (1977).

Karp, R. M., Reducibility among combinatorial problems. In *Complexity of Computer Computations,* R.E. Miller and J.W. Thatcher, eds.. Plenum Press, 85–103 (1972).

Lowrance R., and Wagner R. A., An extension of the string-to-string correction problem. *Journal of the Association for Computing Machinery* **22**:2, 177–183 (1975).

Maier, D., The complexity of some problems on subsequences and supersequences. *Journal of the Association for Computing Machinery* **25**:2, 322–336 (1978).

Maier, D., Personal communication (1978).

Masek, W. J., and Paterson, M. S., A faster algorithm for computing string-edit distances. Technical Report MIT/LCS/TM-105 (1978).

McCreight, E. M., A space-economical suffix-tree construction algorithm. *Journal of the Association for Computing Machinery* **23**:2, 262–272 (1976).

Morris, J. H., and Pratt, V. R., A linear pattern-matching algorithm. TR–40, Computing Center, Univ. of Calif. at Berkeley (1970).

Needleman, S. B., and Wunsch, C. W., A general method applicable to the search for similarities in the amino acid sequence of two proteins. *Journal of Molecular Biology* **48**, 443–453 (1970).

Paterson, M. S., Unpublished manuscript, University of Warwick, England (1974).

Sankoff, D., Matching sequences under deletion/insertion constraints. *Proc. Nat. Acad. Sci. USA* **69**:1, 4–6 (1972).

Szymanski, T. G., A special case of the maximal common subsequence problem. TR–170, Computer Sciences Lab., Princeton University (1975).

Szymanski, T. G., Personal communication (1976).

van Emde Boas, P., An $O(n \log \log n)$ on-line algorithm for the insert-extract min problem. TR 74–221, Dept. of Computer Science, Cornell University (1974).

van Emde Boas, P., Preserving order in a forest in less than logarithmic time. *Proc. 16th Annual Symp. on Foundations of Comp. Sci.,* 75–84 (1975).

Wagner, R. A., On the complexity of the extended string-to-string correction problem. *Proc. Seventh Annual Association for Computing Machinery Symp. on Theory of Computing,* 218–223 (1975).

Wagner, R. A., and Fischer, M. J., The string-to-string correction problem. *Journal of the Association for Computing Machinery* **21**:1, 168–173 (1974).

Weiner, P., Linear pattern-matching algorithms. *Proc. 14th Annual Symp. on Switching and Automata Theory,* 1–11 (1973).

Wong, C. K., and Chandra, A. K., Bounds for the string-editing problem. *Journal of the Association for Computing Machinery* **23**:1, 13–16 (1976).

CHAPTER THIRTEEN

FORMAL-LANGUAGE ERROR CORRECTION

Robert A. Wagner

1. INTRODUCTION

The research that led to identification of the string-to-string correction problem (Wagner and Fischer, 1974) as an area of interest was motivated by the field of computer science called *compiler design*. A *compiler,* in computer science, is a program that translates another program, expressed in a source language, to another language. Formal language theory has developed to the point where it is straightforward to design that part of a compiler which:

1. reads the source program;
2. "parses" the source program (which constitutes a formal proof that the program presented as input in fact belongs to the source language);
3. generates a target-language program "equivalent" to the source program.

Nonetheless, some problem areas remain:

1. Target-language generation, if done in a simple fashion, produces inefficient target programs. Aho, Johnson, and Ullman (1977), together with the designers of BLISS (Wulf *et al.,* 1975) have made significant progress in this area, but they have also uncovered a number of "code-generation" problems that are NP-complete (i.e., probably difficult for a computer to solve quickly).
2. Incorrect source programs constitute a second area of difficulty, one which motivated the notion of formal language-error correction.

Basically, human beings write and keypunch computer programs. This process is almost bound to produce source programs that are not correct in one sense or another. When the compiler or other language-processing system is presented with such "erroneous" input, the human beings who use the system expect considerable help in isolating their errors. Unfortunately conventional formal language theory says little about "erroneous" input programs. Thus, some new theory had to be developed.

2. ERROR PROCESSING

Conventional compilers process erroneous programs, of course. Typically the program is read until its first symbols (its prefix) cannot possibly be the prefix of any "legal" program. The compiler indicates to the program submitter this "first point of error detection," and initiates some kind of "recovery" action. Typically the indication is an *error message,* and the "recovery" action can be *ad hoc,* and hard for the compiler user to understand. Often, "recovery" proceeds by skipping some or all of the input characters that follow the point of error detection, and this action may be accompanied by change in the compiler's internal state, which causes the compiler to act as if some prefix different from the one actually present had been read. Parsing then proceeds on the program *as modified,* and often other error messages are generated, many the result of these "recovery" actions.

The notion of "correction" of an incorrect program can then be expressed as "modify the original program, P, in some way to produce program $S(P)$", where $S(P) \in L$, L being the compiler's source language (a set of strings of symbols over an alphabet, V). The modification action, $S(\cdot)$, applied to P, should be clearly understandable to the compiler user, so that he or she could, with little effort, perform the same modification himself. Also, $S(P)$ should somehow change P as little as possible, so that the compiler user can relate $S(P)$ to P. One obvious candidate for $S(\cdot)$ is the notion of *a sequence of single-token edit operations.* Each edit operation in such a sequence either:

1. Changes one token to another, or
2. Deletes one token from P, or
3. Inserts one token into P.

(A "token" is a lexical unit, usually a *number,* a *word,* or an *operator symbol.*)

Such a modification scheme was explicitly built into the PL/C compiler (Morgan and Wagner, 1971; Conway and Wilcox, 1973), which is particularly helpful to users whose programs contain syntactic errors. That compiler formalizes error recovery, by relating any changes made in the compiler's state to explicit editing operations, performed on P. Each edit operation is accompanied by an error message, and after all "corrections" have been made to a statement, the modified statement is printed on the listing. For example, the listing of an erroneous statement might appear as this:

$$113 \quad A \quad 7,3) = x;$$
$$\text{ERROR 1023:} \quad \text{MISSING (.}$$
$$\text{PL/C USES:} \quad 113 \quad A \quad (7, 3) = x;$$

In effect, PL/C regards errors as single-token "mutations" in an originally correct string of source tokens. PL/C "reverses" each such mutation, printing a message for each mutation it finds, and then reproduces the supposed original to provide feedback to the user.

Problems remain, however. First, there is some ambiguity as to which corrections to make in any erroneous text. Thus, why not this?

$$113 \quad A \quad 7,3) = x;$$
$$\text{ERROR 1022:} \quad \text{tokens} \quad 7,3) \quad \text{deleted}$$
$$\text{PL/C uses:} \quad 113 \quad A \quad = \quad x;$$

Second, there exists the possibility of a "cascade" of error messages, each actually caused by an unintelligent choice of recovery action. Thus, deleting the "7" above, makes the "," illegal, requiring *its* deletion, also.

3. OPTIMAL ERROR CORRECTION

Examples like the one above lead to a mathematical criterion for correction selection:

1. For any two strings **a** and **b**, define $d(\mathbf{a}, \mathbf{b})$ to be the least "weight" of any sequence of single-token edit operations that changes **a** into **b**.
2. Choose as the correction, **c**, for an input string **a**, that string $\mathbf{b} \in L$ such that $d(\mathbf{a}, \mathbf{b})$ is least.

Thus:

$$\mathbf{c} = \text{argument min } \{d(\mathbf{a}, \mathbf{b}) \,|\, \mathbf{b} \in L\}.$$

Note that $d(\mathbf{a}, \mathbf{a}) = 0$, so that the "correction" of an error-free program is itself. Note also that, if each edit operation in the sequence is given a positive weight, and the weight of the sequence is defined as the sum of the weights of its components, minimizing weight tends to minimize the number of edit operations performed in changing **a** to **c**. Interestingly enough, the nonlinear optimization problem of finding **c**, given **a** and L, can be solved efficiently for many classes of languages L.

4. SOME RESULTS

A summary of the known results is as follows:

1. If L consists of a *single string*, **b**, the correction is obvious, but reconstructing a least-weight sequence of edit operations to change **a** into **b** is not. This is the string-to-string correction problem (Wagner and Fischer, 1974), whose applications and variants are discussed throughout this volume. Solution time is $\mathcal{O}(|\mathbf{a}||\mathbf{b}|)$.

2. If L is a *regular language,* with a finite-state automaton recognizer, F, then both the correction string **c** and $d(\mathbf{a}, \mathbf{c})$ can be determined in time $\mathcal{O}((|\mathbf{a}|\ P)$, where F has P arcs in its state diagram. (Equivalently, L's regular grammar has P productions.) (Wagner, 1974)

3. If L can be recognized by a counter-automaton M with N states in the automaton's finite control (and an infinite counter that can be increased or decreased by one, and can be tested for zero on each state transition), then correction time is

$$\mathcal{O}(|\mathbf{a}|^2 N) \quad \text{(Wagner and Seiferas, 1978)}.$$

4. If L is context-free, and its context-free grammar (modified so that no production's right-hand side has more than two symbols) has P productions, then correction time is $\mathcal{O}(|\mathbf{a}|^3 P)$ (Aho and Petersen, 1972; Lyon, 1974).

5. Finally, if L is context-sensitive, the correction problem as defined here becomes NP-complete, and hence probably exponential-time in either $|\mathbf{a}|$, or the "size" of the language's Linear Bounded Automaton recognizer. This last result has not appeared elsewhere in the literature, and is worth sketching here in the context of compiler languages.

5. THE CONTEXT-SENSITIVE CASE AND THE MAXIMUM INDEPENDENT SET PROBLEM

In some sense, compiler source languages are not context-free, although they are context-sensitive. This occurs when the declaration structure of the language is considered part of the measure of syntactic correctness. Thus, consider a special Algol-like language whose syntax is:

⟨program⟩:: = ⟨decl⟩.⟨body⟩
⟨decl⟩:: = ⟨declaration⟩|⟨declaration⟩; ⟨decl⟩
⟨declaration⟩:: = integer ⟨name1⟩|comment ⟨name1⟩

⟨namel⟩: : = ⟨name⟩ | ⟨name⟩, ⟨namel⟩
⟨body⟩: : = ⟨namel⟩

The "semantic" requirement we impose on each legal program is that every ⟨name⟩ occurring within the ⟨body⟩ occur somewhere in an *integer* ⟨namel⟩ within the ⟨decl⟩ section of the program; furthermore, each ⟨name⟩ may occur at most once within *integer* ⟨namel⟩ constructs of the ⟨decl⟩ section.

Correcting a program into this language can now be shown to solve an NP-complete problem, MIS—*Maximum independent set*, on a graph.

First, given a graph $G = (V, A)$ where V is a set of vertices, and $A \subset V \times V$ is a set of undirected arcs, MIS is: Find a subset $N \subset V$ of maximal cardinality such that for every i and every $j \in N$, there is no arc $(i, j) \in A$.

Karp (1972) shows that solving MIS for an arbitrary graph in polynomial time would solve every NP-complete problem in polynomial time. Since this class of problems includes several, such as graph k-coloration, that have been studied unsuccessfully for years, many computer scientists feel that such polynomial-time solutions are impossible. See also Garey and Johnson (1979).

Now, suppose an arbitrary graph $G = (V, A)$ is encoded as follows:

1. A distinct ⟨name⟩ is chosen for each arc in A; let the ⟨name⟩ of arc (i, j) be ⟨i, j⟩.
2. For each vertex v, construct a ⟨namel⟩ containing ⟨v, i⟩, for every i such that [$(v, i) \in A$ and $v < i$] or [$(i, v) \in A$ and $i < v$]. Call this list **v**.
3. Construct a ⟨decl⟩ D consisting of "integer **v**;" strings, concatenated together, for every $v \in V$.

Now the program "D." (with a null body) has a correction in which certain *integer* tokens can be changed to *comment* tokens. A minimal number of such changes would leave as "active" declarations a set corresponding to a maximal independent set of vertices in G.

6. A GRAPH-THEORETICAL FORMULATION

The methods used to solve most of the other string-to-language correction problems can be outlined as follows:

Let $F(i,s) = \min \{d(a_1 \ldots a_i, \mathbf{b}) |$ "string **b**, input to L's recognition automaton, places that automaton M_L into state s"}

When M_L is finite-state, $F(i, s)$ can be calculated recursively as follows:

$F(i, s) = \min \{F(i - 1, t) + G(t, s, a_i)\}$, where $G(t, s, a_i)$ is the smallest cost of any edit sequence that changes character a_i into a string $\mathbf{Q}$ such that M_L, started in state t and given $\mathbf{Q}$ as input, ends in state s.

A somewhat more efficient algorithm can be developed by constructing a graph G^M whose nodes are pairs $\langle i, s \rangle$, and which contains:

1. An "insertion" arc $\lambda \rightarrow b$, of weight $w(\lambda \rightarrow b)$, connecting $\langle i, s \rangle$ to $\langle i, t \rangle$ whenever an arc labelled "b" connects s to t in M_L;
2. A "change" arc $a_i \rightarrow b$, of weight $w(a_i \rightarrow b)$, connecting $\langle i - 1, s \rangle$ to $\langle i, t \rangle$ whenever an arc labelled "b" connects s to t in M_L;
3. A "delete" arc $a_i \rightarrow \lambda$, of weight $w(a_i \rightarrow \lambda)$, connecting $\langle i - 1, s \rangle$ to $\langle i, s \rangle$ for every s.

Then $F(i, s) =$ the least weight in G^M of any directed path from $\langle 0, S_0 \rangle$ to $\langle i, s \rangle$. Thus, the problem becomes a single-origin shortest-distance problem. One efficient solution method for such problems appears in Wagner (1976).

REFERENCES

Aho, A. V., and Petersen, T. G., A minimum-distance error-correcting parser for context-free languages. *SIAM Journal on Computing*, 1, 305–312 (1972).

Aho, A. V., Johnson, S. C., and Ullman, J. D., Code generation for expressions with common subexpressions. *Journal of the Association for Computing Machinery* 24, 146–160 (1977).

Conway, R. W., and Wilcox, T. R., Design and implementation of a diagnostic compiler for PL/I. *Communications of the Association for Computing Machinery* 16, 169–179 (1973).

Garey, M. R., and Johnson, D. B., *Computers and Intractibility, a guide to the theory of NP-completeness*, San Francisco: W. H. Freeman and Co., 1979.

Karp, R. M., Reducibility among combinatorial problems, in R. E. Miller and J. W. Thatcher (eds.), *Complexity of Computer Computations*. New York: Plenum Press, 85–103 (1972).

Lyon, G., Syntax-directed least-errors analysis for context-free languages: a practical approach. *Communications of the Association for Computing Machinery* 17, 3–14 (1974).

Morgan, H. L., and Wagner, R. A., PL/C: The design of a high-performance compiler for PL/I. *Proc. SJCC*, 148–152 (1971).

Wagner, R. A., Order-n correction for regular languages. *Communications of the Association for Computing Machinery* 17, 265–268 (1974).

Wagner, R. A., A shortest-path algorithm for edge-sparse graphs. *Journal of the Association for Computing Machinery* 23, 50–57 (1976).

Wagner, R. A., and Fischer, M. J., The string-to-string correction problem. *Journal of the Association for Computing Machinery* 21, 168–173 (1974).

Wagner, R. A., and Seiferas, J. I., Correcting counter-automation-recognizable languages. *SIAM Journal on Computing;* 7, 357–375 (1978).

Wulf, W. A., Johnsson, R. K., Weinstock, C. B., Hobbs, S. O., and Geschke, C. M., *The Design of an Optimizing Compiler*, New York: American Elsevier, 1975.

CHAPTER FOURTEEN

HOW TO COMPUTE STRING-EDIT DISTANCES QUICKLY

William J. Masek and Michael S. Paterson

1. INTRODUCTION

Wagner and Fischer (1974) presented an algorithm for determining a sequence of edit transformations that changes one string into another. The execution time of their algorithm is proportional to the product of the lengths of the two input strings. The same three types of operations are used here, namely: (1) inserting a character into a string; (2) deleting a character from a string; and (3) replacing one character of a string with another. We present an algorithm with an asymptotically faster execution time, for example $O(n^2/\log n)$ when both strings are of length n, providing that the alphabet for the strings is finite and all edit costs are integral multiples of some real number r.

This algorithm computes an optimal edit sequence for pairs of strings. As a special case, it can compute the longest common subsequence of two strings.

For the infinite-alphabet case, Wong and Chandra (1976) obtained upper and lower bounds proportional to n^2 using a slightly restricted model of computation. Aho, Hirschberg, and Ullman (1976) obtained similar results for the longest-common-subsequence problem. Lowrance and Wagner (1975) extended the string-edit problem to include the operation of interchanging adjacent characters. They developed an $O(n^2)$ algorithm solving their extended problem.

1.1 Basic Definitions

The following notation and conventions will be used.

a	A string of characters over some alphabet $\mathscr{A}$.
$\lvert \mathbf{a} \rvert$	The length of string **a**.
a_n	The nth character of the string **a** ($\lvert a_n \rvert = 1$).
$\mathbf{a}^{i,j}$	The string $a_i \ldots a_j$ ($\lvert \mathbf{a}^{i,j} \rvert = j - i + 1$).
$\mathbf{a}^n$	An abbreviation for $\mathbf{a}^{1,n}$.
λ	The null string, also denoted $\mathbf{a}^0$.

An *edit operation* is a pair $(\mathbf{x}, \mathbf{y}) \neq (\lambda, \lambda)$ of strings of length less than or equal to 1, also denoted as $\mathbf{x} \to \mathbf{y}$. String **b** results from string **a** by the edit operation $\mathbf{x} \to \mathbf{y}$, written "$\mathbf{a} \to \mathbf{b}$ via $\mathbf{x} \to \mathbf{y}$", if $\mathbf{a} = \sigma \mathbf{x} \tau$ and $\mathbf{b} = \sigma \mathbf{y} \tau$ for some strings σ and τ. We call $\mathbf{x} \to \mathbf{y}$ a *replacement* operation if $\mathbf{x} \neq \lambda$ and $\mathbf{y} \neq \lambda$; a *delete* operation if $\mathbf{y} = \lambda$ and an *insert* operation if $\mathbf{x} = \lambda$.

A sequence S of edit operations will be called an *edit sequence*. Let $S = s_1$, $s_2, \ldots, s_m$ be an edit sequence; an S *derivation from* **a** to **b** is a sequence of strings

$$\mathbf{c}_0, \mathbf{c}_1, \ldots, \mathbf{c}_m \qquad \text{such that } \mathbf{a} = \mathbf{c}_0, \ \mathbf{b} = \mathbf{c}_m$$

and, for all $1 \leq i \leq m$,

$$\mathbf{c}_{i-1} \to \mathbf{c}_i \text{ via } s_i.$$

(Note that, in this case, each $\mathbf{c}_i$ represents a complete string, not an individual character.) If there is some S derivation of **a** to **b**, we say S *takes* **a** *to* **b**. *to* **b**.

A *cost function* w is a function assigning a nonnegative real number to each edit operation $\mathbf{x} \to \mathbf{y}$. We define $w(S)$ for any edit sequence $S = s_1, \ldots, s_m$ to be

$$w(S) = \sum_{1 \leq i \leq m} w(s_i).$$

The *edit distance* $d(w, \mathbf{a}, \mathbf{b})$ from string **a** to string **b**, using the cost function w, is defined by

$$d(w, \mathbf{a}, \mathbf{b}) = \min\{w(S) \mid S \text{ is an edit sequence taking } \mathbf{a} \text{ to } \mathbf{b}\}.$$

We may assume that $w(\mathbf{x} \to \mathbf{y}) = d(w, \mathbf{x}, \mathbf{y})$ for all edit operations $\mathbf{x} \to \mathbf{y}$. This leads to no loss of generality, since for any w' we may define a new cost function w by $w(\mathbf{x} \to \mathbf{y}) = d(w', \mathbf{x}, \mathbf{y})$. Then w satisfies the stated property and $d(w', \mathbf{a}, \mathbf{b}) = d(w, \mathbf{a}, \mathbf{b})$ for all strings **a** and **b**.

Let d denote the distance function between the strings **a** and **b** using the cost function w; then we denote $d(w, \mathbf{a}^i, \mathbf{b}^j)$ by $d_{i,j}$. We write the cost of replacing A with B as $R_{A,B}$, the cost of deleting A as D_A, and the cost of inserting A as I_A. We will assume $\lvert \mathbf{a} \rvert \geq \lvert \mathbf{b} \rvert$ throughout.

b

	λ	B	A	B	A	A	A
λ	0	1	2	3	4	5	6
A	1	2	1	2	3	4	5
B	2	1	2	1	2	3	4
a A	3	2	1	2	1	2	3
B	4	3	2	1	2	3	4
B	5	4	3	2	3	4	5
B	6	5	4	3	4	5	6

(a)

b

	λ	B	A	B	A	A	A
λ	0	1	2	3	4	5	6
A	1			2			5
B	2			1			4
a A	3	2	1	2	1	2	3
B	4			1			4
B	5			2			5
B	6	5	4	3	4	5	6

(b)

Edit matrix. Entries are $d_{i,j}$.

Same matrix broken into overlapping 4×4 submatrices.

Figure 1. Computing distances with matrices. The alphabet is {A, B}. Assume $I = D = 1$, $R_{A,B} = R_{B,A} = 2$, and $R_{A,A} = R_{B,B} = 0$.

1.2 Previous results

Wagner and Fischer's matrix-filling algorithm (1974) computes d by constructing an $(|\mathbf{a}| + 1) \times (|\mathbf{b}| + 1)$ *edit-matrix* whose (i, j)th entry is $d_{i,j}$ (Fig. 1(a)). They showed that each internal element of the matrix is determined by three adjacent matrix elements. The *initial vectors* of a matrix are its first row and column. The *final vectors* of a matrix are its last row and column. Theorem 1 describes how they computed the initial vectors of the matrix, and Theorem 2 provides the rule for computing subsequent matrix elements.

Theorem 1. $d_{0,0} = 0$, and for all i, j such that $1 \leq i \leq |\mathbf{a}|$, $1 \leq j \leq |\mathbf{b}|$,

$$d_{i,0} = \sum_{1 \leq r \leq i} D_{a_r} \quad \text{and} \quad d_{0,j} = \sum_{1 \leq r \leq j} I_{b_r}.$$

Theorem 2. For all i, j such that $1 \leq i \leq |\mathbf{a}|$, $1 \leq j \leq |\mathbf{b}|$,

$$d_{i,j} = \min(d_{i-1,j-1} + R_{a_i, b_j}, \quad d_{i-1,j} + D_{a_i}, \quad d_{i,j-1} + I_{b_j}).$$

Each of the $|\mathbf{a}| \cdot |\mathbf{b}|$ internal entries in the edit-matrix for **a** and **b** can thus be computed in constant time, so the construction of the entire matrix can be performed with $\mathcal{O}(|\mathbf{a}| \cdot |\mathbf{b}|)$ elementary steps. Our algorithm reduces the time needed to $\mathcal{O}(|\mathbf{a}| \cdot |\mathbf{b}|/(\min(\log |\mathbf{a}|, |\mathbf{b}|)))$ if the alphabet is finite and the edit costs are restricted.

2. A FASTER ALGORITHM

The transitive closure of a directed graph with n nodes can be easily computed with an $n \times n$ matrix using $O(n^2)$ row operations. Arlazarov, Dinic, Kronrod, and Faradzev (1970) proved that if the matrix was split up into submatrices with a small number of rows and all of the possible computations on submatrices were precomputed, the problem could be solved using $O(n^2/\log n)$ row operations. This algorithm is commonly referred to as the Four Russians' algorithm. Our algorithm applies similar techniques to Wagner and Fischer's edit matrices.

Hopcroft, Paul, and Valiant (1975) provided a generalized version of the Four Russians' technique by showing that every computation performable in $O(n^2)$ steps on a multi-tape Turing machine can be performed in $O(n^2/\log n)$ steps on a unit-cost random-access machine. Since the Wagner–Fischer algorithm can easily be implemented on a multi-tape Turing machine, running in $O(n^2)$ steps, it might appear that our result follows as an immediate corollary. This, however, is not the case, since our method achieves time $O(n^2/\log n)$ on a "charged" random-access machine in which each operation has cost proportional to the size of the operands involved at any step.

The Four Russians' algorithm works faster by splitting the computation into many smaller computations. It computes all possible smaller computations, then puts them together (using some of the small computations many times) to get the larger computation. We follow a similar strategy. First, all possible $(m + 1) \times (m + 1)$ submatrices that can occur in the full matrix are computed for a suitably chosen parameter m; then these submatrices are combined to form the full matrix (like Fig. 1(b)), and the edit cost is computed.

2.1 Computing all the submatrices

Define the (i, j, k) submatrix of the edit matrix d to be the $(k + 1) \times (k + 1)$ submatrix whose upper left-hand entry is (i, j). Figure 1(b) shows the borders of the (i, j, k) submatrices

$$(0, 0, 3), \quad (0, 3, 3), \quad (3, 0, 3) \quad \text{and} \quad (3, 3, 3).$$

It is obvious from Theorem 2 that the values in an (i, j, k) submatrix are determined solely by its initial vectors

$$d_{i, j}, d_{i, j+1}, \quad \ldots, \quad d_{i, j+k},$$

and

$$d_{i,j}, d_{i+1, j}, \quad \ldots, \quad d_{i+k, j},$$

along with its two strings $\mathbf{a}^{i+1,i+k}$ and $\mathbf{b}^{j+1,j+k}$. The first part of our algorithm computes the values for all possible (i, j, m) submatrices that can occur in any edit matrix (using the same alphabet and cost function). It saves each submatrix's final vectors

$$d_{i+m,j+1}, \quad \ldots, \quad d_{i+m,j+m}$$

and

$$d_{i+1,j+m}, \quad \ldots, \quad d_{i+m,j+m},$$

to be used later.

To compute the final vectors for each possible submatrix we must first be able to enumerate the submatrices. We assume that the alphabet is finite, so listing all length m strings is easy. Listing *all* length-m initial vectors, however, may take too long; as we get further into the matrix, the values tend to increase, so listing all initial vectors may become uneconomical. However, under a modest restriction on the costs assigned to edit operations, there are only a finite number of differences between consecutive matrix values for all edit matrices using the same cost function and alphabet. We will operate with these differences instead. Define a *step* to be the difference between any two horizontally or vertically adjacent matrix elements and a *step vector* as a vector of steps. Corollary 1 expresses Theorem 2 in terms of steps.

Corollary 1 (of Theorem 2).

$$d_{i,j} - d_{i+1,j} = \min \begin{cases} R_{a_i,b_j} - (d_{i-1,j} - d_{i-1,j-1}), \\ D_{a_i}, \\ I_{b_j} + (d_{i,j-1} - d_{i-1,j-1}) - (d_{i-1,j} - d_{i-1,j-1}) \end{cases}$$

$$d_{i,j} - d_{i,j-1} = \min \begin{cases} R_{a_i,b_j} - (d_{i,j-1} - d_{i-1,j-1}), \\ D_{a_i} + (d_{i-1,j} - d_{i-1,j-1}) - (d_{i,j-1} - d_{i-1,j-1}), \\ I_{b_j}. \end{cases}$$

Now each (i, j, k) submatrix may be determined by a starting value $d_{i,j}$, two initial step vectors

$$d_{i,j+1} - d_{i,j}, \quad \ldots, \quad d_{i,j+k} - d_{i,j+k-1}$$

and

$$d_{i+1,j} - d_{i,j}, \quad \ldots, \quad d_{i+k,j} - d_{i+k-1,j},$$

along with the two strings $\mathbf{a}^{i+1,j+k}$ and $\mathbf{b}^{j+1,j+k}$. Then our algorithm can compute the final step vectors for each possible submatrix efficiently. To enumerate all possible submatrices, we will enumerate all pairs of length-m strings and all pairs of length-m step vectors.

The initial phase of our algorithm in which all submatrices are computed can now be presented. Assuming some fixed ordering on the alphabet $\mathscr{A}$ and on the finite set of possible step sizes, we enumerate all length-m strings and all length-m step vectors in lexicographic order. Then for each pair of strings $\mathbf{g}$, $\mathbf{h}$ and pair of step vectors F, G, Algorithm Y calculates a submatrix of steps according to Corollary 1. There are two classes of steps to consider, the ones moving horizontally and the ones moving vertically. Therefore our algorithm computes two matrices of steps—T consisting of the vertical steps and U consisting of the horizontal steps. The function Store saves F' and G', the final step vectors of the edit-submatrix determined by $\mathbf{g}, \mathbf{h}, F$, and G so that they can be easily recovered given $\mathbf{g}, \mathbf{h}, F$, and G.

Algorithm Y

for each pair $\mathbf{g}$, $\mathbf{h}$ of strings in $\mathscr{A}^m$ and
      each pair of length-m step vectors F and G

do
    begin
        for $i = 1$ to m do
           begin
               $T(i, 0):= F(i)$;
               $U(0, i):= G(i)$;
           end;
        for $i = 1$ to m do
           for $j = 1$ to m do
              begin

$$T(i, j):= \min \{R_{g_i, h_j} - U(i - 1, j), D_{g_i},$$
$$I_{h_j} + T(i, j - 1) - U(i - 1, j)\};$$
$$U(i, j):= \min \{R_{g_i, h_j} - T(i, j - 1),$$
$$D_{g_i} + U(i - 1, j) - T(i, j - 1), I_{h_j}\};$$

              end;
        $F':= \langle T(1, m), \ldots, T(m, m) \rangle$;
        $G':= \langle U(m, 1), \ldots, U(m, m) \rangle$;
        Store $(F', G', F, G, \mathbf{g}, \mathbf{h})$;
    end;

Algorithm Y takes time $\mathscr{O}(m^2)$ for each submatrix. Assuming that there is a finite number of possible differences between costs of edit instructions, calculating all of the final step vectors takes total time $\mathscr{O}(m^2 c^m) \leq \mathscr{O}(k^m)$ for some k depending only on the number of steps and $\mathscr{A}$, but not on m. Now we observe that the size of steps in a matrix is bounded independently of the strings involved.

Lemma 1. Let $I = \max \{I_A \mid A \in \mathcal{A}\}$, $D = \max\{D_A \mid A \in \mathcal{A}\}$. For all **a**, **b**, i, j such that $1 \le i \le |\mathbf{a}|$, $1 \le j \le |\mathbf{b}|$,

i) $-I \le d_{i,j} - d_{i-1,j} \le D$,
ii) $-D \le d_{i,j} - d_{i,j-1} \le I$.

Let $\Omega = \{D_A \mid A \in \mathcal{A}\} \cup \{I_A \mid A \in \mathcal{A}\} \cup \{R_{A,B} \mid A, B \in \mathcal{A}\}$. The edit cost function is defined to be *sparse* only if there exists some constant r such that every element of Ω is some integral multiple of r. For finite alphabets, cost functions mapping into the integers or the rational numbers are always sparse, while functions mapping into the real numbers may not be. It can be shown that if the set of edit costs is sparse, then there is a finite set of steps occurring in the submatrices independent of the strings we are using for this computation. Hence Algorithm Y is applicable.

Lemma 2. If Ω is sparse, then the set of possible steps in edit matrices is finite.

2.2 Computing the edit distance

The last stage of our algorithm, Algorithm Z, pieces together the (i, j, k) submatrices generated by Algorithm Y, to form the edit matrix of steps. Then the actual edit costs can be calculated by summing the steps along any path to the end. Assume Fetch(F, G, **g**, **h**) returns the pair of final vectors of the submatrix determined by strings **g** and **h**, and initial step vectors F and G. P and Q are matrices of length-m vectors. Graphically, P is the matrix of initial and final column vectors of $(m + 1) \times (m + 1)$ submatrices and Q is the corresponding matrix of row vectors. Define the function Sum(vector) to be the sum of a vector's components. Finally, assume that m divides $|\mathbf{a}|$ and $|\mathbf{b}|$.

Algorithm Z

```
for i = 1 to |a|/m   do P(i, 0) := ⟨D_{a_{(i-1)m+1}}, ..., D_{a_{im}}⟩;
for j = 1 to |b|/m   do Q(0, j) := ⟨I_{b_{(j-1)m+1}}, ..., I_{b_{jm}}⟩;
for i = 1 to |a|/m   do
    for j = 1 to |b|/m   do
        ⟨P(i, j), Q(i, j)⟩ := Fetch [P(i, j − 1), Q(i − 1, j),
                                      a^{(i-1)m+1,im}, b^{(j-1)m+1,jm}];

cost: = 0;

for i = 1 to |a|/m   do cost := cost + Sum(P(i, 0));
for j = 1 to |b|/m   do cost := cost + Sum(Q(|a|/m, j));
```

The timing analysis of Z is straightforward; it requires $\mathcal{O}(|\mathbf{a}| \cdot |\mathbf{b}|/m^2)$ assignments and look-ups of length-m vectors, and hence only $\mathcal{O}(|\mathbf{a}| \cdot |\mathbf{b}|/m)$ basic steps. If we choose $m = \lfloor \min(\log_k |\mathbf{a}|, |\mathbf{b}|)/2 \rfloor$ then the entire algorithm (both Y and Z) runs in time $\mathcal{O}(|\mathbf{a}| \cdot |\mathbf{b}|/m)$. (Algorithm Y runs in time $\mathcal{O}(k^m)$.)

The $\mathcal{O}(|\mathbf{a}| \cdot |\mathbf{b}|/m)$ time bound is still achieved if m does not divide $|\mathbf{a}|$ and $|\mathbf{b}|$. Just pad out $\mathbf{a}$ and $\mathbf{b}$ with a dummy character not in the string, say ϕ, until $|\mathbf{a}|$ and $|\mathbf{b}|$ are multiples of m. Then set $D_\phi = I_\phi = 0$, and for all A in $\mathcal{A}$, $R_{A, \phi} = D_A$ and $R_{\phi, A} = I_A$.

2.3 Edit paths

Algorithms Y and Z describe how to compute the minimum edit cost between any two strings. Now we will explain how to recover a sequence of edit operations that achieves this minimum cost.

It is clear from the Wagner–Fischer algorithm that for any pair of strings $\mathbf{a}$ and $\mathbf{b}$, it is enough to consider edit sequences S with the following properties. Each initial sequence S^r of S edits $\mathbf{a}^i$ to $\mathbf{b}^j$ for some (i, j), and so corresponds to a matrix element. Furthermore, successive elements correspond to a path through the matrix, since for each successive element either i, or j, or both i and j, increase by 1. An *edit path* (Fig. 2) is any such sequence of elements through the matrix, not necessarily starting at the $(0, 0)$ cell. The sequences of elements in an edit-path model sequences of insertions, deletions, and replacements according to their direction, and have costs depending on the symbols involved. The *cost* of an edit path is the sum of the costs of its operations.

Wagner and Fischer described an algorithm for recovering the edit sequence from the edit matrix by working backwards through it. The (i, j)th element of the matrix was originally calculated from the $(i, j - 1)$st, $(i - 1, j)$th, and $(i - 1, j - 1)$st elements along with insert, delete, and replace operations. The characters a_i and b_j are known, so it is easy to decide which operations could have given the value $d_{i,j}$.

The procedure outputs any such operation as the last step of the edit path and remembers the preceding element

$$(i, j - 1), \quad (i - 1, j) \quad \text{or} \quad (i - 1, j - 1).$$

They applied this procedure recursively on the remembered element of each "last step" starting at $(|\mathbf{a}| + 1, |\mathbf{b}| + 1)$, until it reached the starting element $(0, 0)$. There are at most $2n$ steps in any optimal edit path, so this algorithm runs in time $\mathcal{O}(n)$. We can use the same idea to recover the edit path from our sparse edit-matrix by regenerating the $\mathcal{O}(n/\log n) \log n \times \log n$ submatrices crossed by the optimal edit path. This would take time $\mathcal{O}(n \log n)$. Alternatively, if Algorithm Y was modified to save every entry of each submatrix, and

b

		λ	A	B	A	B	A	A
	λ	0	1	2	3	4	5	6
	B	1	2	1	2	3	4	5
	A	2	1	2	1	2	3	4
a	A	3	2	2	2	2	3	4
	B	4	3	2	3	2	3	4
	A	5	4	3	2	3	2	3
	B	6	5	4	3	2	3	4

Figure 2. Paths in matrices. Assume $R_{A,A} = R_{B,B} = 0$, $R_{A,B} = 1$, $R_{B,A} = 2$, and $I = D = 1$. The underlined entries of the matrix form an edit path from $(0, 0)$ to $(5, 5)$ with cost 8. Its operations consist of D, D, I, R, R, R, I.

Algorithm Z saved a pointer to each submatrix whenever it was used, we could do this in linear time on a sparse edit-matrix while using space $\mathcal{O}(n^2/\log n)$.

2.4 Storage Requirements

Algorithm Y saves the initial and final step vectors for each of the submatrices. By our choice of m, there are less than or equal to $\mathcal{O}(k^m) = \mathcal{O}(n^{1/2})$ submatrices to save, so Algorithm Y's storage requirements are $\mathcal{O}(n^{1/2} \log n)$ words.

The linear-time algorithm to construct an optimal edit sequence, which was described in Sec. 2.3, required $\mathcal{O}(n^2/\log n)$ space, but if only the edit distance is required, then Algorithm Z may be made more economical of space by using an observation of Hirschberg (1975). Since the ith row of submatrices depends only on the $(i-1)$st row and the strings **a** and **b**, Algorithm Z can be modified to overwrite the $(i-2)$th row with the ith row. Then Algorithm Z and the whole algorithm would require only linear space.

3. LONGEST COMMON SUBSEQUENCE

Let **u** and **v** be strings. **u** is a subsequence of length n of **v** if there exists $1 \leqq r_1 < \ldots < r_n \leqq |\mathbf{u}|$ such that $u_i = v_{r_i}$. We say **u** is a *longest common subsequence* of **a** and **b** if **u** is a subsequence of both **a** and **b** and there is no longer subsequence of both **a** and **b**.

We may derive $|\mathbf{u}|$ by using the following cost function w:

$$R_{A,B} = \begin{cases} 0 & \text{if } A = B \in \mathcal{A}, \\ 2 & \text{otherwise}; \end{cases}$$
$$D = I = 1.$$

Now $d = |\mathbf{a}| + |\mathbf{b}| - 2|\mathbf{u}|$, or $|\mathbf{u}| = (|\mathbf{a}| + |\mathbf{b}| - d))/2$. The cost function is sparse, so if $|\mathcal{A}|$ is finite, we can compute $|\mathbf{u}|$ in time $\mathcal{O}(|\mathbf{a}| \cdot |\mathbf{b}|/(\min(\log |\mathbf{a}|, |\mathbf{b}|)))$, using Algorithms Y and Z. We can compute the actual string $\mathbf{u}$ using an algorithm similar to our algorithm for recovering edit sequences. The cost function in this section comes from Wagner and Fischer (1974).

4. SPARSENESS IS NECESSARY

The number of steps in all possible $n \times n$ matrices for a given alphabet and cost function has to be finite or very slowly growing, if Algorithm Y is to run efficiently. In this section we demonstrate with an example of a nonsparse cost function that the set of possible steps can grow linearly with the size of the strings. Algorithm Y has an exponential running time for the strings and edit costs we define.

4.1 The Example

Our example gives a nonsparse cost function with an unbounded number of steps. Let $\mathcal{A} = \{A, B, C\}$, then for all $\sigma \in \mathcal{A}$ define the cost function w:

$$R_{\sigma,\sigma} = 0,$$
$$R_{A,B} = R_{B,A} = 1,$$
$$R_{C,B} = R_{C,A} = R_{A,C} = R_{B,C} = \pi,$$
$$I_\sigma = D_\sigma = 5.$$

Then $\mathbf{a}$ and $\mathbf{b}$ are generated as follows. First, let $\mathbf{a}' = BABA\ldots$ and $\mathbf{b}' = ABAB\ldots$. To form $\mathbf{a}$ and $\mathbf{b}$, we replace some characters of $\mathbf{a}'$ and $\mathbf{b}'$ with Cs, so that the number of Cs in $\mathbf{a}^i$ (and Cs in $\mathbf{b}^i$) equals μ_i, where $\mu_{2k} = \mu_{2k+1} = [2k/(2\pi + 1)]$. Note that the Cs can replace characters in even positions only. Figure 3 shows the first 50 characters of $\mathbf{a}$ and $\mathbf{b}$ with the Cs inserted.

$\mathbf{a}^{50} = $ BABABABCBABABABCBABABCBABABABCBABABABCBABABCBABABA
$\mathbf{b}^{50} = $ ABABABACABABABACABABACABABABACABABABACABABACABABAB

Figure 3. The first 50 characters of $\mathbf{a}$ and $\mathbf{b}$.

The edit costs are designed so that optimal paths in the edit matrix contain as many replacements (diagonal steps) as possible. The values in the center diagonal path (the dashed line in Fig. 4) form a nondecreasing integer sequence. The values increase by one every time a C is not encountered and stay constant elsewhere. The values in the two adjacent diagonal paths (the dotted lines in Fig. 4) stay constant until they encounter a C. Each C in either $\mathbf{a}$ or $\mathbf{b}$ increases the cost of the path by π. These paths alone generate about n different steps in an $n \times n$ edit matrix. For a rigorous proof of Theorem 3, see Masek and Paterson (1980).

Theorem 3. Sparseness is a necessary condition for Algorithm Y to run in time $O(k^m)$ on length-m strings and step sequences.

Proof. In our example the edit distances along the main diagonal form an increasing integer sequence, whereas the immediately adjacent diagonals are sequences increasing by multiples of π. Since π is irrational, the number of different step sizes between these diagonals increases linearly with n, the string length. The running time would therefore be about $(kn)^m$ for some k, and no effective use could be made of this preprocessing. □

5. WHEN TO USE THE FASTER ALGORITHM

This algorithm is not practical in all cases. If the strings are short, then Wagner and Fischer's algorithm should be used, but if the strings are long enough our algorithm should be used. In this section we will determine how long the strings must be before our algorithm becomes better for a specific cost function and alphabet.

First, we need to calculate more precisely what the running times for these algorithms are. For Wagner and Fischer's algorithm, if we count assignment statements, we get $2n^2 + 2n + 1$ for its running time. For the running time of Algorithm Y, we will count assignment statements, vector assignments, and uses of the Store function. This yields the running time $|\mathscr{A}|^{2m} |S|^{2m}(2m^2 + 8m)$, where S denotes the set of possible steps in the matrix. For Algorithm Z we will count assignment statements, vector operations, and uses of the Fetch function. This yields $(6n^2/m) + 4n + 1$ for its running time. We must specify the alphabet and the cost function, in order to

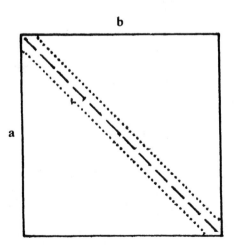

Figure 4. The center and adjacent diagonal paths.

Table 1. Running Times for Computing String-Edit Distances.

String length	Wagner–Fischer Algorithm	Algorithms Y and Z	The Difference
	s	t	$s - t$
200000.	80000400001.	81782333000.	−1781932030.
210000.	88200420001.	89805665000.	−1605244930.
220000.	96800440001.	98189112000.	−1388672000.
230000.	105800460001.	106931478000.	−1131017220.
240000.	115200480001.	116031624000.	−831143940.
250000.	125000500001.	125488475000.	−487974910.
260000.	135200520001.	135301000000.	−100480000.
262000.	137288524001.	137306099000.	−17574912.
262418.	137726938285.	137726955000.	−16384.
262419.	137727987961.	137727961000.	26624.
263000.	138338526001.	138313964000.	24561664.
270000.	145800540001.	145468213000.	332326913.
280000.	156800560001.	155989179000.	811380730.
290000.	168200580001.	166862991000.	1337587710.
300000.	180000600001.	178088780000.	1911820290.

determine $\mathcal{A}$, S, and m. We will use the alphabet $\mathcal{A} = \{A, B\}$ and the cost function specified in Sec. 3 for computing the longest common subsequence. These values yield $2^{2m}3^{2m}(2m^2 + 8m)$ for the running time·of Algorithm Y. If we let $m = (\log n)/6$, then our full algorithm runs in time

$$\frac{36n^2}{\log n} + \frac{n \log^2 n}{18} + \frac{4n \log n}{3} + 4n + 1.$$

We computed the running times for both Wagner and Fischer's algorithm and our algorithm, to find the exact point where our algorithm becomes more efficient. When the strings reach length 262,419, our algorithm becomes better for the alphabet and cost function specified. Table 1 gives some of the running times for the two algorithms.

6. CONCLUSION

We have presented an algorithm for computing the shortest edit distance between two strings of length n in time $\mathcal{O}(n^2/\log n)$. The algorithm works if the alphabet is finite and the domain for the cost function is sparse. Our analysis of an example shows the need for the sparseness condition. We showed that there are cases that violate the sparseness condition when the algorithm does not work. The results in Aho, Hirschberg, and Ullman (1976) show that our algorithm cannot work if the alphabet is infinite. The most important problem remaining is finding a better algorithm for the finite-alphabet case without the sparseness condition.

The question of the complexity of the shortest edit-distance problem for finite alphabets is open. The best lower bound is linear in n (Wong and Chandra, 1976), the upper bound is $O(n^2)$ or $O(n^2/\log n)$, depending on the sparseness condition. This gap seems too large and should be improved.

ACKNOWLEDGMENTS

We are grateful for the interest and guidance during the course of this work given by Ron Rivest, Mike Fischer, Peter Elias, and Albert Meyer. In particular, the observation in Sec. 2 on Hopcroft *et al.* (1975), and the topics addressed in Secs. 2.3 and 2.4 are due to Meyer. This work was supported by the National Science Foundation under the research grant GJ–43–634X, Contract #MCS74–12997 A04.

REFERENCES

Aho, A. V., Hirschberg, D. S., and Ullman, J. D. Bounds on the complexity of the longest common subsequence problem. *Journal of the Association for Computing Machinery* **23**, 1–12 (1976).

Arlazarov, V. L., Dinic, E. A., Kronrod, M. A., and Faradzev, I. A., On economic construction of the transitive closure of a directed graph. *Dokl. Acad. Nauk SSSR* **194**, 487–488 (1970) (in Russian). English translation in *Soviet Math. Dokl.* **11**, 1209–1210 (1975).

Hirschberg, D. S., A linear-space algorithm for computing maximal common subsequences. *Communications of the Association for Computing Machinery* **18**, 341–343 (1975).

Hopcroft, J. E., Paul, W. J., and Valiant, L. G., On time versus space and other related problems. *Proc. 16th Annual Symposium on Foundations of Computer Science,* Berkeley, 57–64 (1975).

Lowrance, R., and Wagner, R. A., An extension of the string-to-string correction problem. *Journal of the Association for Computing Machinery* **22**, 177–183 (1975).

Masek, W. J., and Paterson, M. S., A faster algorithm for computing string-edit distances. *Journal of Computer and System Sciences* **20**(1): 18–31 (1980).

Wagner, R. A., and Fischer, M. J., The string-to-string correction problem. *Journal of the Association for Computing Machinery* **21**, 168–173 (1974).

Wong, C. K., and Chandra, A. K., Bounds for the string-editing problem. *Journal of the Association for Computing Machinery* **23**, 13–16 (1976).

PART FIVE

RANDOM SEQUENCES

I. INTRODUCTION

Many applications of sequence comparison start with the observation or detection of what seems to be a high degree of similarity (or small distance) between two sequences. Often it is important to decide whether this similarity is truly large, or whether it could have occurred by coincidence. The chapters in this part of the volume deal with the expected degree of similarity due to chance, under a simple natural "null hypothesis" for the formation of the sequences. The expected degree of similarity is the first element needed in a statistical test of the null hypothesis.

Following Chvátal and Sankoff (1975), the tradition in this area is to evaluate sequence similarities in terms of $v(\mathbf{a}, \mathbf{b})$, the length of the longest common subsequence of sequences $\mathbf{a}$ and $\mathbf{b}$. For two sequences of length n whose elements are chosen randomly and independently from an alphabet of size k, the expected value of $v(\mathbf{a}, \mathbf{b})$ is asymptotically proportional to n as $n \to \infty$. The results given here pertain to the coefficient of proportionality,

$$c_k = \lim_{n \to \infty} \frac{(\text{expected value of } v(\mathbf{a}, \mathbf{b}))}{n},$$

These results may all be reinterpreted in terms of sequence distances, specifically the simple alignment distance $d(\mathbf{a}, \mathbf{b})$ discussed in Chapter 1, where each insertion and deletion costs 1 and each replacement costs 2 or more. Setting

$$d_k = \lim_{n \to \infty} \frac{(\text{expected value of } d(\mathbf{a}, \mathbf{b}))}{n},$$

and noting that

$$d(\mathbf{a}, \mathbf{b}) = 2(n - v(\mathbf{a}, \mathbf{b})),$$

we have

$$d_k = 2(1 - c_k).$$

Chapter 16, by Deken, sketches the methods that have been used to provide increasingly accurate upper and lower bounds for c_k with particular emphasis on a "waiting times" approach to the upper bound. Chapter 15, by Chvátal and Sankoff, given a compact combinatorial treatment of the upper bound (which Deken has improved within his probabilistic framework). Chapter 17, by Sankoff and Mainville, explores the conjecture that

$$\lim_{k \to \infty} \sqrt{k}\, c_k = 2,$$

based on a connection between the longest-common-subsequence problem and a better-known problem concerning the longest monotone subsequence of a random sequence.

Studying comparisons between random sequences through Monte Carlo simulation is relatively easy (see, for example Chvátal and Sankoff, 1975; Sec. 6 of Chapter 2 by Erickson and Sellers, Chapter 6 by Bradley and Bradley, and Sec. 5 of Chapter 10 by Kruskal and Sankoff), but deriving exact mathematical results seems difficult, and many interesting questions remain unanswered. For example, one question concerns the variance of $v(\mathbf{a}, \mathbf{b})$, which seems to grow surprisingly slowly with n though no mathematical results are known. Another question is the expected value of v when sequence terms are not chosen with equal probabilities among the k elements of the alphabet but instead with unequal probabilities, $p_1, \ldots, p_k$. Mainville (1981) has conjectured that in this case,

$$\lim_{n \to \infty} \frac{(\text{expected value of } v(\mathbf{a}, \mathbf{b}))}{n} < c_{k^*},$$

where k^* is the greatest integer $\leq 1/\Sigma\, p_i^2$.

REFERENCES

Chvátal, V., and Sankoff, D., Longest common subsequences of two random
    sequences. *Journal of Applied Probability* 12:306–315 (1975).
Mainville, S., Comparaisons et autocomparaisons de chaînes finies. Ph.D. thesis,
    Université de Montréal (1981).

CHAPTER FIFTEEN

AN UPPER-BOUND TECHNIQUE FOR LENGTHS OF COMMON SUBSEQUENCES

Václav Chvátal and David Sankoff

1. INTRODUCTION

The purpose of this chapter is to illustrate the application of combinatorial argumentation to sequence-comparison problems, in deriving upper bounds for the expected length of the longest common subsequence of two random k-ary sequences of length n.

Let $\mathbf{a} = a_1, \ldots, a_n$ and $\mathbf{s} = s_1, \ldots, s_m$ be two sequences where $m \leq n$. We say $\mathbf{s}$ is a subsequence of $\mathbf{a}$ if for some $1 \leq i_1 < i_2 < \ldots < i_m \leq n$, we have $a_{i_h} = s_h$ for all $1 \leq h \leq m$.

Let $v(\mathbf{a}, \mathbf{b})$ be the maximum length of any subsequence common to $\mathbf{a}$ and $\mathbf{b}$; that is, $v(\mathbf{a}, \mathbf{b})$ is the maximum value of m such that some $\mathbf{s} = s_1, \ldots, s_m$ is a subsequence of $\mathbf{a}$ and a subsequence of $\mathbf{b}$.

Now, suppose $\mathbf{a}$ and $\mathbf{b}$ are sequences of length n created by random draws from an alphabet of size k (all letters equally likely, and successive draws independent). We will prove that

$$\lim_{n \to \infty} \left[\frac{\text{expected value of } v(\mathbf{a}, \mathbf{b})}{n} \right] \leq V_k,$$

where V_k is easily calculated.

2. UNEXPECTED INVARIANT FOR LENGTH-n SEQUENCES

The first step is to count $F(n, s, k)$, the number of k-ary sequences of length n that contain a fixed sequence s. Note that this formula depends on s only through its length m:

Lemma 1. $F(n, s, k) = \sum\limits_{j=m}^{n} \binom{n}{j}(k-1)^{n-j}.$

Proof. The formula holds trivially if $m = n = 1$. We proceed by induction on n. For any $n > 1$, it is easy to verify the formula for $m = 1$ and $m = n$. For all other m, where $1 < m < n$, let $\mathbf{a}' = a_1, \ldots, a_{n-1}$ and $\mathbf{s}' = s_1, \ldots, s_{m-1}$. Divide the set of length-n sequences that contain s into two subsets: sequences for which $a_n = s_m$, and sequences for which $a_n \neq s_m$. If a is in the first subset, $\mathbf{a}'$ must contain s', so the first subset contains $F(n-1, s', k)$ sequences. If a is in the second subset, $\mathbf{a}'$ must contain s, and a_n can have $k-1$ different values, so the second subset contains $(k-1)F(n-1, s, k)$ sequences. Therefore,

$$F(n, s, k) = F(n-1, s', k) + (k-1)F(n-1, s, k)$$

$$= \sum_{j=m-1}^{n-1} \binom{n-1}{j}(k-1)^{n-j-1} + (k-1)\sum_{j=m}^{n-1}\binom{n-1}{j}(k-1)^{n-j-1},$$

by the induction hypothesis. Using the binomial identity

$$\binom{n-1}{j} + \binom{n-1}{j+1} = \binom{n}{j+1},$$

we then have

$$F(n, s, k) = \sum_{j=m-1}^{n-2}\left[\binom{n-1}{j} + \binom{n-1}{j+1}\right](k-1)^{n-j-1} + 1$$

$$= \sum_{j=m}^{n}\binom{n}{j}(k-1)^{n-j}. \quad\square$$

This result may seem surprising in that the count does not depend on the particular sequence s, but only on its length m. Note that

$$\binom{n}{j}(k-1)^{n-j} \geq \binom{n}{j+1}(k-1)^{n-j-1} \quad \text{if } j \geq \frac{n}{k},$$

so

$$F(n, \text{ s, } k) \le n \binom{n}{m} (k - 1)^{n-m} \quad \text{if } m \ge \frac{n}{k}. \tag{1}$$

3. A BOUND ON THE OCCURRENCES
OF HIGHLY SIMILAR SEQUENCES

The next step is to find an upper bound on the number of pairs of sequences having long common subsequences.

Lemma 2. For sufficiently large n, and $\theta = (m/n) > (1/k)$, the proportion $h_k^{(n)}(\theta)$ of ordered pairs of length-n sequences $(\mathbf{a}, \mathbf{b})$ with $v(\mathbf{a}, \mathbf{b}) \ge \theta n$ is bounded above by $[H_k(\theta)]^{2n}$, where

$$H_k(\theta) = \frac{k^{(\theta/2)-1} (k - 1)^{1-\theta}}{\theta^{\theta} (1 - \theta)^{1-\theta}}.$$

Proof. From Lemma 1, for a fixed s length m, the number of ordered pairs of length-n sequences $(\mathbf{a}, \mathbf{b})$ that both contain s as a subsequence is just $F^2(n, \mathbf{s}, k)$. Then the total number of such triples $(\mathbf{a}, \mathbf{b}, \mathbf{s})$ is

$$G(n, m, k) = \sum_{\mathbf{s}} F^2(n, \mathbf{s}, k),$$

where the summation is over all k^m sequences s of length m. Now $g(n, m, k)$, the number of pairs such that $v(\mathbf{a}, \mathbf{b}) \ge m$, must satisfy

$$g(n, m, k) \le G(n, m, k),$$

since if $(\mathbf{a}, \mathbf{b})$ is counted in $g(n, m, k)$, then at least one and possibly several $(\mathbf{a}, \mathbf{b}, \mathbf{s})$ will be counted in $G(n, m, k)$.

The proportion of all pairs $(\mathbf{a}, \mathbf{b})$ such that $v(\mathbf{a}, \mathbf{b}) \ge m$ is

$$h_k^{(n)} = \frac{g(n, m, k)}{k^{2n}} \le \frac{G(n, m, k)}{k^{2n}}$$

$$= \sum_{\mathbf{s}} \frac{F^2(n, \mathbf{s}, k)}{k^{2n}}$$

$$\le k^{m-2n} \left[n \binom{n}{m} (k - 1)^{n-m} \right]^2$$

for $(m/n) \ge (1/k)$, from (1). Thus, using Stirling's formula,

$$\lim_{n \to \infty} \left[h_k^{(n)}(\theta) \right]^{\frac{1}{n}} \leq k^{\theta-2}(k-1)^{2(1-\theta)} \lim_{n \to \infty} \left[n \binom{n}{m} \right]^{2/n}$$

$$= \frac{k^{\theta-2} (k-1)^{2(1-\theta)}}{\theta^{2\theta} (1-\theta)^{2(1-\theta)}}$$

$$= [H_k(\theta)]^2. \ \square$$

The key properties of H_k are established in the following:

Lemma 3. There is a unique solution of $H_k(\theta) = 1$ in the interval $[1/k, 1)$. Denoting this solution by V_k, then $H_k(\theta) < 1$ for $\theta > V_k$.

Proof. Note first that $H_k(\theta) > 0$ in $[1/k, 1)$; then

$$H_k\left(\frac{1}{k}\right) = k^{1/2k} > 1, \qquad \lim_{\theta \to 1} H_k(\theta) = k^{-1/2} < 1,$$

and

$$\frac{d}{d\theta} H_k(\theta) = H_k(\theta) \ \log\left[\frac{(1-\theta)k^{1/2}}{\theta(k-1)} \right].$$

Since the second factor of the derivative changes sign once from positive to negative when θ passes through the solution to $(1-\theta)/\theta = (k-1)/k^{1/2}$, we conclude that H_k first increases and then decreases as θ goes from $1/k$ to 1. Hence in this interval there is a unique solution, which we denote by V_k, of

$$H_k(\theta) = 1,$$

and $H_k(\theta) < 1$ for $\theta > V_k$. $\square$

4. CALCULATING EXPECTED VALUES

We can now prove our main result.

Theorem. For $k \geq 2$

$$\lim_{n \to \infty} \left[\frac{\text{expected value of } v(\mathbf{a}, \mathbf{b})}{n} \right] \leq V_k.$$

Proof. For any $\varepsilon > 0$ such that $V_k + \varepsilon < 1$, we divide the k^{2n} pairs of sequences into two parts, those which have $0 \leq v(\mathbf{a}, \mathbf{b}) \leq V_k + \varepsilon)n$, and those that have $(V_k + \varepsilon)n < v(\mathbf{a}, \mathbf{b}) \leq n$. The expected value of $v(\mathbf{a}, \mathbf{b})$ is less than

$$n \left(V_k + \varepsilon \right) \left[1 - h_k^{(n)} (V_k + \varepsilon) \right] + n h_k^{(n)} (V_k + \varepsilon)$$

$$\leq n \left(V_k + \varepsilon + \left[H_k(V_K + \varepsilon) \right]^{2n} \right), \quad \text{by Lemma 2.}$$

By Lemma 3, we have $H_k(V_k + \varepsilon) < 1$, so that

$$\lim_{n \to \infty} \left[\frac{\text{expected value of } v(\mathbf{a}, \mathbf{b})}{n} \right] \leq V_k + \varepsilon.$$

(The existence of the limit is guaranteed by the monotone-increasing nature of the expected value of $v(\mathbf{a}, \mathbf{b})/n$ (Chvåtal and Sankoff, 1975a)). $\square$

These arguments, drawn from Chvåtal and Sankoff (1975b), have since been refined within a probabilistic framework by Deken (Chapter 16) to achieve sharper bounds on expected common-subsequence length.

REFERENCES

Chvåtal, V., and Sankoff, D., Longest common subsequences of two random sequences. *Journal of Applied Probability* **12**, 306–315 (1975a).
Chvåtal, V., and Sankoff, D., Longest common subsequences of two random sequences. *Technical Report STAN-CS-75-477,* Stanford University, Computer Science Department (1975b).

CHAPTER SIXTEEN

PROBABILISTIC BEHAVIOR OF LONGEST-COMMON-SUBSEQUENCE LENGTH

Joseph Deken

1. PROBABILISTIC MODELS FOR RANDOM SEQUENCES

The length of a longest common subsequence of two sequences can be thought of as a measure of how "close" the sequences are to each other, and in this context it is natural to ask "How close is close?" For example, are the sequences much closer than two randomly generated sequences would be? There are many possible models for random sequences. One may suppose that all letters appear independently on both sequences, and have equal probability, or that they appear independently but perhaps with different probabilities for different letters. Alternatively there may be a fixed set of letters for each sequence, and the observed sequences may be obtained by permuting these letters randomly.

For a fairly broad class of models for random sequences (cf. Deken, 1979), it can be shown that the length of a longest common subsequence, divided by the total sequence length, approaches a constant as the lengths of the random sequences become large. In particular, this fractional length, averaged over all random sequences, will approach some constant c_k that is a function of the random model used and the number k of letters in the alphabet. The constants c_k are difficult to determine except in trivial cases, and the only results available for finite alphabets are upper and lower bounds.

Lower bounds for the constants c_k may produced by constructing an algorithm that takes two sequences as input and produces a common subsequence as output. If the algorithm is simple enough, its probabilistic behavior can be determined; and since the common subsequence produced is, by definition, no longer than the longest common subsequence, the limit

Table 1. Bounds for c_k

Alphabet size	Lower bound	Upper bound	Alphabet size	Lower bound	Upper bound
2	0.7615	0.8602 (0.8575)	9	0.4032	0.5599
3	0.6153	0.7769	10	0.3965	0.5406
4	0.5454	0.7181	11	0.3719	0.5234
5	0.5061	0.6733	12	0.3589	0.5079
6	0.4716	0.6373	13	0.3473	0.4942
7	0.4450	0.6074	14	0.3368	0.4811
8	0.4223	0.5819	15	0.3273	0.4650

obtained furnishes a lower bound for c_k. Upper bounds may be derived by counting methods, since the probability that the two sequences contain any *particular* subsequence is usually easily computed. Examples of these upper- and lower-bound techniques may be found in the papers by Chvátal and Sankoff (1975) and by Deken (1979).

Table 1 gives upper and lower bounds for the constants c_k derived by these methods, when the sequences are completely independent, with all letters equally likely. The corresponding lower-bound algorithms have been described in detail (cf. Deken, 1979) elsewhere, and will not be restated here. The counting method used to derive the upper bounds is of some independent interest, and so is described below. The bounds in the table are believed to be, at least temporarily, the best known.

2. UPPER BOUNDS

Upper bounds for the constants c_k are derived here by an overcounting method: Suppose we wish to show that $c_k < p$. We compute the number $N(p, n)$ of pairs of sequences of length n containing in common a particular sequence **s** of length $\lceil pn \rceil$. The total number of pairs of sequences containing a common subsequence of length greater than or equal to $\lceil pn \rceil$ is not greater than $c(p, n) = k^{\lceil pn \rceil} N(p, n)$, and if the ratio $(c(p, n)/k^{2n})$ approaches 0 as n increases, then $c_k \leq p$. This method is essentially that used by Chvátal and Sankoff (1975). Our efforts here are focused on reducing the overcounting caused by the fact that, in general, one pair of sequences may contain many common subsequences of length $\lceil pn \rceil$.

Consider a sequence $\mathbf{s} = (s_1, s_2, \ldots, s_{\lceil pn \rceil})$. The probability that a sequence of length n contains **s** may be expressed as the probability that the sum of "waiting times" to collect $\lceil pn \rceil$ digits on an infinite random sequence does not exceed $n \cong \lceil pn \rceil / p$. Since the expected waiting time to collect a digit is k, and the values of p of interest will be $> 1/k$, this probability may be bounded by an argument due to Chernoff (1952). Squaring this bound gives an upper bound to the probability that two independent sequences both contain **s**.

It is convenient to think of examining the pair of sequences left to right to determine whether they contain the subsequence **s**. For $i = 1, 2, \ldots, [pn]$, we scan the "top" sequence until the desired digit s_i of **s** appears, then switch to the "bottom" sequence and search for s_i there. A particular pair of sequences is *marked* by the subsequence **s** if this process can be completed without searching more than n digits on either sequence. Suppose, however, that in collecting s_i we examine a segment **v** of $n_1 > 0$ digits different from s_i on the top sequence and a segment **v'** of $n_2 > 0$ digits different from s_i on the bottom sequence. In this case, there may be a digit $r \neq s_i$ contained by both **v** and **v'** (e.g., in the binary alphabet, there *must* be a match). In this case, we will not mark the pair of sequences by **s**, since if both sequences contain **s**, they also contain the sequence **s'** formed by inserting the digit r before s_i. In a natural way, we say that the sequence **s** is *preceded at i*, by **s'**. The probability that a sequence is not preceded is the product of the probabilities that it is not preceded at $i = 1$, $2, \ldots, [pn]$; that is, $m(k)^{[pn]}$, where $m(k)$ is the probability (as a function of alphabet size k) that no match occurs on initial segments of a random sequence before the appearance of a "1" on both sequences. New upper bounds are thus obtained by bounding the probability $c'(k, p, n)$ that the sum of the collection times for **s** does not exceed n on either the top or bottom sequence and **s** is not preceded. Functions $m(k, p)$ are derived below such that $c'(k, p, n) \leq m(k, p)^{[pn]}$. This gives the bounds

$$c_k \leq \inf\{p : \lim_{n \to \infty} k^{[pn]} c'(k, p, n) = 0\} \leq \inf\{p : k m(k, p) < 1\}.$$

3. UPPER BOUNDS FOR $c'(k, p, n)$

In order to use Chernoff's inequality, we compute the generating function

$$\phi_k(t) = \sum_{i=0}^{\infty} P_i(k) t^{i+2},$$

where $P_i(k)$ is the probability that i digits in total occur on top and bottom sequences before the occurrence of "1" on each sequence, and no other digit is common to the two initial segments before the "1"s. In fact, we have the following formula:

$$\phi_k(t) = \sum_{\alpha=0}^{k-1} \binom{k-1}{\alpha} \sum_{\beta=0}^{\alpha} \binom{\alpha}{\beta} (-1)^\beta \frac{t^2}{(k - t(\alpha - \beta))(k - t(k - 1 - \alpha))}.$$

In probabilistic terms, $(\phi_k(e^t))/(\phi_k(1))$ is the moment-generating function of the sum of the number of digits in the two initial segments, including the "1"'s, conditional on no other matches, so that the desired inequality

$$c'(k, p, n) \le m(k, p)^{\lfloor pn \rfloor}, \quad \text{with } m(k, p) = \inf_{t>0} \{t^{-2/p} \phi_k(t)\}$$

follows from Chernoff's inequality and the fact that if the maximum of two positive numbers is less than n, then their sum is less than $2n$.

4. MORE EXTENSIVE PRECEDENCE RULES

To further improve the upper bounds, we consider blocks of digits of length b. A subsequence s of length $\lceil cn \rceil$ is then considered to be made up of $\lceil cn/b \rceil$ blocks. Precedence rules are formulated so that s is not preceded if and only if it is not preceded at any block $i = 1, 2, \ldots, \lceil cn/b \rceil$. A sequence a of length n is marked by s if and only if s is not preceded on a and the total number of collections (summed over top and bottom sequences and all blocks) for s does not exceed $2n$.

The upper bounds given above were obtained by numerical solution of the equation

$$km(k, p) = 1,$$

and are accurate to four decimal places. The additional upper bound (0.8575) for the binary case was obtained by utilizing an extended precedence rule with block size $b = 8$. Increasing b causes a geometric increase in the number of terms to be calculated for the moment-generating function, and the improvement in the bound seems to be small.

REFERENCES

Chernoff, H., A measure of asymptotic efficiency for tests of a hypothesis based on the sum of observations, *Annals of Mathematical Statistics* **23**, 493–507 (1952).

Chvátal, V., and Sankoff, D., *Longest Common Subsequences of Random Sequences,* Department of Computer Science, Stanford University CS–75–477 (1975).

Deken, J., Some limit results for longest common subsequences, *Discrete Mathematics* **26**, 17–31 (1979).

CHAPTER SEVENTEEN

COMMON SUBSEQUENCES AND MONOTONE SUBSEQUENCES

David Sankoff and Sylvie Mainville

1. COMMON SUBSEQUENCES

Let $\mathbf{a} = (a_1, a_2, \ldots, a_n)$, $\mathbf{b} = (b_1, b_2, \ldots, b_n)$ be two sequences of positive integers not greater than k. If $1 \le i_1 < i_2 < \ldots < i_m \le n$ and $1 \le j_1 < j_2 < \ldots < j_m \le n$ and $a_{i_h} = b_{j_h}$ for $h = 1, \ldots, m$, we say that $\mathbf{a}$ and $\mathbf{b}$ have a common subsequence of length m. Let $v(\mathbf{a}, \mathbf{b})$ be the length of the longest common subsequence of the two sequences $\mathbf{a}$ and $\mathbf{b}$. Let $V_{n,k}$ be the mean value of $v(\mathbf{a}, \mathbf{b})$ over all possible sequences $(\mathbf{a}, \mathbf{b})$.

It is known that as $n \to \infty$, $V_{n,k}/n$ approaches, in a monotone increasing way, a constant, c_k (Chvátal and Sankoff, 1975). We conjecture here that as $k \to \infty$, $\sqrt{kc_k} \to 2$.

Let $\varepsilon > 0$. Can we find a k_0 such that for all $k > k_0$, $|\sqrt{kc_k} - 2| < \varepsilon$? Because

$$\left| \lim_{m \to \infty} \frac{V_{m,k}}{m} - \frac{2}{\sqrt{k}} \right| \le \left| \lim_{m \to \infty} \frac{V_{m,k}}{m} - \frac{V_{n,k}}{n} \right| + \left| \frac{V_{n,k}}{n} - \frac{2}{\sqrt{k}} \right|$$

for all k and n, it suffices to find an $n = n(k)$ for which each term on the right side is less than $\varepsilon/2\sqrt{k}$. We will find a range of values of n for which the second term converges this fast, but we cannot as yet confirm that this range includes values that ensure that $V_{n(k),k}/n(k)$ converges quickly enough to c_k.

Specifically, we shall prove that $V_{n,k}/n$ converges to $2/\sqrt{k}$ if $n^2/k \to \infty$ while $n/k \to 0$. We first show, by a simple probabilistic argument, that we can neglect the set of terms in **a** equal to more than one term of **b**. Actually the probability of an a_i matching at most one b_j is equal to

$$\left[1 - \frac{1}{k}\right]^n + \frac{n}{k}\left[1 - \frac{1}{k}\right]^{n-1} = \left[1 + \frac{n}{k-1}\right]\left[1 - \frac{1}{k}\right]^n.$$

which is of the order of $1 - \frac{1}{2}n^2/k^2$. The expected number of terms of **a** matching exactly one term of **b** is

$$N = \frac{n^2}{k}\left[1 - \frac{1}{k}\right]^{n-1}.$$

If the ith **a**-term matches b_j we set $x_i = j$. That is, we define a sequence x_1, $x_2, \ldots, x_N$ as the subscripts of the **b**-terms matched by successive **a**-terms, ignoring that some **a**-terms may match more than one **b**-term.

2. MONOTONE SUBSEQUENCES

The x_i are independent, identically distributed random variables, and hence satisfy the conditions of Hammersley's (1972) conjecture, as partly clarified by Logan and Shepp (1977) and finally proved by Veršhik and Kerov (1977):

> If $x_1, x_2, \ldots, x_N$ is a sequence of independent, identically distributed random variables, and L_N is the length of a longest nondecreasing subsequence, then $L_N/\sqrt{N}$ converges in probability to 2 as $N \to \infty$. The pth absolute moment of L_N also converges to $2\sqrt{N}$ for any p satisfying $0 < p < \infty$.

Applying this result to our sequence $x_1, x_2, \ldots, x_N$ produces a monotone subsequence of expected length

$$2\sqrt{\frac{n^2}{k}\left[1 - \frac{1}{k}\right]^{n-1}} = \frac{2n}{\sqrt{k}} + \delta,$$

where δ is of order of n/k. This subsequence corresponds to a longest common subsequence of **b** and those terms of **a** that match at most one term of **b**. How much better could we do by including in our considerations those terms of **a** that match two or more terms of **b**? Let this improvement be γ, an addition of at least zero and at most one to the length of our common subsequence for each such multi-matching term. From the above calculation γ is of the order of $n \cdot n^2/k^2$. Then

$$\frac{V_{n,k}}{n} = \frac{2}{\sqrt{k}} + \frac{(\delta + \gamma)}{n}.$$

Thus we can be assured that

$$\left| \frac{V_{n,k}}{n} - \frac{2}{\sqrt{k}} \right| < \frac{\varepsilon}{2\sqrt{k}}$$

as long as $n = o(k)$. However it is not proved that this allows n to increase fast enough for

$$\left| c_k - \frac{V_{n(k),k}}{n(k)} \right| < \frac{\varepsilon}{2\sqrt{k}} .$$

REFERENCES

Chvátal, V., and Sankoff, D., Longest common subsequences of two random sequences. *Journal of Applied Probability* **12**, 306–315 (1975).

Hammersley, J. M., A few seedlings of research. *Proc. Sixth Berkeley Symp. Math. Stat. and Prob.* **1**, 345–394 (1972).

Logan, B. F., and Shepp, L. A., A variational problem for random Young tableaux. *Advances in Mathematics* **26**, 206–222 (1977).

Veršhik, A. M., and Kerov, S. V., Asymptotics of the Plancherel measure of the symmetric group and the limiting form of Young tables. *Soviet Math. Dokl.* **18**, 527–531 (1977).

AUTHOR INDEX

SUBJECT INDEX

A

#, *see* dummy
− (blank character), 14, 57
  see also null character
λ, 14, 216, 336, 338
  see also null character
∅, 14, 255, 266, 344
  see also null character
⟨/, *see* alphabet
A, *see* adenine
acceptance-rejection problem, 35
accessibility, in secondary structure, 98, 100
acoustic units, in speech recognition, 166
adenine (A), 49, 93
algorithm, 211–214
  allowing string insertion and deletion,
    296–298
  basic, 23–30, 57, 146–148, 196, 258, 266,
    326–327, 337–339
  biased against nondiagonal arcs, 176
  CELLAR, 216, 224–229
  for comparing networks, 269–274
  for distance between trees, 246–247, 251
  for *G*-based distances, 303–306
  incorporating deletion-insertion constraints,
    282–290
  for interval resemblance, global
    and local, 60–68
  KQW, 46–48
  level-building, 274–276
  for local alignments, 294–296
  for minimizing distance, 20
  for optimizing interior nodes of tree,
    260–263

  for partial homology, 294–296
  for RNA secondary structure, 94, 104–119
  for similar portions of two sequences,
    293–296
  for special types of input string, 328
  SS, 60
  for string-to-language distance, 335
  SU, 67–68, 70, 85
  time-warping, 145–157
  for tree distance (length of tree), 258–260
  weighted average distance between pairs,
    195
  Y and Z, 342–344
  see also dynamic programming, recurrence
    equation
alignment, 10–18, 266
  distance, 21, 307–308
  generalized, 33, 301
  local (partial homology), 292–296
  metric, 57–59
    diagram of, 72–73, 82, 86, 88
    local, 73
  satisfying constraints on deletions and
    insertions, 282, 285, 288
  shortest, 26, 97
  and stacked pairs, 105–106
  standard format, 71
  with string insertion or deletion, 298
  of three sequences, 33
  tree, 255–258
  valence, 71–74
    total, 72
  see also correspondence, homology, listing,
    matching, trace